T0231117

FLAMMABILITY
HANDBOOK
FOR
PLASTICS

FLAMMABILITY HANDBOOK FOR PLASTICS

Fifth Edition

Carlos J. Hilado
PRODUCT SAFETY CORPORATION

CRC Press
Taylor & Francis Group
Boca Raton London New York

CRC Press is an imprint of the
Taylor & Francis Group, an **informa** business

Flammability Handbook for Plastics
a TECHNOMIC publication

Published in the Western Hemisphere by
Technomic Publishing Company, Inc.
851 New Holland Avenue, Box 3535
Lancaster, Pennsylvania 17604 U.S.A.

Distributed in the Rest of the World by
Technomic Publishing AG
Missionsstrasse 44
CH-4055 Basel, Switzerland

Main entry under title:
  Flammability Handbook for Plastics, Fifth Edition

A Technomic Publishing Company book
Bibliography: p.
Includes index p. 311

Library of Congress Catalog Card No. 98-60209
ISBN No. 1-56676-651-6

To Connie

For their assistance in studying this broad field of knowledge and in preparing this Fifth Edition of the <u>Flammability Handbook for Plastics,</u> the author is grateful to Vytenis Babrauskas of Fire Science and Technology, Bode Buckley of American Society for Testing and Materials, Loren M. Caudill of E. I. du Pont de Nemours and Company, J. Thomas Chapin of Lucent Technologies, Gordon H. Damant of Inter-City Testing and Consulting Corporation of California, Richard G. Hill of Federal Aviation Administration, James R. Hoover of E. I. du Pont de Nemours and Company, April Horner of Federal Aviation Administration, James D. Innes of Flame Retardants Associates, Melvyn A. Kohudic of Technomic Publishing Company, Matti Kokkala of Technical Research Centre of Finland, Jeffrey J. Lear of ARCO Chemical Company, Richard S. Magee of New Jersey Institute of Technology, William S. Metes of Underwriters Laboratories, Gordon L. Nelson of Florida Institute of Technology, James C. Norris of Oak Ridge National Laboratory, John H. Nosse of ICBO Evaluation Service, Deborah M. Oates of Underwriters Laboratories, William C. Page of Dow Corning Corporation, Roy F. Pask of BASF Corporation, James M. Peterson of Boeing Airplane Company, John E. Roberts of Underwriters Laboratories of Canada, Bjorn Sundstrom of Swedish National Testing and Research Institute, James J. Urban of Underwriters Laboratories, Kay M. Villa of American Textile Manufacturers Institute, Thomas R. Washer of Dow Corning Corporation, Robert H. White of USDA Forest Products Laboratory, and David W. Yarbrough of Tennessee Technological University.

The assistance of Sheila D. Byers in computer searches and manuscript preparation is gratefully acknowledged.

The author is indebted to Connie L. Runyan, without whose encouragement this book might never have been written.

ABS	acrylonitrile-butadiene-styrene
ACS	acrylonitrile-chlorinated polyethylene-styrene terpolymer
ASA	acrylic-styrene-acrylonitrile polymer
BEO	brominated epoxy oligomer
BPA	bisphenol A
CHDM	cyclohexanedimethanol
CPVC	chlorinated polyvinyl chloride
CTFE	chlorotrifluoroethylene polymer
DMT	dimethyl terephthalate
EAA	ethylene-acrylic acid polymer
EBA	ethylene-butyl acrylate polymer
EEA	ethylene-ethyl acrylate polymer
EG	ethylene glycol
EMA	ethylene-methyl acrylate polymer
EMAA	ethylene-methacrylic acid polymer
EnBA	ethylene-normal butyl acrylate polymer
EPS	expandable polystyrene
ETCFE	ethylene-chlorotrifluoroethylene copolymer
ETFE	ethylene-tetrafluoroethylene copolymer
ETP	engineering thermoplastics
EVA	ethylene-vinyl acetate polymer
EVOH	ethylene-vinyl alcohol polymer
FEP	fluorinated ethylene propylene copolymer
GPPS	general purpose polystyrene (crystal)
LCP	liquid crystal polymer

HMW-HDPE high-molecular-weight high-density polyethylene
HDPE   high-density polyethylene
HIPS   high impact polystyrene

LDPE   low-density polyethylene
LLDPE   linear low-density polyethylene

MDI   diphenylmethane diisocyanate
MIPS   medium impact polystyrene
MPE   metallocene polyethylene
MSW   municipal solid waste

NDA   2,6-napththalene dicarboxylic acid
NDC   dimethyl ester of 2,6-napththalene dicarboxylic acid

PA   polyamide
PAI   polyamide-imide
PBI   polybenzimidazole
PBT   polybutylene terephthalate
PC   polycarbonate
PCDD   polychlorinated dioxin
PCDF   polychlorinated furan
PCTA   copolymer of CHDM and PTA
PDF   packaging derived fuel
PE   polyethylene
PEBA   polyether block polyamide or polyester block polyamide
PEEK   polyetheretherketone
PEI   polyetherimide
PEK   polyetherketone
PEN   polyethylene naphthalate
PES   polyethersulfone
PET   polyethylene terephthalate
PETG   PET modified with CHDM
PFA   perfluoroalkoxy polymer
PMDI   polymeric MDI
PP   polypropylene
PPE   polyphenylene ether
PPO   polyphenylene oxide
PPS   polyphenylene sulfide or polyphenyl sulfone
PS   polystyrene
PSU   polysulfone
PTA   terephthalic acid
PTFE   polytetrafluoroethylene
PUR   polyurethane
PVC   polyvinyl chloride
PVDC   polyvinylidene chloride

PVDF	polyvinylidene fluoride
PVF	polyvinyl fluoride
RIM	reaction injection molding
SAN	styrene-acrylonitrile polymer
SMA	styrene-maleic anhydride polymer
TDI	toluene diisocyanate
TEO	thermoplastic elastomeric olefin
TPE	thermoplastic elastomer
TPP	triphenyl phosphate
TPU	thermoplastic polyurethane
TPV	thermoplastic vulcanizate
UHMWPE	ultrahigh-molecular-weight polyethylene
ULDPE	ultralow-density polyethylene
VOC	volatile organic compound

AATCC	American Association of Textile Chemists and Colorists (US)
ABNT	Associacao Brasileira de Normas Tecnicas (Brazil)
ACIPLAST	Asociacion Costarricense de la Industria del Plastico (Costa Rica)
ACOPLAS-TICOS	Asociacion Colombiana de Industrias Plasticas (Colombia)
AENOR	Asociacion Espanola de Normalizacion y Certificacion (Spain)
AFMA	American Fiber Manufacturers Association (US)
AFMA	American Furniture Manufacturers Association (US)
AFNOR	Association Francaise de Normalisation (France)
AFPR	Association of Foam Packaging Recyclers (US)
AHAM	Association of Home Appliance Manufacturers (US)
AISI	American Iron and Steel Institute (US)
AITM	Airbus Industrie Test Method (Europe)
AMTRAK	National Railroad Passenger Corporation (US)
ANIPAC	Asociacion Nacional de Industrias del Plastico A.C. (Mexico)
ANSI	American National Standards Institute US)
APC	American Plastics Council (US)
APINDO	Asosiasi Industri Plastik Indonesia (Indonesia)
APIP	Associacao Portuguesa de Industria de Plasticos (Portugal)
APME	Association of Plastics Manufacturers in Europe
AS	Australian Standard (Australia)
ASEAN	Association of Southeast Asian Nations
ASEP	American Society of Electroplated Plastics (US)
ASEPLAS	Asociacion Ecuatoriana de Plasticos (Ecuador)
ASHRAE	American Society of Heating, Refrigerating, and Air-Conditioning Engineers (US)
ASIPLA	Asociacion de Industriales del Plastico de Chile (Chile)
ASME	American Society of Mechanical Engineers (US)
ASP	American Society for Plasticulture (US)
ASTM	American Society for Testing and Materials (US)
ATMI	American Textile Manufacturers Institute (US)
ATT	American Telephone and Telegraph (US)
AVIPLA	Asociacion Venezolana de Industrias Plasticas (Venezuela)

BART	Bay Area Rapid Transit (US)
BIFMA	Business and Institutional Furniture Manufacturers Association (US)
BNSI	Barbados National Standards Institute (Barbados)
BOCA	Building Officials and Code Administrators International (US)
BPA	Bonneville Power Administration (US)
BS	British Standard (UK)
BSI	British Standards Institute (UK)
CAA	Civil Aviation Authority (UK)
CABO	Council of American Building Officials (US)
CACM	Central America Common Market
CAMI	Civil Aeromedical Institute (FAA) (US)
CANACIN-TRA	Camara Nacional de la Industria de la Transformacion (Mexico)
CARICOM	Caribbean Common Market
CEFIC	European Chemical Industry Council
CEGB	Central Electricity Generating Board (UK)
CEN	Comite Europeen de Normalisation (European Committee for Standardization)
CENELEC	Comite Europeen de Normalisation Electrotechnique
CFFA	Chemical Fabrics and Film Association (US)
CFR	Code of Federal Regulations (US)
CIMA	Cellulose Insulation Manufacturers Association (US)
CNET	Centre National d'Etudes et de Telecommunications (France)
COGUANOR	Comision Guatemalteca de Normas (Guatemala)
COPANT	Comision Panamericana de Normas Technicas
COVENIN	Comision Venezolana de Normas Industriales (Venezuela)
CPAI	Canvas Product Association International (US)
CPI	chemical process industry
CPSC	Consumer Product Safety Commission (US)
CSA	Canadian Standards Association (Canada)
DGN	Direccion General de Normas (Mexico)
DIGENOR	Direccion General de Normas y Sistemas de Calidad (Dominican Republic)
DIN	Deutsches Institut fur Normung (Germany)
DOC	Department of Commerce (US)
DOD	Department of Defense (US)
DOE	Department of Energy (US)
DOT	Department of Transportation (US)
EC	European Community or European Commission
ECM	European Common Market
EFRA	European Flame Retardants Association
EFTA	European Free Trade Association
EN	Norme Europeenne (European Standard)

EPA	Environmental Protection Agency (US)
EPIC	Environment and Plastics Institute of Canada (Canada)
ERMA	European Resin Manufacturers Association
EU	European Union
FAA	Federal Aviation Administration (US)
FAR	Federal Aviation Regulations (US)
FCC	Federal Communications Commission (US)
FHA	Federal Highway Administration (US)
FHA	Federal Housing Administration (US)
FM	Factory Mutual (US)
FPA	Flexible Packaging Association (US)
FPL	Forest Products Laboratory (USDA) (US)
FRA	Federal Railroad Administration (US)
FRCA	Fire Retardant Chemicals Association (US)
FTA	Federal Transit Administration (US)
FTAA	Free Trade Area of the Americas
FTC	Federal Trade Commission (US)
FTMS	Federal Test Method Standard (US)
G3	Group of Three (Colombia, Mexico, Venezuela)
GATT	General Agreement on Tariffs and Trade
GDP	gross domestic product
GNBS	Guyana National Bureau of Standards (Guyana)
GRENADA	Grenada Bureau of Standards (Grenada)
GSA	General Services Administration (US)
HHS	Department of Health and Human Services (US)
HPI	hydrocarbon processing industry
HUD	Department of Housing and Urban Development (US)
HVI	Home Ventilating Institute (US)
IAPD	International Association of Plastics Distributors (US)
IBN	Institut Belge de Normalisation (Belgium)
IBNORCA	Instituto Boliviano de Normalizacion y Calidad (Bolivia)
ICAITI	Instituto Centroamericano de Investigaciones y Tecnologia (Central America)
ICBO	International Conference of Building Officials (US)
ICFI	International Fire Code Institute (US)
ICONTEC	Instituto Colombiano de Normas Technicas y Certificacion (Colombia)
IEC	International Electrotechnical Commission
IEEE	Institute of Electrical and Electronic Engineers (US)
INDECOPI	Instituto Nacional de Defensa de la Competencia y de la Proteccion a la Propiedad Intelectual (Peru)
INEN	Instituto Ecuatoriano de Normalizacion (Ecuador)
INN	Instituto Nacional de Normalizacion (Chile)

INTECO   Instituto de Normas Tecnicas de Costa Rica (Costa Rica)
INTN   Instituto Nacional de Tecnologia y Normalizacion (Paraguay)
IPQ   Instituto Potugues da Qualidade (Portugal)
IRAM   Instituto Argentino de Normalizacion (Argentina)
ISO   International Standardization Organization

JAA   Joint Airworthiness Authorities (Europe)
JAR   Joint Airworthiness Requirements (Europe)
JBS   Jamaica Bureau of Standards (Jamaica)
JCS   Japanese Cable Standard (Japan)
JIS   Japanese Industrial Standard (Japan)
JSA   Japanese Standards Association (Japan)

KVS   Kunststoff-Verband Schweiz (Swiss Plastics Association) (Switzerland)

LBL   Lawrence Berkeley Laboratory (now LBNL) (US)
LBNL   Lawrence Berkeley National Laboratory (US)
LLL   Lawrence Livermore Laboratory (now LLNL) (US)
LLNL   Lawrence Livermore National Laboratory (US)

MESA   Mine Enforcement and Safety Administration (US)
MHI   Manufactured Home Institute (US)
MIL   Military Standard (US)
MITI   Ministry of International Trade and Industry (Japan)
MPMA   Malaysian Plastic Munfacturers Association (Malaysia)
MSHA   Mine Safety and Health Administration (US)
MTA   New York Metropolitan Transit Authority (US)

NAFTA   North American Free Trade Agreement
NAPCOR   National Association for Plastic Container Recovery (US)
NAHB   National Association of Home Builders (US)
NAIMA   North American Insulation Manufacturers Association (US)
NAS   Norsk Allmennstandardisering (Norway)
NASA   National Aeronautics and Space Administration (US)
NBL   Norges Branntekniske Laboratorium (Norway)
NBS   National Bureau of Standards (now NIST) (US)
NC   Oficina Nacional de Normalizacion (Cuba)
NEC   National Electrical Code (US)
NEMA   National Electrical Manufacturers Association (US)
NFPA   National Fire Protection Association (US)
NHTSA   National Highway Traffic Safety Administration (US)
NHTSB   National Highway Transportation Safety Board (US)
NIBS   National Institute of Building Sciences (USA)
NIOSH   National Institute of Occupational Safety and Health (US)
NIST   National Institute of Standards and Technology (formerly NBS) (US)
NKB   Nordiska Kommiten for Byggbestammelser (Nordic countries)

NMAB	National Materials Advisory Board (US)
NNI	Netherlands National Institute (Netherlands)
NRC	National Recycling Coalition (US)
NRC	National Research Council (US)
NTSB	National Traffic Safety Board (US)
NTT	Nippon Telegraph and Telephone (Japan)
OAS	Organization of American States
OPEC	Organization of Petroleum Exporting Countries
ORNL	Oak Ridge National Laboratory (US)
PACIA	Plastics and Chemicals Industries Association (Australia)
PANYNJ	Port Authority of New York and New Jersey (US)
PAGEV	Turk Plastik Sanayiceleri Araslirma Gelistirme ve Egitim Vakfi (Turkey)
PASC	Pacific Area Standards Congress
PBA	Plastic Bag Association (US)
PFA	Polyurethane Foam Association (US)
PIA	Plastics Institute of America (US)
PINZ	Plastics Institute of New Zealand (New Zealand)
PPI	Plastic Pipe Institute (US)
PPI	Polymer Processing Institute (US)
PRF	Packaging Research Foundation (formerly Plastics Recycling Foundation) (US)
RIMA	Reflective Insulation Manufacturers Association (US)
RMA	Rubber Manufacturers Association (US)
SAA	Standards Association of Australia (Australia)
SAE	Society of Automotive Engineers (US)
SAMPE	Society for the Advancement of Material and Process Engineering (US)
SBCC	Southern Building Code Congress International (US)
SCC	Standards Council of Canada (Canada)
SEPTA	Southeastern Pennsylvania Transit Authority (US)
SIPA	Structural Insulated Panel Association (US)
SNZ	Standards New Zealand (New Zealand)
SP	Statens Provningsanstalt (Sweden)
SPE	Society of Plastics Engineers (US)
SPI	Society of the Plastics Industry (US)
SPMP	Syndicat des Producteurs de Matieres Plastiques (France)
TBT	Agreement on Technical Barriers to Trade
TTBS	Trinidad and Tobago Bureau of Standards (Trinidad and Tobago)
TVA	Tennessee Valley Administration (US)
UBC	Uniform Building Code (US)
UFAC	Upholstered Furniture Action Council (US)

UL	Underwriters Laboratories (US)
ULC	Underwriters Laboratories of Canada (Canada)
UMTA	Urban Mass Transit Administration (US)
UN	United Nations
UNI	Ente Nazionale Italiano de Unificazione (Italy)
UNIT	Instituto Uruguayo de Normas Tecnicas (Uruguay)
USAF	U.S. Air Force (US)
USCG	U.S. Coast Guard (US)
USDA	U.S. Department of Agriculture (US)
USMC	U.S. Marine Corps (US)
USN	U.S. Navy (US)
UTE	Union Technique de l'Electricite (France)
VDE	Verband Deutscher Elektrotechniker (Germany)
VKF	Vereinigung Kantonaler Feuerversicherungen (Switzerland)
VUPS	Vyzkumny Ustav Pozemnich Staveb (Czechoslovakia)
WMTA	Washington Metropolitan Transit Authority (US)
WTO	World Trade Organization

The general class of materials known as plastics has become the most versatile and the most widely useful class of materials known to man. The range of processing and performance characteristics, and of combinations of these characteristics, made available through creative research and judicious design, has brought these materials into almost every conceivable application. Many of these applications involve a significant possibility of exposure to fire.

A plastic is defined as a material which contains as an essential ingredient an organic substance of large molecular weight, is sold in its finished state, and at some stage in its manufacture or its processing into finished articles, can be shaped by flow. Because the essential ingredient of a plastic material is an organic substance, combustion can occur under sufficiently severe exposure to heat and oxygen. The severity of exposure required to produce combustion, and the results of the combustion process, vary as widely as the materials themselves. Because overall fire hazard is a function, not only of the fire performance of the material, but also of the degree of exposure of the fire involved, and the possibility of occurrence and the nature of the fire involved, all of these factors must be considered in determining the degree of inherent fire resistance which the material must possess in order to be satisfactory for a particular application.

Considerable effort has gone into the study of various aspects of flammability and of various plastic materials, so that these materials which are proving so useful to man will always be used in ways which will not compromise his safety. The task is a continuing one, because the family of plastics continues to grow, and, along with it, its variety of applications. Some of these future applications can not even be conceived of at the present time. The needs of man and his society are changing, and with them the factors that affect his safety, comfort, and convenience.

Because of population growth and demographic changes, conditions in the plastics market are changing rapidly and generating pressures for changes in many requirements affecting plastics. These pressures may tend to make acceptance requirements more reasonable, but the requirement of adequate safety for human life and property will never be abolished, and government involvement is making certain that safety will not be neglected.

The plastics market is an international market. An effort is made to address the international market as well as the United States market.

A flammability handbook for plastics must necessarily involve a variety of sciences and technologies spread across the whole spectrum of human knowledge, and it is impossible to discuss all the subjects in great depth. Any details extracted for attention are brought out because they are believed to be significant to the overall effort to make plastics as useful and as safe as humanly possible.

CHAPTER 1

Materials for the Plastics Industry

Materials for the plastics industry fall into the general class of materials known as synthetic polymers: organic substances of high molecular weight, man-made from repeating units of lesser molecular weight called monomers.

Synthetic polymers are divided into three groups on the basis of the industry in which they are used: plastics, synthetic fibers, and synthetic rubber. Plastics are used in the plastics industry, synthetic fibers are used in the fiber industry, and synthetic rubber is used in the rubber industry. The fibers industry uses significant amounts of natural polymers such as cotton and wool, and the rubber industry uses significant amounts of natural rubber. The plastics industry uses the largest portion of the synthetic polymers produced in the United States, as shown in Table 1.1.

The market for plastics is a world market. Estimated plastics consumption by region and country is shown in Table 1.2.

Plastics are produced all over the world. Estimated plastics production by region is shown in Table 1.3.

Some major markets for plastics are shown in Table 1.4.

SECTION 1.1. PHYSICAL CLASSIFICATION

Plastics or polymers are available in a variety of physical shapes. The physical form in which a plastic is present has a significant influence on its flammability characteristics. In some cases physical structure is much more important than chemical structure where fire behavior is concerned. For this reason, physical classification will be discussed first. Six basic physical groupings will be considered: formed, filled, film, foam, fiber, and fines. These groupings are listed in increasing order of increasing susceptibility to flame.

1. Formed, the general grouping which includes rods, blocks, castings, extrusions, and moldings, covers structures in which there is a significant thickness and mass of polymer, so that heat transfer to and heat absorption by adjacent material is a function of the

1

characteristics of the polymer itself. Formed plastics generally have the lowest surface to volume ratio and tend to be the least flammable among the physical classifications.

2. Filled, the general grouping which includes reinforced plastics and filled plastics, covers structures in which solid materials other than the polymer under discussion are more or less uniformly distributed throughout the material, to a large enough extent that heat transfer to and heat absorption by adjacent material is a function of both the characteristics of the polymer and the characteristics of the filler or reinforcing material.

3. Film, the general grouping which includes films, sheeting, coatings, and adhesives, covers structures in which thickness is very small relative to length and width. Film is strictly defined as sheeting having nominal thicknesses not greater than 0.010 inch, but the term is used loosely here to simplify classification. The small thickness results in relatively little mass per unit area, so the material is essentially exposed or unexposed surface, and heat transfer to and heat absorption by adjacent material is essentially a function of the characteristics of the adjacent material or materials. In the case of adhesives, solids are the adjacent materials on both sides. In the case of coatings, the adjacent materials are a solid on one side and a gas, liquid, or solid on the other side. In the case of self-supporting sheeting, the adjacent material on either side could be a gas, liquid, or solid.

4. Foam, the general grouping which includes cellular plastics, structural foam, and honeycomb structures, covers structures in which gaseous materials are more or less uniformly distributed throughout the polymer. These are characterized by a relatively low mass per unit volume, high surface area per unit volume, and low thermal conductivity. Heat transfer to and heat absorption by adjacent material is a function of the characteristics of the polymer, the characteristics of the contained gas, and the manner of distribution of polymer and gas.

5. Fiber, the general grouping which includes fibers, filaments, fabrics, and textiles, covers structures in which the polymer is present in the form of essentially continuous cylindrical rods of very small diameter, arranged by weaving or other methods into more or less uniform matrices of varying characteristics. Heat transfer to and heat absorption by adjacent material is a function of the characteristics of the polymer, the characteristics of the matrix, and the characteristics of the intervening material, which may be air in most but not all cases.

6. Fines, the general grouping which includes powders and dusts, covers materials which are finely divided into small particles and are characterized by high surface area per unit volume or unit mass, and rather complete exposure to the surrounding atmosphere. Fines are the most susceptible among the physical classifications and under certain conditions can produce dust explosions.

SECTION 1.2. COMMERCIAL CLASSIFICATION

Polymers or plastics can be divided into three groups on a commercial basis: tonnage plastics, specialty plastics, and venture plastics.

Tonnage plastics are characterized by high demand and high volume (hence many plastics in this group are discussed in terms on tons as well as pounds), high capital intensity, relatively low R&D costs, emphasis on process development, relatively little need for knowledge on end-use applications, great need for vertical integration in raw materials, and small product differentiation. Growth is generally slow, and prices are generally less than $1.00 per lb. Low density polyethylene (LDPE), high density polyethylene (HDPE), polyvinyl chloride (PVC), polypropylene, and polystyrene are tonnage plastics on the basis of their high volume, as shown in Table 1.3.

Specialty plastics are characterized by lower volume, lower capital intensity, moderately higher R&D costs, greater need for knowledge of end-use applications, less need for a raw material base, and greater product differentiation. Growth can vary, and prices can range from $1.00 to $4.00 and more per lb.

Venture plastics are characterized by low demand but promising potential, low volume but relatively high sales revenues, low capital intensity, high R&D costs, need for knowledge of end-use applications, and definite product differentiation. Growth can vary, and prices can range from $3.00 to $30 and more per lb.

SECTION 1.3. FUNCTIONAL CLASSIFICATION

Polymers or plastics can be divided into three groups on the basis of function: commodity plastics, engineering plastics, and high-performance plastics. This functional classification is often but not always similar to the commercial classification. Tonnage plastics are often commodity plastics, specialty plastics are often engineering plastics, and venture plastics are often high-performance plastics. The two classifications, however, are not always identical. For example, thermoplastic polyesters are a tonnage plastic, but part of the thermoplastic polyesters were in the form of engineering grades. Polypropylene, polystyrene, and acrylonitrile-butadiene-styrene (ABS) are tonnage plastics, but upgraded polypropylene, polystyrene, and ABS are competing as engineering plastics.

Traditionally, performance and specifications were used to distinguish engineering plastics from commodity plastics. Engineering plastics were broadly defined as everything above commodity plastics. As engineering plastics grew in importance, the workhorse engineering plastics like polycarbonate and polybutylene terephthalate (PBT) were gradually distinguished from more specialized engineering plastics like polyphenylene sulfide and from plastics developed to meet higher performance requirements. It is therefore appropriate to divide plastics on the basis of function into commodity plastics, engineering plastics, and high-performance plastics.

SECTION 1.4. CHEMICAL CLASSIFICATION

Polymers have traditionally been divided into two general groups: thermoplastic and thermosetting.

Thermoplastic polymers are capable of being repeatedly softened by increase of temperature and hardened by decrease of temperature, the change upon heating being physical rather than chemical.

Thermosetting polymers are capable of being changed into a substantially infusible or insoluble product when cured by application of heat or chemical means.

The division between thermoplastic and thermosetting polymers is now less clearly defined, because polymers are manufactured and modified and blended to give an increasing range of properties. There are thermoplastic and thermosetting polyurethanes, thermoplastic and thermosetting polyesters, and thermoplastic and thermosetting polyimides. There are thermoplastic elastomers, which bring together the traditionally separate disciplines of thermoplastics and thermoset rubber.

The major types of thermoplastic materials are olefin, including polyethylene (PE) and polypropylene (PP); vinyl, including polyvinyl chloride (PVC), polystyrene (PS), acrylonitrile-butadiene-styrene (ABS), and acrylic; cellulosic, including cellulose nitrate, acetate, butyrate, and propionate; and aromatic, including polycarbonate (PC), phenoxy, polysulfone, and polyimide. Other thermoplastic materials are polyamide (nylon), acetal, and thermoplastic polyester.

The major types of thermosetting materials are polyurethane; phenolic (including phenol-formaldehyde); amino (melamine-formaldehyde and urea-formaldehyde); epoxy; and unsaturated polyester.

Estimated production by region of the different types of plastics is shown in Table 1.3.

Some major markets for the different types of plastics are shown in Table 1.4.

SECTION 1.5. OLEFIN POLYMERS

Olefin polymers are the plastics produced in the largest quantity, and are produced all over the world. Olefin polymer capacities for some countries are shown in Table 1.5.

Polyethylene (PE) is produced commercially from ethylene monomer, either alone to produce homopolymer grades, or with alpha olefin comonomers such as butene, hexene, 4-methyl-1-pentene, and octene to produce copolymer grades.

Polyethylene is divided into two general types: low-density polyethylene (LDPE), including conventional high-pressure polyethylene and therefore sometimes called high-pressure polyethylene, and high-density polyethylene (HDPE), sometimes called low-pressure polyethylene.

Polyethylene is also divided into linear polyethylene, which includes a range of densities from 0.880 to 0.965 g/cc, and branched polyethylene, which is primarily LDPE with densities from 0.910 to 0.955 g/cc.

Linear polyethylene includes the following types of polyethylene:

1. ULDPE, ultralow-density polyethylene, densities from 0.880 to 0.915 g/cc
2. LLDPE, linear low-density polyethylene, densities from 0.916 to 0.940 g/cc
3. HDPE, high-density polyethylene, molecular weights from 50,000 to 250,000, densities from 0.941 to 0.965 g/cc
4. HMW-HDPE, high-molecular-weight high-density polyethylene, molecular weights from 250,000 to 1 million, densities from 0.940 to 0.960 g/cc
5. UHMWPE, ultrahigh-molecular-weight polyethylene, molecular weights over 1.5 million

There is great versatility possible in physical properties and processing characteristics, depending on the molecular weight distribution, extent and type of branching, crosslinking, and crystallinity. The wide variety of applications which can result from this versatility is illustrated by the following list of examples: thin-walled containers by thermoforming, injection molding, or blow molding; pipe; wire coating; paper coatings; floor polishes; textile treatment; films and sheeting; blister packaging; and toys.

LDPE applications include general packaging, bread bags, frozen foods, kraft paper, milk and food cartons, toys, cable jacketing and insulation, and drug packaging. ULDPE applications include meat, cheese and poultry packaging, stretch film, shrink film, frozen food packaging, disposable gloves, medical packaging and film, and heat seal layers.

LLDPE applications include trash liners, newspaper bags, industrial liners, garment bags, wire and cable insulation and jacketing, pipe and tubing.

HDPE applications include food containers, housewares, caps and closures, pails, crates, totes, pipe and profiles, grocery bags, can liners, merchandise bags, bag lines, wire and cable insulation, tubing and pipe, 55-gal drums, gasoline tanks, bottles, and toys.

HMW-HDPE applications include grocery sacks, can liners, pressure pipe, sewer and industrial pipe, shipping containers, municipal refuse containers, and pond liners.

UHMWPE applications include gears, impellers, sliding surfaces, and abrasion-resistant and chemical-resistant surfaces.

Total polyethylene sales in the United States in 1996 were 35,811 million lb.

Polypropylene (PP) is produced commercially from propylene monomer, either alone to produce homopolymer grades, or with comonomers such as ethylene to produce copolymer grades. It has the lowest density (0.90 to 0.91 g/cc) of the major commercially available thermoplastics, and exhibits high yield strength and rigidity, exceptional flex life, good surface hardness, resistance to most chemicals, and excellent dielectric properties.

Polypropylene exists in three forms: isotactic, syndiotactic, and atactic. Isotactic PP is the principal form used commercially.

Applications include interior trim parts for automobiles, bucket seat backs, heater ducts, dishwasher racks, ice cube trays, television cabinets and components, wire coating, coaxial cable coating, fibers, pipe, hospital equipment, housewares, luggage, film, shrink packaging, decorative ribbon, brush bristles, indoor-outdoor carpeting, furniture, office equipment, battery cases, carpet backing, and tamper-evident closures.

Total polypropylene sales in the United States in 1996 were 12,183 million lb.

Polybutylene is produced commercially by polymerization of 1-butene, either alone to produce homopolymer grades, or with comonomers such as ethylene to produce copolymer grades. It exhibits low cold flow, flexibility, resistance to stress cracking, and chemical stability.

Applications include well pipe, water service tubing, collapsible irrigation tubing, heavy duty packaging, cold- and hot-water plumbing, underfloor hydronic heating, heat pump piping, food packaging, and hot-fill containers.

Polymethylpentene is produced by polymerization of 4-methyl-pentene-1, and is characterized by extremely low density (0.83 g/cc, close to the theoretical minimum for thermoplastics), high light transmission, high melting point, and excellent electrical properties.

Applications include hospital and laboratory ware, encapsulation of electronic relays, piping in milking machines, buttons, frozen dinner trays, dielectrics for coaxial cable connections, wire covering, cook-in containers, and paper coating.

Polyallomers are a class of olefin block copolymers which exhibit higher crystallinity than earlier copolymers of the olefins. Commercially available polyallomers are of the propylene-ethylene type, and offer high stiffness and high impact, exceptional flow properties, and resistance to flexing fatigue.

Applications include heat sterilizable containers, typewriter cases, integral hinge applications, and cowl panels.

Ethylene copolymers are produced by polymerizing ethylene with various comonomers to obtain the desired properties and processing characteristics. The important ethylene copolymers include:

EVA, ethylene-vinyl acetate
EnBA, ethylene-normal butyl acrylate
EVOH, ethylene-vinyl alcohol
EMA, ethylene-methyl acrylate

EEA, ethylene-ethyl acrylate
EBA, ethylene-butyl acrylate
EAA, ethylene-acrylic acid
EMAA, ethylene-methacrylic acid

Ethylene-vinyl acetate (EVA) copolymers are produced by polymerizing ethylene with vinyl acetate, and exhibit excellent low temperature flexibility, high impact strength, low elastic moduli, good resilience, good clarity, and dielectric sealability. Applications include medical tubing, disposable syringes, construction film, shower curtains, cattle ear tags, pool table bumpers, automobile mudflaps, meat and poultry wrap, ice bags, adhesives, carpet backing, shoes, wire and cable.

Ethylene-vinyl alcohol (EVOH) resins are hydolyzed copolymers of ethylene and vinyl acetate. They are outstanding as barriers to gases. EVOH applications include food packaging for products such as ketchup, barbecue sauce, mayonnaise, relish, and processed meats.

Ethylene-normal butyl acrylate (EnBA) copolymer is produced by polymerizing ethylene with normal butyl acrylate. EnBA applications include flexible packaging and extrusion coating applications.

Ethylene-methyl acrylate (EMA) copolymer is produced by polymerizing ethylene with methyl acrylate. EMA applications include food packaging and soft squeeze toys.

Ethylene-ethyl acrylate (EEA) copolymers are produced by polymerizing ethylene with ethyl acrylate. They are among the toughest and most flexible of the polyolefins. EEA applications include small rubberlike parts, gasketing, bumpers, disposable gloves, diaper liners, hospital sheeting, and microchip packaging.

Ethylene-acrylic acid (EAA) and ethylene-methacrylic acid (EMAA) copolymers are produced by copolymerizing ethylene with acrylic or methacrylic acid. Applications include flexible packaging of food, coated paperboard, toothpaste tubes, and bonded cable sheath.

Ionomers are ion-linked ethylene-methacrylic acid interpolymers. They are characterized by toughness, flexibility, high transparency, and chemical resistance. Applications include housewares, toys, safety shields, tool handles, containers, film, sheet, coatings, food packaging, skin packaging for hardware and electronic products, golf balls, ski boots, and boat bumpers.

Chlorinated polyethylene is produced by chemical substitution of chlorine on linear polyethylene. Variations of the molecular weight of the polyethylene feed stock, chlorine content, and chlorine placement can produce a wide range of characteristics. Applications include flexible film and sheet, electrical insulation, cable jacketing, and molded goods.

SECTION 1.6. VINYL POLYMERS

Vinyl chloride polymers and copolymers are produced by polymerization of vinyl chloride, with or without comonomers such as vinyl acetate.

Polyvinyl chloride (PVC) may be the most versatile of all plastics because of its blending capability with plasticizers and other materials. Polyvinyl chloride can be prepared by a variety of processes, and can be formulated for numerous applications.

Over 90% of PVC resins are suspension resins, and 5-7% are dispersion resins.

Applications include automobile seat covers, moldings and floor mats, shower curtains, kitchen utensils, baby pants, meat wrap, adhesives, house siding, weatherstripping, flooring, pipe, records, raincoats, toys, paper and textile coating, gloves, and medical tubing. Vinyl foam is used in seat cushions and marine products such as buoys.

Total sales of polyvinyl chloride in the United States in 1996 were 13,388 million lb.

Polyvinyl acetate is prepared by polymerization of vinyl acetate.

Polyvinyl alcohol is prepared by hydrolysis of polyvinyl acetate and is used for water-soluble packaging.

Polyvinyl butyral is prepared from polyvinyl alcohol and butyraldehyde, and is used in safety glass interlayers and table tops.

Chlorinated polyvinyl chloride (CPVC) is produced by post-chlorination of polyvinyl chloride resin. Applications include the following: residential pipe and fittings, and industrial liquid handling.

Polyvinylidene chloride (PVDC) is produced by the polymerization of vinylidene chloride, either alone to produce homopolymer or with comonomers such as vinyl chloride, acrylates, and nitriles to produce copolymers. Applications include food packaging and household wrap.

Polystyrene (PS) is manufactured by the polymerization of styrene monomer, and exhibits good physical and dielectric properties, chemical resistance, dimensional stability, and versatility in coloring. Crystal polystyrene is a clear non-crystalline homopolymer of styrene. Impact polystyrene is produced by introducing elastomers such as polybutadiene into the polystyrene matrix for impact modification. Polystyrene is modified with elastomers to provide impact resistance, and copolymerized with acrylonitrile to provide hardness, rigidity, tensile strength, and chemical resistance.

Polystyrene can be divided into crystal or general-purpose polystyrene (GPPS), and rubber modified medium-impact polystyrene (MIPS) and high-impact polystyrene (HIPS). Applications include housewares, toys, closure caps, luggage, appliance housings, automotive parts, lighting diffusers, illuminated interior signs, and, for modified polystyrene, business machine shells, telephones, and packaging.

Polystyrene foam can be divided into two general types: extruded polystyrene foam board and sheet, containing chlorofluorocarbons as blowing agents, and expandable polystyrene (EPS), crystal polystyrene containing pentane as a blowing agent. Polystyrene foam applications include building insulation, cushioned shipping containers, and wall panels.

Total sales of polystyrene in the United States in 1996 were 6059 million lb.

Styrene-butadiene polymers are produced by polymerizing styrene and 1,3-butadiene to produce block copolymers known for transparency and high impact strength. Applications include single-service disposable packaging, centrifuge tubes, toys, office supplies, tool handles, and shrink wrap.

Styrene-butadiene elastomers are produced from styrene and polybutadiene, and combine the end-use properties of vulcanized elastomers with the processing advantages of thermoplastics. They exhibit good resilience, elastic recovery, tensile properties, frictional properties, and chemical resistance. Applications include footwear, toy wheels, swim fins, aircraft oxygen masks, stair treads, bumpers, suction cups, and pipe seals.

Styrene-acrylonitrile (SAN) resins are random linear copolymers of styrene and acrylonitrile. Applications include refrigerator shelves, blender bowls, cosmetic packaging, safety glazing, chair backs and shells, and battery cases.

Styrene-maleic anhydride (SMA) copolymers are made by copolymerization of styrene and maleic anhydride. Applications include instrument panels and components, interior trim parts, and parts of small appliances.

Acrylonitrile-butadiene-styrene (ABS) copolymers are manufactured from the basic building blocks of styrene, acrylonitrile, and polybutadiene. ABS is not a random terpolymer of acrylonitrile, butadiene, and styrene, but a mixture of styrene-acrylonitrile (SAN) copolymer with SAN-grafted polybutadiene rubber. ABS is actually one of the first commercially successful polymer alloys. A wide range of performance and processing characteristics can be obtained by varying the composition and the polymerization process.

Applications include housings and cases for appliances, refrigerator door and inner liners, luggage, automotive air ducts, housings for lamps and turn signals, pipe and fittings in residential sanitary systems, safety helmets, housings for electronic components, machine gun grips and gun stocks, trays and tote boxes, boat hulls and decks, automotive interior trim components, and business machine housings.

Total sales of ABS in the United States in 1996 were 1427 million lb.

ACS is a terpolymer of acrylonitrile, chlorinated polyethylene, and styrene. Applications include housings and parts of office machines and home appliances.

Acrylic-styrene-acrylonitrile (ASA) polymers are characterized by excellent weatherability. Applications include home and commercial siding, truck tops, and recreational vehicle parts.

Acrylic polymers are produced by the polymerization of acrylates such as methyl methacrylate, ethyl acrylate, and acrylonitrile. Polymethyl methacrylate offers light transmittance, color stability, toughness, dimensional stability, abrasion resistance, and weather stability.

Applications include outdoor signs, interior and exterior lighting, automotive lenses, aircraft canopies, architectural panels, safety glazing, insulated skylights, and sanitaryware.

Total sales of acrylics in the United States in 1996 were 556 million lb.

SECTION 1.7. ENGINEERING THERMOPLASTICS

Engineering thermoplastics are synthetic polymers which provide an improved balance of properties for a variety of demanding applications. These materials are generally characterized by improved mechanical properties, high temperature capability, and flame resistance, and offer improvements in processing flexibility, freedom of design, and overall assembly economics over metal, ceramics, and glass.

Engineering plastics are a functional class of plastics, but are almost a chemical class because they include many if not most of the high temperature performance thermoplastics. Some types of thermoplastics, such as acrylonitrile-butadiene-styrene (ABS) and thermoplastic polyesters, are available in both engineering and non-engineering grades.

Engineering thermoplastics can be divided into several groups: polyesters, polyamides, polyimides, ketone-based polymers, polyacetals, sulfur-containing polymers, and fluoropolymers.

The most important polyesters include thermoplastic polyesters and polycarbonates.

Section 1.7.1. Thermoplastic Polyesters

Thermoplastic polyesters are produced by the polyesterification reaction between a glycol and a dibasic acid.

Polyethylene terephthalate (PET) is prepared by the reaction of either terephthalic acid (PTA) or dimethyl terephthalate (DMT) with ethylene glycol (EG). Applications include beverage bottles, dual-ovenable food trays, and food packaging. Various copolymers of PET broaden the range of PET applications.

PETG, which is PET modified with cyclohexanedimethanol (CHDM), has improved melt strength for extrusion-blow molding. It is used in small blow-molded containers.

PCTA copolyester, a copolymer of CHDM and PTA with another acid substituted for a portion of the PTA, is used in extruded amorphous film and sheeting for packaging.

Polyethylene napththalate (PEN) is the homopolymer of dimethyl-2,6-naphthalene dicarboxylate (NDC) and ethylene glycol. NDC is a dimethyl ester of 2,6-naphthalene dicarboxylic acid (NDA). PEN has thermal, mechanical, chemical, and dielectric properties which are generally superior to those of PET.

Applications include film, recording tape, and circuit boards.

Polybutylene terephthalate (PBT) is obtained by the condensation reaction of dimethyl terephthalate and 1,4-butanediol.

Applications include automotive exterior body parts, motor housings, fuse cases, and computer keyboards.

Polybutylene naphthalate (PBN) is obtained by the reaction of dimethyl-2,6-naphthalene dicarboxylate (NDC) and 1,4-butanediol.

Applications include connectors, switches, coil bobbins, fuel sensors, fuel tanks, lines, and hoses.

Total sales of thermoplastic polyesters in the United States in 1996 were 3687 million lb.

Section 1.7.2. Polycarbonates

Polycarbonates (PC) are polyesters of carbonic acid, and are prepared by three major processes: transesterification, phosgenation of aromatic dihydroxy compounds, and interfacial condensation.

The most commercially significant class of polycarbonates is based on bisphenol A (BPA). This polycarbonate is prepared by the catalytic reaction of bisphenol A with carbonyl chloride.

Polycarbonates offer high temperature capability, high impact strength, high stiffness, good dimensional stability, high creep resistance, good electrical properties, low water absorption, high transparency, and good stain resistance.

Applications include electric knife handles, housewares, appliance housings, electrical components, food vending machines, glazing, outdoor lighting, automotive parts, returnable milk containers, blood processing equipment, tractor cab parts, mine lighting guards, and business machine housings.

Total sales of polycarbonates in the United States in 1996 were 804 million lb.

Section 1.7.3. Other Polyesters

Polyarylate polymers are aromatic polyesters. Applications include automotive headlight housings and window trim, printed wiring boards, and housings and enclosures for appliances and business machines.

Liquid crystal polymers (LCP) are aromatic copolyesters which exhibit a highly ordered or liquid crystalline structure in solution and molten states. They are based on compounds such as terephthalic acid, isophthalic acid, p,p'-dihydroxybiphenyl, p-hydrobenzoic acid, and hydroxynaphthoic acid. Applications include electronic components and automotive components.

Section 1.7.4. Polyethers

Polyphenylene ether (PPE), also known as polyphenylene oxide (PPO), is produced by oxidative coupling of substituted phenols. The homopolymer is based on 2,6-dimethyl-phenol, and the copolymer contains 2,6-dimethyl-phenol and 2,3,6-trimethyl-phenol. PPE or PPO polymers are usually available commercially only as blends or alloys with polystyrene.

Section 1.7.5. Polyamides

Polyamides or nylons are prepared by two methods:

1. Condensation polymerization of a diamine with a dibasic acid
2. Polymerization of amino acids

The nylon homopolymers prepared by condensation polymerization of a diamine with a dibasic acid include:

1. Nylon 6/6 or poly(hexamethylene adipamide), from hexamethylene diamine and adipic acid
2. Nylon 6/10, from hexamethylene diamine and sebacic acid
3. Nylon 6/12, from hexamethylene diamine and dodecanedioic acid
4. Nylon 6/9
5. Nylon 4/6

The numbers refer to the number of carbon atoms in the diamine and the number of carbon atoms in the dibasic acid used in the reaction.

The nylon homopolymers prepared by polymerization of amino acids include:

1. Nylon 6 or polycaprolactam, from epsilon-caprolactam

2. Nylon 11, from 11-amino-undecanoic acid
3. Nylon 12, from 12-amino-dodecanoic acid

The number refers to the number of carbon atoms in the starting compound.

These materials are characterized by outstanding resistance to repeated impact and fatigue, low friction coefficient, excellent abrasion resistance, and adequate electrical properties.

Applications include food packaging, sausage casings, cable jacketing, braided tubing for automotive applications, appliance parts, floats, cams and gears, electrical connectors, wire jackets, moving machine parts, and sporting goods.

Total sales of nylons in the United States in 1996 were 1143 million lb.

Section 1.7.6. Polyimides

Thermoplastic polyimides are produced by the polycondensation reaction of an aromatic dianhydride with an aromatic diamine. They offer superior performance at higher temperatures. Applications include bearing materials, missile wire cable, high-temperature structural adhesives, and multilayer printed circuit boards.

Polyamide-imide (PAI) is the condensation polymer of trimellitic anhydride and various aromatic diamines, and is characterized by high strength and good impact resistance. Applications include high-performance electrical/electronic parts, textile equipment, pumps, valves, turbines, compressor and generator parts, and jet engine components.

Polyetherimides are polymers that contain regular repeating ether and imide linkages. The repeating aromatic imide units are connected by aromatic ether units. They are also characterized by high temperature capability, high mechanical strength, and good electrical properties that are stable across a wide range of frequencies and temperatures. Applications include flame-resistant aircraft interiors, heat-resistant automotive parts, commercial and military connectors, and printed circuit boards.

Polybenzimidazoles (PBI) are produced by reactions of compounds such as 3,3'-diaminobenzidine and diphenylisophthalate or isophthalamide. They are designed for high-temperature applications such as glass-reinforced composites and adhesives.

Section 1.7.7. Polyketones

The aromatic polyketones are a family of engineering thermoplastics combining ketone and aromatic moieties.

Polyetherketone (PEK) and polyaryletheretherketone (PEEK) are high-temperature-resistant aromatic ketone-based thermoplastics.

Applications include compressor plates, valve seats, pump impellers, aircraft fairings, interiors, radomes, ducting, wire coating, and semiconductor wafer carriers.

Section 1.7.8. Polyacetals

Polyacetals are produced by polymerization or aldehydes, and exhibit strength, stiffness, creep resistance, toughness, chemical resistance, and electrical properties. Acetal homopolymers, such as polyformaldehyde, are used in business machine cams, gears, and printing wheels, automotive parts, ball cocks and shower heads in plumbing, furniture casters, buckles and clothing fitments, electric clocks, and lawn mower wheels. Acetal copolymers, such as those based on trioxane, are used in gears, cams, housings, fender extensions, valves and valve stems, tub assemblies, and toys.

Total sales of acetals in the United States in 1996 were 345 million lb.

Section 1.7.9. Sulfur-Containing Polymers

Polyphenylene sulfide (PPS) is a crystalline aromatic polymer where repeating benzene rings are para-substituted with sulfur atom linkages. It exhibits high temperature stability, flame resistance, and chemical resistance. Applications include electrical sockets and connectors, component encapsulation, switch and relay components, microwave oven components, and computer disk drives.

Sulfone-based polymers consist of phenylene units linked by different chemical groups: isopropylidene, ether, and sulfone. Polysulfone (PSU) contains aliphatic isopropylidene linkages. Polyethersulfone (PES) and polyphenylsulfone (PPS) are wholly aromatic. They are characterized by thermal stability, oxidation resistance, chemical resistance, and rigidity. Applications include electrical and electronic parts, aircraft interiors, battery cases, food processing equipment, medical instrumentation, and chemical processing equipment.

Section 1.7.10. Fluorine-Containing Polymers

Fluoropolymers are paraffinic polymers which have some or all of the hydrogen replaced by fluorine. They include:

PTFE, polytetrafluoroethylene
PVF, polyvinyl fluoride
PVDF, polyvinylidene fluoride
FEP, fluorinated ethylene propylene copolymer
CTFE, chlorotrifluoroethylene polymer
ETFE, ethylene-tetrafluoroethylene copolymer
PFA, perfluoroalkoxy polymer

Polyetetrafluoroethylene (PTFE) is a completely fluorinated polymer manufactured by polymerization of tetrafluoroethylene. Applications include coatings and linings.

Polyvinyl fluoride (PVF) has been commercially available only as film which has good abrasion and staining resistance.

Polyvinylidene fluoride (PVDF) is produced by polymerization of vinylidene fluoride.

Fluorinated ethylene-propylene (FEP) polymer is produced by copolymerization of tetrafluoroethylene and hexafluoropropylene. Applications include aircraft hookup wire, plenum cable, fire alarm cable, and pipe linings.

Chlorotrifluoroethylene (CTFE) polymer is produced by polymerization of chlorotrifluoroethylene. Applications include hookup wire insulation, flexible cable jacketing, and film.

Ethylene-tetrafluoroethylene (ETFE) copolymer is a predominantly 1:1 alternating copolymer of ethylene and tetrafluoroethylene. Applications include aircraft hookup wire, computer back panels, and components for chemical process equipment.

Ethylene-chlorotrifluoroethylene (ETCFE) copolymer is a predominantly 1:1 alternating copolymer of ethylene and chlorotrifluoroethylene. Applications include plenum cable, fire alarm cable, optical fiber jacketing, and pipe linings.

SECTION 1.8. OTHER THERMOPLASTICS

Cellulosic polymers are produced by chemical modification of cellulose. Cellulose nitrate is prepared by direct nitration of cellulose, and offers dimensional stability, low water absorption, and toughness. Unfortunately, it is flammable and lacks stability to heat and sunlight. Applications include personal accessories, toilet articles, and various industrial items.

Cellulose acetate is produced by reaction of cellulose with acetic acid and acetic anhydride. To produce cellulose acetate butyrate and cellulose acetate propionate, the acetic anhyride is partly replaced with butyric anhydride and propionic anhydride, respectively. Cellulose esters offer the advantageous combination of being both tough and transparent, and are used in film, sheet, and packaging applications.

Ethyl cellulose is produced by reaction of cellulose with caustic to form alkali cellulose, which then reacts with ethyl chloride to form ethyl cellulose. It offers superior toughness at low temperatures. Applications include football helmets, tool handles, electric appliance parts, and flashlight cases.

SECTION 1.9. THERMOSETTING POLYMERS

Polyurethanes are produced by the reaction of polyisocyanates with compounds containing reactive hydrogen atoms. For most commonly used polyurethanes, the compounds containing reactive hydrogen atoms are polyols. Polyurethanes have found commercial use in flexible, semi-flexible, and rigid foams, in coatings, elastomers, and adhesives, and in spandex fibers.

Total sales of polyurethanes in the United States in 1996 were 4256 million lb.

Flexible polyurethane foams use toluene diisocyanate (TDI) in conventional grades and blend diphenylmethane diisocyanate (MDI) with TDI in some high-resiliency grades. They are used in furniture, bedding, and transportation seating.

Rigid polyurethane foams are usually made from polymeric MDI and high functionality polyols. A significant fraction are actually rigid polyisocyanurate foams. They are used in construction, appliances, and transportation, primarily as thermal insulation.

Reaction injection molding (RIM) polyurethane elastomers are used largely in transportation.

Polyurethane coatings offer toughness, hardness, mar-resistance, flexibility, chemical resistance, and resistance to abrasion. Polyurethane elastomers are used in textures for specialty upholstering and luggage, automotive fuel tanks, collapsible fuel storage cells, and wire and cable jacketing. Polyurethane adhesives are used in shoe construction.

Phenolic resins are prepared by the condensation of a phenol and aldehyde, phenol-formaldehyde being the most common combination. They offer low cost and satisfactory all-around properties. Applications include appliance parts, telephone handsets, washing machine agitators, pump impellers, switches, oven parts, resistor casings, particle board, buttons, coatings, adhesives, and plywood bonding.

Amino resins are prepared by the reaction of amines, or compounds bearing amino groups, with aldehydes. Urea-formaldehyde and melamine-formaldehyde resins are the most commonly used. They offer moisture and solvent resistance, electrical properties, heat resistance, and mar resistance. Applications include electrical wiring devices, dinnerware, industrial laminates, coatings, adhesives, plywood, boat hulls, flooring, and furniture.

Epoxy resins are divided into three types:

1. Conventional epoxy, prepared by reacting epichlorohydrin with a polyhydroxy compound such as bisphenol-A
2. Epoxidized novolacs, such as epoxy cresol novolac and epoxy phenol novolac, prepared in a similar manner
3. Cycloaliphatic epoxy, prepared by the peracetic acid epoxidation of cyclic olefins

Epoxy resins offer toughness, high bond strength, and resistance to moisture and chemicals. Applications include adhesives, coatings, waterproofing membranes, and pipe fittings.

Polyesters are produced by the reaction of dihydric alcohols (glycols) and dicarboxylic acids. Polyesters are classified as saturated or unsaturated, depending on the absence or presence of reactive double bonds in the linear polymer. Saturated polyesters, such as ethylene glycol terephthalate, find their greatest use in the production of fibers and film.

Unsaturated polyester resins are condensation polymers which in their simplest forms are reaction products of a glycol, such as propylene glycol, with a dibasic acid, such as fumaric acid. Unsaturated polyesters are used principally with fibrous reinforcement for fabricating a wide variety of articles, fibrous glass being the reinforcement most generally used. By design of the polyester molecule, properties such as fire retardance, corrosion resistance, and electrical properties can be achieved. Applications include cultured marble and onyx, sanitaryware, tub shower units, building facades, roof panels, skylights, awnings, wall sidings, boat hulls, aircraft components, buttons, truck bodies, advertising displays, and structural lay-ups.

Total sales of unsaturated polyester in the United States in 1996 were 1579 million lb.

Alkyds are a class of polyesters which were developed primarily for high performance electronics and electrical systems. Alkyd molding compounds are combinations of liquid polyester resins (mixtures of unsaturated polyester and monomer) and various fillers.

Allyl resins are esters of organic acids, such as the diallyl esters of phthalic, isophthalic, maleic, and chlorendic acids. They offer high temperature stability, chemical and moisture resistance, and good molding performance. Applications include electrical and electronic parts, impregnated glass cloth, decorative laminates, coatings, and sealants.

Bismaleimides are prepared by the condensation reaction of a diamine and maleic anhydride. Applications include printed circuit boards and high-performance structural composites.

Furan resins are prepared by the condensation polymerization of furfuryl alcohol. They offer chemical resistance and high char yield. Applications include chemical-resistant coatings, mortars, and grouts, adhesives, and laminates.

Silicone polymers are polymers with a backbone of alternating atoms of silicon and oxygen, with organic groups attached to the silicon atoms. They are produced by the hydrolysis of organochlorosilanes. They offer thermal stability, surface properties, electrical properties, and inertness.

Applications include mold release agents, defoamers, heat transfer fluids, electrical and electronic components, paper coatings, surfactants, curtain wall sealants, and wire insulation.

SECTION 1.10. BLENDS AND ALLOYS

The terms blends and alloys tend to be used interchangeably in the plastics industry, although there are no doubt various strict definitions for these terms. Blends and alloys are prepared to obtain performance or processing characteristics which can not be obtained economically, or at all, from individual polymers.

Acrylonitrile-butadiene-styrene (ABS) is actually one of the first commercially successful polymer alloys, because it is a mixture of styrene-acrylonitrile (SAN) copolymer with SAN-grafted polybutadiene rubber. The SAN graft provides good adhesion between the otherwise incompatible matrix and rubber phases, but it may make ABS a terpolymer rather than a blend or alloy, depending on definition.

Blends of polycarbonate (PC) with ABS are widely used. Total sales of PC/ABS blends in the United States in 1996 were 160 million lb.

Most blends and alloys are proprietary compositions. Among the widely known are blends or alloys of polyphenylene ether (PPE) or polyphenylene oxide (PPO), usually with polystyrene. Applications include automotive instrument panels and seat backs, business machines, and small appliances.

SECTION 1.11. THERMOPLASTIC ELASTOMERS

A thermoplastic elastomer (TPE) is a hybrid material with the performance properties of a conventional thermoset rubber and the ability to be molded or extruded as if it were a rigid thermoplastic. TPEs bring together the traditionally separate disciplines of rubber and thermoplastics.

The plastics industry has become very familiar with the processing of TPEs into useful rubber-like articles, but the rubber industry is much more familiar with their marketing. The growth rate of TPEs during the 1990s has been three to five times that of either the rubber or plastics industry. The TPE market in North America was approximately 420,000 metric tons in 1995.

The advantages of thermoplastic elastomers over thermoset rubbers include:

1. Lower fabrication costs
2. Shorter processing times
3. Little or no compounding required
4. In-process scrap fully recyclable
5. Lower consumption of energy
6. Processing by means of blow molding, thermoforming, and heat welding
7. Lower costs for quality control
8. High-speed, automated fabrication and assembly
9. Lower unit costs (more parts per pound) due to lower density

The disadvantages of thermoplastic elastomers relative to thermoset rubber include:

1. Unfamiliarity of many rubber processors with technology and equipment
2. Drying normally required before processing
3. Volume production needed in order to economically replace existing rubber parts

The chemistry and morphology of TPEs are the basis for the unusual and useful properties and for their generic classification. A TPE consists of two (or more) intermingled polymer systems, each of which has its own phase and softening temperature (T_s). These systems consist of a hard thermoplastic phase and a soft elastomeric phase. The useful temperature range of a TPE is in the region where the soft phase is above and the hard phase below their respective T_s.

There are six generic classes of TPEs:

A. BLOCK COPOLYMERS
    1. Styrene block copolymer
    2. Copolyester (COP)
    3. Thermoplastic polyurethane (TPU)
    4. Polyether block polyamide or polyester block polyamide (PEBA)
B. ELASTOMER/THERMOPLASTIC COMPOSITIONS
    5. Thermoplastic elastomeric olefin (TEO)
    6. Thermoplastic vulcanizate (TPV)

A styrene block copolymer, or styrenic TPE, is a block copolymer of styrene joined with either a diene such as butadiene or isoprene or with an olefin pair such as ethylene-propylene or ethylene-butylene. The hard thermoplastic phase is an aggregate of polystyrene blocks, and the soft elastomer phase is an aggregate of rubberlike polydiene or olefin pair blocks.

A copolyester (COP) is a block copolymer composed of alternating hard and soft segments, polyalkylene terephthalate and polyalkylene ether, respectively.

A thermoplastic polyurethane (TPU) is a block copolymer with hard and soft blocks, analogous to the styrenics and copolyesters. They are prepared by the melt polymerization of a low molecular weight glycol with a diisocyanate and a macroglycol. The soft segments consist of either a polyester or polyether macroglycol. The hard segments result from the reaction of the low molecular weight glycol with diisocyanate, which binds them to the soft segments.

A polyether block polyamide or polyester block polyamide (PEBA) is a block copolymer in which the hard segment is a polyamide and the soft segment is either an aliphatic polyether or aliphatic polyester.

A thermoplastic elastomeric olefin (TEO) is a simple blend of two polymer systems, each with its own phase. The harder polymer system is a polyolefin, commonly polypropylene (PP) or polyvinyl chloride (PVC). The softer polymer system is an elastomer, ethylene-propylene rubber with PP or nitrile-butadiene with PVC, with little or no crosslinking. The continuous phase is normally that of the polymer present in greatest amount, with the thermoplastic polyolefin having the greater tendency to be continuous.

A thermoplastic vulcanizate (TPV) is a simple blend of two polymer systems, each with its own phase. The harder polymer system is a polyolefin, commonly polypropylene (PP) or polyvinyl chloride (PVC). The softer polymer system is an elastomer, ethylene-propylene rubber with PP or nitrile-butadiene with PVC, highly crosslinked and finely divided in a continuous matrix of polyolefin.

The thermoplastic elastomers are listed in order of increasing level of performance, from low to high:

1. Styrene block copolymer
2. Thermoplastic elastomeric olefin (TEO)
3. Thermoplastic vulcanizate (TPV)
4. Thermoplastic polyurethane (TPU)
5. Copolyester (COP)

The thermoset rubbers are listed in order of increasing level of performance, from low to high:

1. Styrene-butadiene rubber (SBR)
2. Natural rubber
3. Butyl
4. EPDM
5. Polychloroprene
6. Chlorosulfonated polyethylene
7. Nitrile
8. Epichlorohydrin
9. Acrylate
10. Fluoroelastomer
11. Silicone

Pneumatic tires, which consume approximately 54% of the rubber used in the world, are not a suitable market for TPEs. The remaining 46% of the rubber market offers a massive opportunity for penetration by TPEs, primarily thermoset rubber replacement in non-tire applications.

TPEs are used in automotive applications such as bumpers, rubstrips, dashboard trim, and under-the-hood air ducts, building applications such as expansion joints, window glazing, plumbing, and door and window seals, electrical applications such as primary insulation and jacketing materials, and assembled devices such as household appliances and business machines.

SECTION 1.12. CONDUCTIVE PLASTICS

One of the traditional advantages of plastics over metals has been the much lower electrical conductivity of plastics, which made them outstanding as electrical insulation. Electrically conductive plastics have been developed to satisfy the demands of many applications.

Static electricity is electrical energy at rest on a surface. It is generally created by the rubbing together and separating of two surfaces, one of which is usually non-conductive. This generation of static electricity by friction is called triboelectrification.

Electrostatic discharge (ESD) is the sudden discharge of electrostatic potential from one body to another. Perhaps the most common example is the shock received when touching a metal door after walking across a carpeted floor.

ESD can cause serious problems in many environments. It can damage or destroy sensitive electronic components, erase or alter magnetic media, and set off explosions or fires in flammable environments.

ESD can be controlled with materials such as conductive plastics, which do not generate high levels of charge, dissipate charges before they accumulate to dangerous levels, or provide electrostatic shielding that prevents charges from reaching the sensitive product.

Electromagnetic interference (EMI) is electrical energy, either electromagnetic or in the radio frequency (RF) range, that is radiated by computer circuits, radio transmitters, fluorescent lamps, color TV oscillators, electric motors, automotive ignition coils, overhead power lines, lightning, TV games, and many other sources. Radio frequency interference (FRI) is interference in the radio frequency range.

One of the most familiar examples of EMI/RFI is the unwanted operation of garage door openers.

EMI/RFI can cause serious problems in many environments. It can corrupt data in large-scale computer systems, cause inaccurate readings and output in aircraft guidance systems, and interrupt the function of medical devices such as pacemakers.

The major weapon against EMI/RFI is shielding of electronic circuits. Proper shielding can prevent products from emitting electromagnetic or radio frequency energy to susceptible equipment, and can protect susceptible equipment from the effects of externally radiated EMI/RFI.

An electrically conductive thermoplastic material is a resin that has been modified with electrically conductive additives including powdered carbon black, carbon fibers, metal fibers, metal-coated carbon fibers, and metal powders. These materials can be used in a variety of applications, particularly in ESD control and EMI/RFI protection.

The advantages of conductive thermoplastics over other materials such as metals for ESD protection or EMI/RFI shielding include:

1. Usually easier and less expensive fabrication of finished parts
2. Finished parts that are lighter in weight, easier to handle, and less costly to ship
3. Finished parts less subject to denting, chipping, and scratching and thus often have more consistent electrical performance
4. Colors inherent in the material rather than added as a secondary operation

Conductive thermoplastics fall into three groups:

1. ESD protective materials
2. EMI/RFI protective materials
3. Electrically conducting materials

ESD protective materials include:

1. Plastics compounded with chemical antistats such as fatty amine derivatives, which require at least 15% relative humidity for their performance. They are not permanent and are used mainly in short-term packaging such as short-term packaging of printed circuit boards and semiconductors.
2. Plastics compounded with permanent conductive additives

EMI/RFI protective materials are used primarily for shielding against emission or reception of EMI and RFI. Traditionally, shielding has been accomplished by encasing sensitive electronic parts in metal housings or by using metallic coatings on the inside of plastic housings. Thermoplastic compounds with appropriate shielding additives are cost effective alternatives in many applications because of their ability to take on complex shapes and maintain tight tolerances. They are most cost effective when used in small housings that demand close tolerances and high shielding effectiveness.

Electrically conducting materials can be used as conducting elements in electrical circuits when current loading is low and when the alternative would be an expensive, complex, hard-to-fabricate metal part.

REFERENCES

Advanced Elastomer Systems, Akron, Ohio, product literature received September 1997

Amoco Polymers, Alpharetta, Georgia, product literature received September 1997

BASF Corporation, Mount Olive, New Jersey, product literature received October 1997

Bayer Corporation, Pittsburgh, Pennsylvania, product literature received September 1997

B.F. Goodrich Specialty Chemicals, Cleveland, Ohio, product literature received September 1997

Cyro Industries, Rockaway, New Jersey, product literature received October 1997

Dow Chemical Company, Midland, Michigan, product literature received September 1997

General Electric Plastics, Pittsfield, Massachusetts, product literature received September and October 1997

Hilado, C. J., "Flammability Handbook for Plastics", 4th Ed., Technomic Publishing Co., Lancaster, Pennsylvania (1990)

Hitech Polymers, Hebron, Kentucky, product literature received October 1997

Hoechst Aktiengesellschaft, Frankfurt, Germany, product literature received September 1997

Japan Synthetic Rubber Company, Tokyo, Japan, product literature received October 1997

Millenium Petrochemicals, Cincinnati, Ohio, product literature received October 1997

Modern Plastics (January 1997)

Modern Plastics Encyclopedia '96, McGraw-Hill, New York (1996)

Nova Chemicals, Calgary, Alberta, Canada, product literature received October 1997

Plastics Engineering Company, Sheboygan, Wisconsin, product literature received October 1997

RTP Company, Winona, Minnesota, product literature received September 1997

Unichem Products, Ridgefield, New Jersey, product literature received September 1997

Table 1.1. Production of Polymers in the United States.

Material	Production in 1988, Billion lb
Plastics	46.15
Synthetic fibers	9.10
Synthetic rubber	5.13

Table 1.2. Estimated Plastics Consumption by Region and Country.

Region	1000 Metric Tons	
	1994	2000
North America	35,840	43,690
United States	32,070	39,000
Canada	2,580	3,190
Mexico	1,190	1,500
South America	3,800	4,950
Argentina	460	640
Brazil	1,760	2,350
Other South America	1,580	1,960
Asia and Oceania	27,250	36,900
Australia	1,100	1,250
China	2,750	4,500
Japan	12,550	15,200
South Korea	3,920	5,780
Taiwan	3,450	4,800
Other Asia/Oceania	3,480	5,370
Western Europe	31,850	38,600
France	4,200	5,030
Germany	9,100	10,800
Italy	4,600	5,500
Netherlands	985	1,200
Spain	2,290	2,950
United Kingdom	3,550	4,390
Other Western Europe	7,125	8,730
Eastern Europe	7,240	8,420
Former Czechoslovakia	910	1,080
Hungary	390	510
Poland	680	870
Romania	300	360
Former Soviet Union	4,400	5,000
Other Eastern Europe	560	600
Africa and Mideast	1,900	2,700
World plastics consumption	107,880	135,260

Reference: Modern Plastics Encyclopedia, 1996.

24

Table 1.3. Estimated Plastics Production by Region.

	1000 Tonnes	
Region	**1994**	**2000**
All Resins		
North America	37,315	46,860
South America	3,830	5,650
Asia and Oceania	28,620	38,200
Western Europe	31,065	36,820
Eastern Europe	6,575	7,620
Africa and Mideast	2,100	2,840
World total	109,505	137,990
LDPE/LLDPE Resins		
North America	7,120	8,330
South America	1,130	1,750
Asia and Oceania	4,500	6,300
Western Europe	6,195	7,300
Eastern Europe	1,390	1,600
Africa and Mideast	700	885
World total	21,035	26,165
HDPE Resins		
North America	5,840	7,350
South America	470	650
Asia and Oceania	3,050	4,100
Western Europe	2,730	3,300
Eastern Europe	290	360
Africa and Mideast	270	375
World total	12,650	16,135
PP Resins		
North America	4,765	6,700
South America	670	1,050
Asia and Oceania	4,650	6,700
Western Europe	4,500	5,650
Eastern Europe	575	720
Africa and Mideast	200	330
World total	15,360	21,150
PVC Resins		
North America	5,650	7,030
South America	840	1,260
Asia and Oceania	5,750	7,320
Western Europe	5,170	6,070
Eastern Europe	1,250	1,450
Africa and Mideast	450	620
World total	19,110	23,750

continued

Table 1.3 (continued). Estimated Plastics Production by Region.

Region	1000 Tonnes	
	1994	2000
PS Resins		
North America	2,840	3,350
South America	310	430
Asia and Oceania	3,250	4,180
Western Europe	2,620	3,100
Eastern Europe	520	650
Africa and Mideast	160	190
World total	9,700	11,900
Other Thermoplastic Resins		
North America	5,600	7,400
South America	120	150
Asia and Oceania	4,300	5,400
Western Europe	4,720	5,500
Eastern Europe	540	650
Africa and Mideast	45	70
World total	15,325	19,170
Thermoset Resins		
North America	5,500	6,700
South America	290	360
Asia and Oceania	3,120	4,205
Western Europe	5,130	5,900
Eastern Europe	2,010	2,190
Africa and Mideast	275	365
World total	16,325	19,720

Reference: Modern Plastics Encyclopedia, 1996.

Table 1.4. Some Major Markets for Plastics

United States	Million lb.	
	1995	1996
LDPE and copolymer	9,651	11,141
LLDPE	10,830	11,274
HDPE	11,920	13,396
Polypropylene	10,644	12,183
Polyvinyl chloride	12,032	13,388
Polystyrene	5,799	6,059
ABS	1,463	1,427
Polyurethane	4,030	4,256
Polycarbonate	753	804
PC/ABS blends	149	160
Acrylic	546	556
Nylons	1,040	1,143
Acetals	330	345
Polyethylene terephthalate (PET)	2,967	3,397
Polybutylene terephthalate (PBT)	272	290
Polyester, unsaturated	1,569	1,579

Canada	Million lb.	
	1995	1996
LDPE	1,408	1,256
HDPE	759	693
Polypropylene	790	825
Polyvinyl chloride	1,005	1,186

Western Europe	Million lb.	
	1995	1996
LDPE	9,944	9,953
LLDPE	3,137	3,394
HDPE	7,847	8,159
Polypropylene	10,978	11,715
Polystyrene	4,120	4,199
Expandable polystyrene	1,445	1,399
Acrylonitrile-butadiene-styrene	1,221	1,162
Vinyls	11,762	11,945
Nylons	1,000	1,020
Polycarbonate	448	459
Polycarbonate/ABS blends	136	145
Acrylic	548	519
Acetals	275	273
Polyethylene terephthalate (PET)	1,938	2,103
Polybutylene terephthalate (PBT)	157	169
Polyester, unsaturated	948	923

continued

Table 1.4 (continued). Some Major Markets for Plastics

Japan	Million lb.	
	1995	1996
LDPE and LLDPE	3,274	3,285
HDPE	2,251	2,265
Polypropylene	4,893	4,904
Polystyrene	2,258	2,276
Acrylonitrile-butadiene-styrene (ABS)	990	983
Polyvinyl chloride (PVC)	4,073	4,079
Polycarbonate	256	269
Nylon 6	223	229
Nylon 66	161	174
Nylon 11 and 12	20	20
Acrylic	353	357
Polyacetal	179	181
Polyethylene terephthalate (PET)	957	1,025
Polybutylene terephthalate (PBT)	112	115
Polyester, unsaturated	525	527
Reinforced polyester	1,008	1,021

Reference: Modern Plastics, January 1997.

Table 1.5. Olefin Polymer Capacities for Some Countries.

Country	Million lb., 1996			
	LDPE	LLDPE	HDPE	PP
United States	11,141	11,274	13,396	12,183
Canada	3,113	—	1,639	699
Mexico	825	—	418	462
Argentina	286	264	132	374
Brazil	1,496	1,122	968	1,716
Chile	88	—	—	165
Colombia	121	132	66	396
Venezuela	264	660	220	165
Australia	297	198	352	572
China	2,266	2,145	1,309	1,958
India	416	1,078	638	770
Indonesia	—	990	220	924
Japan	2,278	1,519	2,736	5,506
Malaysia	—	528	—	264
Singapore	319	—	352	440
South Korea	1,397	1,628	2,046	3,740
Taiwan	528	264	440	1.034
Thailand	550	132	880	792

Reference: Modern Plastics, January 1997.

CHAPTER 2

Decomposition, Combustion, and Propagation

The entire process of thermal decomposition, combustion, and fire propagation is perhaps best described by considering it on three scales: micro, macro, and mass. The process can be described on the micro scale by discussing the behavior of the polymer molecule; on the macro scale by discussing the behavior of a unit mass of material, such as one gram; and on the mass scale by discussing the behavior of a complete system such as a room or structure.

SECTION 2.1. THE BURNING PROCESS ON A MICRO SCALE

The behavior of the polymer molecule with increasing temperature can be divided into five stages.

Stage I. <u>Heating</u>. Heat from an external source is applied to the polymer, progressively raising its temperature. There is relatively little change in physical properties.

Stage II. <u>Transition</u>. Within a relatively narrow temperature range, called the glass transition temperature, the polymer changes from a relatively hard and brittle material to a viscous or rubbery condition. The mechanical properties, and some thermal properties, change rapidly above this temperature. In general, the load-bearing capabilities of the material decrease above this temperature (Table 2.1).

The transition stage would tend to be more difficult to define for blends and alloys of polymers than for individual polymers. One purpose of blends and alloys is to provide some control over the transition stage.

Stage III. <u>Degradation</u>. The polymer chain is no stronger than its weakest link, and the temperature of initial degradation is usually the temperature at which the least thermally stable bonds fail. The bulk of the polymer may be stable, but the failure of the weakest bonds often produces results such as discoloration. Degradation may be of two types: thermal-nonoxidative degradation, in the absence of oxygen, and thermal-oxidative degradation, influenced by heat and oxygen.

29

The characteristics of the polymer which are important in this stage are:

1. Decomposition temperature of the least thermally stable bonds
2. Proportion of the least stable bonds in the polymer
3. Latent heat of decomposition of the least stable bonds

Endothermic decomposition absorbs heat and slows the temperature rise, while exothermic decomposition supplies heat and accelerates the temperature rise.

Stage IV. Decomposition. The majority of the bonds existing in the polymer reach the failure point, and the polymer mass itself changes. The polymer may exhibit behavior ranging from disintegration, with complete loss of physical integrity, to rearrangement to give different properties, with little weight loss. An example of the former is the complete depolymerization of some polymers, such as polyformaldehyde and poly(methyl methacrylate), to yield the original monomers, in these cases formaldehyde and methyl methacrylate, respectively. An example of the latter is the thermal modification of polyacrylonitrile to yield "black orlon".

The degradation and decomposition stages can be separated only when the failure temperature of the least stable bonds is significantly lower than the decomposition temperature of the majority of the bonds in the polymer. When a polymer contains a variety of bonds with an almost continuous spectrum of decomposition temperatures, the degradation and decomposition stages tend to merge into a single stage.

The decomposition products vary with polymer composition, temperature level, rate of temperature rise, endotherms and exotherms, and rate of volatiles evolution. Differential thermal analysis (DTA), thermogravimetric analysis (TGA), and time-of-flight mass spectrometer analysis indicate that relatively small changes in chemical composition, temperature, and heating rate can drastically change the nature of the decomposition products. It is obvious that actual fires involve varying and changing temperatures well above the normal decomposition temperature of many polymers. The products obtained in careful laboratory decompositions will therefore not necessarily be the same as those produced at actual fire temperatures and under actual fire conditions. Even laboratory studies cover a range of decomposition temperatures, rather than a specific level considered to be the decomposition temperature (Table 2.2).

The decomposition of the polymer produces two types of materials: the polymer chain residue, which continues to provide some structural integrity, if such a residue is obtained, and polymer fragments which are highly vulnerable to oxidation. Excess of the heated carbonaceous residue to oxygen can result in glowing of the char, but flaming combustion generally occurs in the gas phase near the polymer residue, and involves gaseous and finely divided solid material.

The decomposition stage is significantly affected by the following characteristics of the polymer:

1. Decomposition temperature of the various bonds comprising the bulk of the polymer.
2. Latent heat of decomposition of the various bonds. If the decomposition is on the average endothermic, heat must be supplied from other sources to continue decomposition. If the decomposition is on the average exothermic, the decomposition can maintain itself.
3. Decomposition behavior, or the manner in which the polymer decomposes. This involves the relative amounts of combustible and noncombustible fragments.

Stage V. Oxidation. At a high enough temperature and in the presence of sufficient oxygen, the oxidation of the polymer fragments proceeds rapidly enough to produce heat and flame in the gas phase, and perhaps glowing in the solid residue.

SECTION 2.2. THE BURNING PROCESS ON A MACRO SCALE

Consideration of the burning process on the macro scale differs from that on the micro scale in that the basic polymer alone is considered on the micro scale, while the plastic material, including fillers or blowing agents, is considered on the macro scale.

The burning of an individual unit mass of material, such as one gram, can be considered as occurring in five stages:

Stage I. Heating. Heat from an external source is applied to the material, progressively raising its temperature. The external supply of heat may come from direct exposure to flame (radiation and convection) in the case of material at an exposed surface, from heat transfer from hot fire gases (conduction and convection), or from a hot solid mass (conduction) in the case of an adhesive behind exposed material or a plastic covered by a protective coating or covering.

The rate of temperature rise is a function of the flow rate of applied heat, of the temperature differential, and of the following characteristics of the material:

1. Specific heat, or the amount of heat required to raise the temperature of the unit mass by one unit of temperature measurement. Materials with high specific heats increase more slowly in temperature than materials with low specific heats (Table 2.3).
2. Thermal conductivity, or the rate at which heat flows through a given thickness of material under a given temperature differential. A high thermal conductivity means that heat is transferred to an adjacent unit mass more rapidly than would be the case with a low thermal conductivity (Table 2.4).
3. Latent heat of fusion (melting), vaporization, or other changes which may occur in the material during the heating process.

Stage II. Decomposition. The material reaches its decomposition temperature, and begins to evolve one or more of the following types of products:

1. Combustible gases, or gases which will burn in the presence of air. This group would include methane, ethane, ethylene, formaldehyde, acetone, and carbon monoxide.
2. Noncombustible gases, or gases which normally would not burn in the presence of air. This group would include carbon dioxide, hydrogen chloride, hydrogen bromide, and water vapor.
3. Liquids, usually partially decomposed polymer and high-molecular-weight organic compounds
4. Solids, usually carbonaceous residue, char, or ash
5. Entrained solid particles, or polymer fragments, which appear as smoke

Since in most cases ignition and combustion occur in the gas phase, it is obvious that complete elimination of combustible gases would effectively preclude burning. In most cases, unfortunately, this is impossible, because most organic materials cannot be reduced to a highly carbonaceous residue without releasing some volatile compounds containing hydrogen.

Noncombustible gases are of course preferred to combustible gases from the viewpoint of burning. The evolution of any gases, however, results in expansion of the system, into adjacent masses or into the adjacent atmosphere, or both, and by disrupting the chemical and physical structure of the material can increase the degree of exposure to destructive temperatures. Some gases also present the additional disadvantages of irritation or toxicity.

Liquids may not be as combustible as gases because their latent heat of vaporization is needed to transfer them to the gas phase. However, they present possible hazards in spreading the unit mass so that it affects other unit masses other than those originally adjacent to it, and they have the potential for being converted into gases through a temperature rise and change in state (vaporization).

The solid residue is the most desirable product of decomposition, because it helps preserve structural integrity and protect adjacent unit masses from decomposition, and impedes the mixing of air with combustible gases. The extent of weight loss on decomposition is therefore rightly used as one indication of resistance to fire.

Decomposition does physical violence to the polymer in varying degrees. Fragments of the polymer can be broken off from the mass and entrained in moving gases. These fragments appear as smoke particles, and, if swept into the flame itself, as incandescence.

The decomposition stage is significantly affected by the following characteristics of the material:

1. Temperature of initial decomposition, or the lowest temperature at which decomposition occurs. If the temperature of the unit mass never reaches this level, decomposition does not occur and the burning process is terminated before ignition can occur. Where two polymers of equal specific heat and equal thermal

conductivity are exposed to heat at a surface, the extent to which decomposition proceeds into the mass is largely a function of the initial decomposition temperature.

2. Latent heat of decomposition, or the heat absorbed or released during decomposition. If decomposition is exothermic, the heat released increases the rate of temperature rise. If decomposition is endothermic, the heat absorbed is removed from the available supply, slowing the heating process.

3. Decomposition behavior, or the manner in which the material decomposes. This involves the relative amounts of combustible and noncombustible gases, liquids, solid residue, and solid particles, as well as the sequence of the phase changes.

Stage III. Ignition. The combustible gases ignite in the presence of sufficient oxygen or oxidizing agent, and combustion begins. Ignition depends on the presence of an external source of ignition, such as flames or sparks, and the temperature and composition of the combined gas phase.

The ignition stage is significantly affected by the following characteristics of the material:

1. Flash ignition temperature, or the temperature at which the gases evolved from the material can be ignited by a spark or flame. This is generally higher than the initial decomposition temperature (otherwise there would be no gases to ignite). Flash-ignition temperatures of various polymers are shown in Table 2.5.

2. Self-ignition or autoignition temperature, or the temperature at which reactions within the material become self-sustaining the point of ignition. This is generally higher than the flash ignition temperature, because more energy is required to initiate self-sustaining decomposition than externally-sustained decomposition. One notable exception is cellulose nitrate. Autoignition temperatures of various polymers are shown in Table 2.5.

3. Limiting oxygen concentration, or the minimum level of oxygen required to sustain ignition and combustion. For normal use, a material can be considered self-extinguishing if it cannot continue burning with less than 21 per cent oxygen, and non-igniting if it cannot be ignited with less than 21 per cent oxygen. Oxygen index for various polymers are shown in Table 2.6. The conditions under which a material appears "non-igniting" or "self-extinguishing," however, vary with the severity of the exposure.

Increasing the oxygen concentration in the gas phase increases the probability of ignition. Autoignition temperature generally decreases with increasing oxygen concentration. Autoignition temperature at elevated oxygen pressure are shown in Table 2.7.

Stage IV. Combustion. Combustion of the unit mass generates a certain amount of combustion heat (see Table 2.8). The heat of combustion raises the temperature of the gaseous products of combustion and of the noncombustible gases, which increases heat transfer by conduction; expansion of the heated gases increases heat transfer by convection. Heating the entrained solid particles to incandescence increases heat transfer by radiation, and heating the solid residue increases heat transfer by conduction.

On the small scale of the unit mass, this stage represents full-scale or fully developed burning. Once the burning process has reached this stage, extinguishment, rather than inhibition, becomes critical. The most important characteristic of the material in this stage is heat of combustion, the amount of heat released by the combustion of a unit mass. The net heat of combustion is the heat released by the combustion reactions, reduced by the amount of heat required to bring the unit mass from its initial stage to the combustion stage. If the net heat of combustion is negative, then an external heat supply is required to continue combustion. If the net heat of combustion is positive, then the unit mass is making available an excess of heat to increase the exposure of an adjacent unit mass.

Stage V. Propagation. For propagation to occur, the net heat of combustion of a unit mass, decreased by the heat lost to the surroundings and increased by heat supplied from external sources such as an adjacent fire, must be sufficient to bring an adjacent unit mass to the combustion stage. Where the initial unit mass is at an exposed surface, an adjacent unit mass at the surface is more readily brought to the combustion stage, because the material at the surface is exposed to the external fire source, while the material in the interior is shielded from external heat by the solid residue of the initial unit mass, and in addition dissipates heat to material deeper into the interior. For this reason, propagation is often treated as a surface phenomenon.

For plastic materials which are applied on or as an exposed surface over substantial areas, surface flame spread is a realistic measure of propagation.

Where the plastic material is not continuously supplied for significant distances, or where it is not applied on or as an exposed surface, surface flame spread becomes less realistic a measure of fire propagation than other flammability characteristics such as heat contribution and combustion products of the same material, and ease of ignition of other materials in the vicinity.

SECTION 2.3. THE BURNING PROCESS ON A MASS SCALE

The burning process in a system, such as a room containing articles associated with typical occupancy, can be considered as occurring in five stages, analogous, but on a large scale, to the stages described for the unit mass.

    I. Initial fire
    II. Fire build-up
    III. Flashover
     IV. Fully developed fire
     V. Fire propagation

Stage I. <u>Initial Fire</u>. The original fire source in a system such as a room can vary widely. It may be a burning match, a cigarette dropped on a fabric, an electrical fire from overheated wiring, or a burning door or window ignited by fire in an adjacent room or structure. In almost all cases, the plastic material in the system is not the source of the initial fire. In many cases, however, the plastic material in the system may be the first item to be ignited by the original fire source.

In the initial fire stage, the following properties of the plastic material are relevant:

1. Decomposition temperature and behavior. This characteristic is particularly important if the plastic material is relied upon for structural capabilities (see Table 2.2).
2. Ease of ignition. This characteristic is measured by properties such as flash ignition temperature (if the material is exposed to direct flame), self-ignition temperature (if the material is exposed to heat without flame), and limiting oxygen concentration (if the material is exposed to air mixed with combustion gases from the initial fire).
3. Extent of exposure. Plastic materials presenting an exposed surface are more vulnerable to fire than plastic materials shielded by a coating or covering. Plastics near the ceiling, where the heated gases tend to accumulate, and more vulnerable than plastics near the floor.
4. Extent of involvement. Plastics present in large quantities in the system, such as wall and ceiling material, are more significant than those present in relatively small amounts, such as buttons and utensil handles.

Stage II. <u>Fire Build-Up</u>. The heat from the initial fire accumulates in the system, raising the temperature of the material in the system by conduction, convection, and radiation. A limited amount of fire spread may occur in this stage. In the fire build-up stage, the following properties of the plastic materials involved are relevant:

1. Ease of ignition. Easily ignited materials can contribute to fire build-up.
2. Surface flammability. This is an important characteristic if the material is exposed over significantly large surface areas, such as interior finishes and wall coverings, but may be irrelevant where the material is present at widely separated locations, or in small units.
3. Heat contribution. If the material is ignited, the generation of large amounts of heat during combustion will contribute to fire build-up.
4. Smoke production. The evolution of dense smoke hinders the escape of occupants and the location of the fire for extinguishing efforts.
5. Extent of exposure. This is relevant only if the material is exposed to fire in this stage.
6. Extent of involvement. If the material is present in a small enough quantity, it can not make a significant contribution to fire build-up.
7. Fire gases. Highly toxic combustion gases can present a hazard to the occupants.

Stage III. <u>Flashover</u>. The point of flashover is the point at which most of the combustible materials in the system reach their ignition temperatures at essentially the same time, so that the combustible materials seem to burst into flame almost simultaneously. The following properties of the plastic material are relevant in this stage:

1. Ease of ignition. The temperatures present in the system in the flashover stage are probably higher than the ignition temperatures of most plastic materials, so that relative ease of ignition is academic except in the case of high-temperature polymers.
2. Surface flammability. Although much of the heat transfer leading to flashover occurs through radiation, surface flammability can contribute significantly to flashover.
3. Heat contribution. The intensity of flashover is influenced by the heat contribution of the materials in the system.
4. Extent of exposure. The intensity of flashover is influenced by the amount of material exposed.
5. Extent of involvement. The intensity of flashover is influenced by the amount of material involved.

Stage IV. Fully Developed Fire. Essentially all of the combustible materials in the system contribute to the fire in this stage, and fire damage is essentially total for these materials. Once the burning process has reached this stage, containment, rather than extinguishment, becomes critical. The most important factor is the total heat contribution of the materials involved. In the fully developed fire stage, the following properties of the plastic materials involved are relevant:

1. Extent of involvement. The maximum heat contribution is a function of the unit heat of combustion and the total amount of material. The actual heat contribution is a function of the unit heat of combustion and the actual amount of material involved.
2. Heat contribution. The intensity of the fully developed fire is influenced by the heat contribution of the materials in the system.
3. Smoke production
4. Fire gases

Surface flammability is irrelevant in this stage, because fire has reached all combustible surfaces. Extent of exposure is essentially total.

Stage V. Fire Propagation. The total heat contribution of the combustible materials in a system, reduced by the heat required to bring the system to the fully developed fire stage, can provide the initial fire for adjacent systems if the boundaries of the system fail to contain the fire. For this reason, fire endurance is an important characteristic of the system boundaries. A "fire wall" is exactly what the term implies: a wall placed to serve as a barrier to fire propagation.

In the fire propagation stage, the following properties of plastic materials are relevant:

1. Fire endurance. If and when the boundaries of the system fail to contain the fire determines if and when the fire propagates to adjacent systems.
2. Extent of involvement. The rapidity of fire propagation is influenced by the amount of material involved in the original system.
3. Heat contribution. The rapidity of fire propagation is influenced by the heat contribution of the material involved in the original system.
4. Smoke production
5. Fire gases

SECTION 2.4. COMBUSTION AND PEOPLE

Some results of the burning process are obvious threats to human life, wherever occupants are present within the burning system and in adjacent systems, but their relative importance can vary with the conditions of each individual fire. Because there can be such wide variety in types of plastic materials and in their extent of exposure and involvement, the hazard of each situation must be judged on an individual basis.

The results of the burning process which can threaten human life can be summarized as the following: (1) oxygen depletion, (2) flame, (3) heat, (4) toxic gases, (5) smoke, and (6) structural strength reduction.

1. <u>Oxygen Depletion</u>. The average human being is accustomed to operating satisfactorily with the usual level of about 21 per cent oxygen in the atmosphere. When the oxygen level falls to 17 per cent, muscular skill is diminished because of a phenomenon called anoxia. At oxygen levels of 10 to 14 per cent, a man is still conscious but may exhibit faulty judgement, not obvious to himself. At oxygen levels of 6 to 8 per cent, breathing ceases, and death by asphyxiation occurs in 6 to 8 minutes. These results are summarized in Table 2.9. The excitement and exercise occasioned by a fire tend to increase the oxygen demands of the body, and oxygen deficiency systems may appear at higher oxygen levels than would be the case without excitement or exercise.

An oxygen concentration of 10 per cent is considered the minimum level for survival. Whether this level is reached, and how rapidly it is reached, varies with each fire, and with location within the system, because it is affected by the concentration of combustibles, rate of burning, volume of the system, and rate of ventilation.

2. <u>Flame</u>. Burns can be caused by direct contact with flames or by heat radiated from flames. Burns can result if skin temperature is held above 150 deg. F. for one second. Flame temperatures and their radiant heat may prove immediately or eventually fatal.

3. <u>Heat</u>. Unlike direct flame, heat can be a hazard to occupants of adjacent systems in addition to occupants of the burning system. Hot air and gases from a fire, ignoring any effects of oxygen depletion or toxicity, can cause burns, heat exhaustion, dehydration, and blockage of the respiratory tract (edema).

A breathing level temperature of 300 deg. F. is considered to be the maximum value for survival. Breathing level, the distance above the floor, is considered to be about 5 feet, although the 4 foot level is sometimes used when children comprise a significant fraction of the occupants. Temperatures above 150 deg. F. are considered untenable, and temperatures in this range can hold back firefighters and keep occupants from entering passages leading to exits.

4. Fire Gases. The toxicity of some gaseous products of combustion is well known, but the concentration of these gases in an actual fire is not well known, even from simulated conditions in laboratory experiments. What gaseous products are formed is determined by the chemical composition of the material, the amount of available oxygen, the temperature, and the rate of temperature change.

The hazard of toxic fire gases is often psychological and sometimes overemphasized. Fire fatalities from the inhalation of hot air and gases are a significant proportion of fire deaths. Experimental fires in typical structures have shown that in many cases the minimum concentration of oxygen for survival, or the maximum breathing level temperature for survival, was reached before any toxic gases attained a lethal concentration.

Because plastic materials are based on carbon-containing polymers, the gaseous products of combustion include carbon dioxide and carbon monoxide. Carbon dioxide is not generally considered toxic, but it can present a hazard in two ways. First, it represents oxygen depletion. In the complete absence of ventilation, a certain volume of carbon dioxide represents removal of an equal volume of oxygen. Second, it overstimulates the respiratory system, causing an abnormally high intake of other gases, resulting in toxic or lethal effects which would not otherwise have been attained. A carbon dioxide concentration of 1.8 per cent is said to increase speed and depth of breathing 50 per cent, and a 2.5 per cent concentration is said to increase breathing rate 100 per cent. The response of humans to carbon dioxide is summarized in Table 2.10.

Carbon monoxide ranks first as a cause of fire deaths because it is always formed by an uncontrollable accidental fire, in which oxygen concentrations and availability are novel ideals for complete oxidation to carbon dioxide. The maximum concentration of carbon monoxide for survival is considered to be 1.28 per cent. At this or higher levels, a person will become unconscious after two or three breaths and probably die in one to three minutes. The response of humans to various levels of carbon monoxide is summarized in Table 2.11. Variables such as physical condition and amount of exertion affect the amount of carbon monoxide which can be tolerated.

Carbon monoxide combines with the hemoglobin in the blood to form carboxyhemoglobin, displacing oxygen in the blood and leading to anoxia and death. The response of humans to carboxyhemoglobin is summarized in Table 2.12. The response of animals to various levels of carbon oxides is summarized in Table 2.13.

The gaseous products of combustion from polymers containing carbon, hydrogen, and oxygen may include, in addition to carbon dioxide, carbon monoxide, and water, other gases which contain carbon, hydrogen, and oxygen. Gases which contain carbon and hydrogen may include butane and other aliphatic hydrocarbons, and benzene and other aromatic hydrocarbons. Gases which contain carbon, hydrogen, and oxygen may include acetone and other ketones, formaldehyde and other aldehydes, acetic acid and other acids, and ethyl acetate and other esters.

Many polymers contain other elements in addition to carbon, hydrogen, and oxygen, and their gaseous products of combustion may include other gases in addition to gases containing only carbon, hydrogen, and oxygen. The gaseous products of combustion from polymers containing sulfur may include sulfur dioxide and hydrogen sulfide. The gaseous products of combustion from polymers containing nitrogen may include ammonia, hydrogen cyanide, and nitrogen dioxide. The gaseous products of combustion from polymers containing chlorine may include hydrogen chloride.

Some gases such as sulfur dioxide and ammonia are extremely irritating at concentrations well below lethal levels, and a person will normally leave the burning system or don protective breathing apparatus, if he is able, before suffering serious effects. Hydrogen sulfide is identified by its "rotten egg" odor, but continued exposure to concentrations above 0.2 per cent may paralyze the sense of smell. Hydrogen cyanide has a characteristic bitter almond odor which may be masked by other odors. Nitrogen dioxide is identified by its reddish-brown color, but it tends to anesthetize the throat and its toxic effect is often delayed. Fatalities may occur hours or days after exposure although the exposed persons may show no immediate ill effects.

The response of humans to various toxic gases is summarized in Table 2.14. The response of animals to some of these gases is summarized in Table 2.15.

5. <u>Smoke</u>. The principal hazard of smoke is that it hinders the escape of occupants and the entry of firefighters seeking to locate and extinguish the fire. It can contribute to panic conditions because of its blinding and irritating effects. In many cases, smoke reaches untenable levels in exitways before temperature reaches untenable levels.

The amount of light absorption measured by photometer methods is not necessarily indicative of the degree of sight obscuration. The size and distribution of smoke particles affect the optical absorption level at which sight obscuration is essentially total, as does the degree of irritation to the observer. Smoke density is influenced by the rate of burning and degree of ventilation, and is generally in inverse proportion to the degree of ventilation.

6. <u>Structural Strength Reduction</u>. The failure of structural components through heat damage or burning can present a serious hazard. Perhaps the most dramatic examples are the collapse of weakened floors under the weight of firefighters, and the collapse of walls and roofs on persons beneath them.

Table 2.1. Glass Transition Temperature for Various Polymers.

Polymer	Glass Transition Temperature Tg, °C
Polyethylene (PE)	−125
Polypropylene (PP)	−20
Polybutylene	−20 to −34
Polybutadiene	−85
Poly (4-methyl-1-pentene)	29
Polyvinyl chloride (PVC)	80
Chlorinated polyvinyl chloride (CPVC)	115 to 135
Polyvinylidene chloride (PVDC)	−18
Polystyrene (PS)	100 to 104
Polymethyl methacrylate (PMMA)	50
Polyethylene terephthalate (PET)	70 to 74
Polybutylene terephthalate (PBT)	40 to 52
Polycarbonate, bisphenol A	149
Polyphenylene ether/oxide (PPE/PPO)	218
Polycaprolactam (nylon 6)	75
Polyhexamethylene adipamide (nylon 6/6)	57
Polyhexamethylene sebacamide (nylon 6/10)	50
Polyimide, thermoplastic	280 to 340
Polyamide-imide (PAI)	270 to 285
Polyetherimide (PEI)	215
Polyetherketone (PEK)	205
Polyformaldehyde	−85
Polyacetaldehyde	−30
Polysulfone	190
Polyethersulfone (PES)	221
Polyarylsulfone	220
Polytetrafluoroethylene (PTFE)	−113, 127
Polyvinyl fluoride (PVF)	−20
Polyvinylidene fluoride (PVDF)	−45
Polychlorotrifluoroethylene (CTFE)	45
Polypropylene oxide	−75
Poly (dimethyl siloxane)	−123

Table 2.2. Decomposition Temperature for Various Polymers.

Polymer	Decomposition Temperature Range, °C
Polyethylene (PE)	335 to 450
Polypropylene (PP)	328 to 410
Polyisobutene	288 to 425
Polyvinyl chloride (PVC)	200 to 300
Polyvinylidene chloride (PVDC)	225 to 275
Polyvinyl acetate	213 to 325
Polyvinyl alcohol	250
Polyvinyl butyral	300 to 325
Polystyrene (PS)	285 to 440
Styrene-butadiene	327 to 430
Polymethyl methacrylate (PMMA)	170 to 300
Polyacrylonitrile	250 to 280
Polyethylene terephthalate (PET)	283 to 306
Polycarbonate (PC)	420 to 620
Poly-p-xylylene	420 to 465
Liquid crystal polymer (LCP)	560 to 567
Nylon 6 and nylon 6/6	310 to 380
Polyformaldehyde	222
Polytetrafluoroethylene (PTFE)	508 to 538
Polyvinyl fluoride (PVF)	372 to 480
Polyvinylidene fluoride (PVDF)	400 to 475
Polychlorotrifluoroethylene (CTFE)	347 to 418
Cellulose triacetate	250 to 310
Polyethylene oxide	324 to 363
Polypropylene oxide	270 to 355

Table 2.3. Specific Heat for Various Materials.

Material	Specific Heat, cal/g-°C
Polyethylene (PE)	0.55
Ethylene-vinyl acetate copolymer (EVA)	0.55
Ethylene-ethyl acrylate copolymer (EEA)	0.55
Polypropylene (PP)	0.46
Polypropylene, rubber modified	0.5
Ethylene methacrylic acid ionomer	0.55
Polytetrafluoroethylene (PTFE)	0.25
Fluorinated ethylene-propylene (FEP)	0.28
Polychlorotrifluoroethylene (CTFE)	0.22
Polybutylene	0.45
Methylpentene polymer	0.52
Polyvinyl chloride (PVC)	0.20 to 0.28
Polyvinylidene chloride (PVDC)	0.32
Polyvinylidene fluoride (PVDF)	0.33
Polystyrene (PS)	0.32
Polystyrene, 20–30% glass filled	0.23 to 0.27
Styrene-acrylonitrile copolymers (SAN)	0.32 to 0.34
Styrene-butadiene thermoplastic elastomers	0.45 to 0.50
Acrylonitrile-butadiene-styrene (ABS)	0.30 to 0.40
Polymethyl methacrylate (PMMA)	0.35
Cellulose nitrate	0.30 to 0.40
Cellulose acetate	0.30 to 0.50
Cellulose acetate butyrate	0.30 to 0.40
Cellulose propionate	0.30 to 0.40
Ethyl cellulose	0.30 to 0.75
Acetal	0.35
Nylon 6	0.38
Nylon 6/6	0.40
Nylon 6/10	0.40
Nylon 6, 20–40% glass filled	0.30 to 0.35
Nylon 11	0.58
Nylon 11, glass filled	0.42
Polycarbonate (PC)	0.30
Phenoxy	0.40
Polysulfone	0.31
Polyphenylene oxide, modified	0.32
Polyimide, aromatic	0.27
Phenol-formaldehyde	0.38 to 0.42
Phenolic, asbestos filler	0.30
Melamine-formaldehyde, cellulose filler	0.40
Urea-formaldehyde, cellulose filler	0.40
Epoxy	0.25
Epoxy, silica filled	0.20 to 0.27
Polyester, chopped glass filler	0.25
Polyurethane	0.40 to 0.45
Allyl	0.26 to 0.55
Silicone, glass fiber filled	0.24 to 0.30

Table 2.4. Thermal Conductivity of Various Materials.

Material	Thermal Conductivity, 10^{-4} cal/sec-cm² (°C/cm)
Polyethylene, low density	8.0
Polyethylene, medium density	8.0 to 10.0
Polyethylene, high density	8.0 to 12.4
Polypropylene (PP)	2.8
Polypropylene, rubber modified	3.0 to 4.0
Ethylene methacrylic acid ionomer	5.8
Methylpentene polymer	4.0
Polytetrafluoroethylene (PTFE)	6.0
Fluorinated ethylene-propylene (FEP)	5.9 to 6.0
Polychlorotrifluoroethylene (CTFE)	4.7 to 6.0
Polyvinyl chloride (PVC)	3.0 to 7.0
Polyvinylidene chloride (PVDC)	3.0
Polyvinylidene fluoride (PVDF)	3.0
Polystyrene (PS)	1.9 to 3.3
Styrene-acrylonitrile copolymers (SAN)	2.9 to 3.0
Styrene-butadiene thermoplastic elastomers	3.6
Acrylonitrile-butadiene-styrene (ABS)	4.5 to 8.0
Polymethyl methacrylate (PMMA)	4.0 to 6.0
Cellulose nitrate	5.5
Cellulose acetate	4.0 to 8.0
Cellulose acetate butyrate	4.0 to 8.0
Cellulose propionate	4.0 to 8.0
Ethyl cellulose	3.8 to 7.0
Acetal homopolymer	1.6 to 5.5
Acetal copolymer	5.5
Nylon 6	5.9
Nylon 6/6	5.8
Nylon 6/10	5.2
Nylon 6, 20–40% glass filled	5.2
Nylon 11	7.0
Nylon 11, glass filled	8.8
Polycarbonate (PC)	4.6
Polycarbonate, 10–40% glass filled	2.5 to 5.2
Phenoxy	4.2
Polyphenylene oxide	4.5
Polyphenylene oxide, 30% glass filled	3.4
Polyphenylene oxide, modified	5.2
Polyphenylene oxide, modified, 20–30% glass	3.8
Chlorinated polyether	3.1

continued

Table 2.4 (continued). Thermal Conductivity of Various Materials.

Material	Thermal Conductivity, 10^{-4} cal/sec-cm² (°C/cm)
Phenol-formaldehyde	3.0 to 6.0
Phenolic, asbestos filler	8.4
Melamine-formaldehyde, asbestos filler	13 to 17
Melamine-formaldehyde, cellulose filler	7 to 10
Urea-formaldehyde, cellulose filler	7 to 10
Epoxy	4 to 5
Epoxy, silica filled	10 to 20
Polyester	4
Polyester, glass fiber filled	10 to 16
Alkyd, glass filled	15 to 25
Polyurethane	1.5 to 7.4
Allyl	4.8 to 5.0
Allyl, glass filled	5 to 15
Silicone	3.5 to 7.5
Silicone, glass filled	7.5

Table 2.5. Ignition Temperatures of Various Polymers.

Polymer	Flash-Ignition Temperature, °C	Self-Ignition Temperature, °C
Polyethylene (PE)	341 to 357	349
Polypropylene (PP), fiber		570
Polyvinyl chloride (PVC)	391	454
Polyvinyl chloride-acetate	320 to 340	435 to 557
Polyvinylidene chloride (PVDC)	532	532
Polystyrene (PS)	345 to 360	488 to 496
Styrene-acrylonitrile (SAN)	366	454
Acrylonitrile-butadiene-styrene (ABS)		466
Styrene-methyl methacrylate	329	485
Polymethyl methacrylate (PMMA)	280 to 300	450 to 462
Acrylic, fiber		560
Polycarbonate (PC)	375 to 467	477 to 580
Nylon	421	424
Nylon 6/6, fiber		532
Polyetherimide (PEI)	520	535
Polyethersulfone (PES)	560	560
Polytetrafluoroethylene (PTFE)		530
Cellulose nitrate	141	141
Cellulose acetate	305	475
Cellulose triacetate, fiber		540
Ethyl cellulose	291	296
Polyurethane, polyether rigid foam	310	416
Phenolic, glass fiber laminate	520 to 540	571 to 580
Melamine, glass fiber laminate	475 to 500	623 to 645
Polyester, glass fiber laminate	346 to 399	483 to 488
Silicone, glass fiber laminate	490 to 527	550 to 564
Wool	200	
Wood	220 to 264	260 to 416
Cotton	230 to 266	254

Table 2.6. Oxygen Index for Various Polymers.

Polymer	Oxygen Index, pct
Polyethylene (PE)	17.3 to 30.2
Polypropylene (PP)	17.0 to 29.2
Polybutadiene	18.3
Chlorinated polyethylene	21.1
Polyallomer	17
Polyvinyl chloride (PVC)	20.6 to 80.7
Polyvinyl alcohol	22.5
Polyvinylidene chloride (PVDC)	60
Polystyrene (PS)	17.0 to 23.5
Styrene-acrylonitrile (SAN)	18 to 28
Acrylonitrile-butadiene-styrene (ABS)	18 to 39
Acrylic	16.7 to 25.1
Polyethylene terephthalate (PET)	20 to 40
Polybutylene terephthalate (PBT)	24 to 32
Polycarbonate (PC)	21.3 to 44
Liquid crystal polymer (LCP)	35 to 50
Polyphenylene ether/oxide (PPE/PPO)	24 to 33
Nylon 6	23 to 28
Nylon 6/6	21 to 38
Nylon 6/10	25 to 28
Nylon 6/12	25
Polyimide	36.5
Polyamide-imide	43
Polyetherimide (PEI)	47
Polybenzimidazole (PBI)	40.6
Polyacetal	14.7 to 16.2
Polysulfone	30 to 51
Polyethersulfone (PES)	37 to 42
Polytetrafluoroethylene (PTFE)	95
Polyvinyl fluoride (PVF)	22.6
Polyvinylidene fluoride (PVDF)	43.7
Fluorinated ethylene propylene (FEP)	95
Polychlorotrifluoroethylene (CTFE)	83 to 95
Ethylene-tetrafluoroethylene (ETFE)	30
Ethylene-chlorotrifluoroethylene (ECTFE)	60
Cellulose acetate	16.8 to 27
Cellulose butyrate	18.8 to 19.9
Cellulose acetate butyrate	18 to 20
Phenolic	18 to 66
Epoxy	18.3 to 49
Polyester, unsaturated	20 to 60
Alkyd	29 to 63.4

continued

Table 2.6 (continued). Oxygen Index for Various Polymers.

Polymer	Oxygen Index, pct
Polypropylene, fabric	18.6
Acrylic, fabric	18.2 to 19.6
Modacrylic, fabric	26.7 to 29.8
Wool, fabric	23.8 to 25.2
Nylon, fabric	20.1
Aromatic polyamide, fabric	26.7 to 28.2
Cellulose acetate, fabric	18.6
Cellulose triacetate, fabric	18.4
Cotton, fabric	18.6 to 27.3
Rayon, fabric	18.9 to 19.7
Polyester, fabric	20.6 to 21.0
Polybutadiene, rubber	17.1
Styrene-butadiene rubber (SBR)	16.9 to 19.0
Polychloroprene, rubber	26.3
Chlorosulfonated polyethylene, rubber	25.1
Silicone, rubber	25.8 to 39.2
Natural rubber	17.2
Wood	22.4 to 24.6
Cardboard	24.7
Fiber board, particle board	22.1 to 24.5
Plywood	25.4 to 73.6

Table 2.7. Autoignition Temperature of Various Polymers at Elevated Oxygen Pressure of 10.3 MPa.

Polymer	Autoignition Temperature, °C
Plastics	
Polytetrafluoroethylene (PTFE)	434
Copolymer of tetrafluoroethylene and perfluoro (propyl vinyl ether)	424
Polychlorotrifluoethylene (PCTFE)	388
Polyetherimide	385
Copolymer of hexafluoropropylene and tetrafluoroethylene	378
Polyethersulfone (PES)	373
Polyphenylene oxide/polystyrene blend	348
Polyimide with 15 wt% graphite fiber	343
Polyetheretherketone (PEEK)	305
Polycarbonate (PC)	286
Polyphenylene sulfide (PPS)	285
Polyvinylidene fluoride (PVDF)	268
Polyamide (nylon 6/6)	259
Acrylonitrile-butadiene-styrene (ABS)	243
Copolymer of tetrafluoethylene and ethylene (ETFE)	243
Polyvinyl chloride (PVC)	239
Polyvinyl fluoride (PVF)	222
Polyethylene terephthalate (PET)	181
Polymethylene oxide	178
Polyethylene (PE)	176
Polypropylene (PP)	174
Copolymer of ethylene and chlorotrifluoroethylene (ECTFE)	171
Rubbers	
Copolymer of tetrafluoroethylene and perfluoro(methyl vinyl ether)	355
Copolymer of vinylidene fluoride and hexafluoropropylene	302–322
Polysiloxane (silicone rubber)	262
Polychloroprene (neoprene)	258
Copolymer of tetrafluoroethylene and propylene	254
Copolymer of isobutylene and isoprene	208
Polyurethane rubber	181
Copolymer of butadiene and acrylonitrile (nitrile rubber)	173
EPDM rubber	159

Ref. Hshieh, F.-Y., Stoltzfus, J. M., Beeson, H. D., "Note: Autoignition Temperature of Selected Polymers at Elevated Oxygen Pressure and Their Heat of Combustion," Fire and Materials, Vol. 20, 301–303 (1996).

Table 2.8. Thermochemical Properties of Various Materials.

	Heat of Combustion, Kcal/g·mol	Heating Value, Btu/lb	Stoichiometric Flame Temperature	
			°C	°F
Polyethylene, high density	−312.5	20,050	2120	3850
Polyethylene, low density	−312.0	20,020	2120	3850
Ethylene/propylene copolymer, 69/31	−360.8	20,270	2120	3850
Polypropylene, isotactic, syndiotactic	−468.3	20,030	2120	3850
Polypropylene, atactic	−467.8	20,010	2120	3850
Poly-1-butene, isotactic	−625.4	20,060	2120	3850
Polyisobutylene	−628.2	20,150	2130	3870
Poly-1-pentene, isotactic	−780.8	20,040	2120	3850
Poly-3-methyl-1-butene, isotactic	−780.2	20,030	2120	3850
Poly-4-methyl-1-pentene, isotactic	−935.7	20,010	2120	3850
Poly-1,4-butadiene, atactic	−584.3	19,440	2220	4020
Polytetrafluoroethylene	+8.01	−144	—	—
Polychlorotrifluoroethylene	−31.2	482	320	615
Polyvinyl chloride	−268.0	7,720	1960	3550
Polyvinylidene chloride	−232.4	4,315	1840	3340
Polyvinyl fluoride	−238.8	9,180	1710	3100
Polyvinylidene fluoride	−140.3	3,940	1090	2000
Polystyrene, isotactic	−1033	17,850	2210	4010
Polystyrene, atactic, crystal	−1034	17,870	2210	4010
Poly-alpha-methylstyrene	−1196	18,220	2210	4010
Butadiene/styrene (8.58%) copolymer	−604.2	19,300	2220	4020
Butadiene/styrene (25.5%) copolymer	−650.5	19,010	2220	4020
Butadiene/acrylonitrile (37%) copolymer	−512.6	17,180	2190	3970
Polyoxymethylene	−121.4	7,280	2050	3750
Polyoxytrimethylene	−437.6	13,560	2130	3860
Polyethylene oxide	−280.6	11,470	2120	3850
Polypropylene oxide, 27% isotactic	−432.7	13,410	2100	3810
Polypropylene oxide, 100% atactic	−432.3	13,400	2120	3810
Chlorinated polyether	−660.7	7,673	1990	3610
Polycarbonate	−1880	13,310	2190	3980
Polyphenylene oxide	−993.8	14,880	2200	3990
Polypropene sulfone	−462.7	7,850	1970	3570
Poly-1-butene sulfone	−618.9	9,290	2000	3640
Poly-1-hexene sulfone	−913.5	11,310	2040	3710
Polymethyl methacrylate	−637.7	11,470	2070	3760
Phenol-formaldehyde (1:1)	−1496	12,000	1860	3380
Urea-formaldehyde (1:2)	−358.8	7,680	1950	3540
Melamine-formaldehyde (1:3)	−749.3	8,310	1990	3610

Table 2.8. Thermochemical Properties of Various Materials.

	Heat of Combustion, Kcal/g·mol	Heating Value, Btu/lb	Stoichiometric Flame Temperature	
			°C	°F
Polyurethane, ester-based	−743.1	10,180	2100	3810
Polyester, unsaturated	−723.1	12,810	2250	3910
Epoxy, bisphenol A	−1700	14,430	2220	4030
Polyacenaphthalene	−1429	16,900	2230	4040
Polycarbon suboxide	−224.1	5,940	2260	3910
Polytetrahydrofuran	−592.7	14,790	2120	3850
Polyvinyl alcohol	−263.2	10,760	1980	3600
Poly-beta-propiolactone	−333.3	8,330	2075	3770
Polynitroethylene	−278.7	6,870	2670	4830
Polyacrylonitrile	−408.6	13,860	1860	3380
Cellulose	−1011	7,520	—	—
Paper	—	7,590	—	—
Woodflour	—	8,520	—	—
Wood	—	8,835	—	—
Chipboard (90% woodflour, 10% resin)	—	8,715	—	—
Bituminous coal, med. volatile, W. Va.	—	15,178	—	—
Lignite, Texas	—	11,084	—	—
Peat, Minn.	—	9,057	—	—
Oil shale	—	6,300	—	—
No. 1 fuel oil	—	19,800	—	—
No. 6 fuel oil	—	18,300	—	—

Table 2.9. Response of Humans to Various Concentrations of Oxygen.

Concentration, percent	Symptoms
21	Normal concentration in air
17	Respiration volume increased, muscular coordination diminished, more effort required for attention and clear thinking
12 to 15	Shortness of breath, headache, dizziness, quickened pulse, quick fatigue upon exertion, loss of muscular coordination for skilled movements
10 to 14	Faulty judgement
10 to 12	Nausea and vomiting, exertion impossible, paralysis of motion
6 to 8	Collapse and unconsciousness, but rapid treatment can prevent death
6 or below	Death in 6 to 8 minutes
2 to 3	Death in 45 seconds

Table 2.10. Response of Humans to Various Concentrations of Carbon Dioxide.

Concentration, ppm	Symptoms
250 to 350	Normal concentration in air
900 to 5000	No effect
18,000	Ventilation increased by 50 percent
25,000	Ventilation increased by 100 percent
30,000	Weakly narcotic, decreasing acuity of hearing, increase in pulse and blood pressure
40,000	Ventilation increased by 300 percent, headache, weakness
50,000	Symptoms of poisoning after 30 minutes, headache, dizziness, sweating
80,000	Dizziness, stupor, unconsciousness
90,000	Distinct dyspnoea, loss of blood pressure, congestion, death within 4 hours
100,000	Headaches and dizziness
120,000	Immediate unconsciousness, death in minutes
200,000	Narcosis, immediate unconsciousness, death by suffocation

Table 2.11. Response of Humans to Various Concentrations of Carbon Monoxide.

Concentration, ppm	Symptoms
100	No poisoning symptoms even for longer periods of time, allowable for several hours
200	Headache after 2 to 3 hours, collapse after 4 to 5 hours
300	Headache after 1.5 hours, distinct poisoning after 2 to 3 hours, collapse after 3 hours
400	Distinct poisoning, frontal headache, and nausea after 1 to 2 hours, collapse after 2 hours, death after 3 to 4 hours
500	Hallucinations felt after 30 to 120 minutes
800	Collapse after 1 hour, death after 2 hours
1000	Difficulty in ambulation, death after 2 hours
1500	Death after 1 hour
2000	Death after 45 minutes
3000	Death after 30 minutes
8000 or above	Immediate death by suffocation
12,800	Unconsciousness after 2 to 3 breaths, death in 1 to 2 minutes

Table 2.12. **Response of Humans to Various Concentrations of Carboxyhemoglobin.**

Concentration, percent	Symptoms
0 to 10	No signs or symptoms
10 to 20	Tightness across forehead, possible slight headache, dilation of cutaneous blood vessels
20 to 30	Headache, throbbing in temples
30 to 40	Severe headache, weakness, dizziness, dimness of vision, nausea, vomiting, collapse
40 to 50	Same as above, increase in pulse and breathing rate, greater possibility of collapse, asphyxiation
50 to 60	Same as above, coma, intermittent convulsions, and Cheyne-Stokes respiration
60 to 70	Coma, intermittent convulsions, depressed heart action and respiratory rate, possible death
70 to 80	Weak pulse, slowing of respiration leading to death within hours
80 to 90	Death in less than 1 hour
90 to 100	Death in a few minutes

Table 2.13. **Response of Animals to Various Concentrations of Carbon Oxides.**

Carbon Oxide and Concentration, ppm	Effects
Carbon dioxide	
400,000	Lethal to mice after 4 hours
Carbon monoxide	
1250	Lethal to mice after 4 hours
1500	25% COHb in mice after 5 minutes
2100	25% COHb in rats after 5 minutes
4670	LC_{50} for rats in 60 minutes
5000	Minimum lethal dose for rats in 30 minutes
5500	LC_{50} for rats in 30 minutes
6100	LC_{50} for rats in 20 minutes
8800	LC_{50} for rats in 10 minutes

Table 2.14. Response of Humans to Various Concentrations of Some Materials.

Compound and Concentration, ppm	Symptoms
Acetaldehyde	
0.07 to 0.21	Threshold of odor
25 to 50	Transient slight irritation of eyes after 15 minutes
134	Slight irritation of respiratory tract after 30 minutes
200	Irritation of nose and throat
Acetic acid	
20 to 30	Apparently no danger to workers exposed for 7 to 12 years
60	Slight irritation of respiratory tract, stomach, and skin
800 to 1200	Cannot be tolerated for more than 3 minutes
Acetone	
0.5 to 1000	Threshold of odor, quick adaptation
2000	Apparently no symptoms on workers exposed over many years
5000	Induces first narcotic symptoms
9300	Irritation of throat after 5 minutes
Acetonitrile	
40	Odor detectible, no symptoms
80	No symptoms after 4 hours
160	Slight feeling of bronchial tightness after 4 hours
Acrolein	
0.805	Lachrymation, irritation of mucous membranes
1.0	Immediately detectible, irritation
5.5	Intense irritation
10 and over	Lethal in a short time
24	Unbearable
Ammonia (NH_3)	
1 to 50	Detectable odor
57 to 72	Respiration not significantly changed
96	Slight irritation of nose, throat, and eyes
100	Working possible, adaptation
200	Irritation of the mucous membranes
500 to 1000	Strong irritation of upper respiratory tract
over 1000	Fatal
Benzene	
500	Slight irritation
1500 to 4000	Dangerous to life after several hours
8000	Fatal after 30 to 60 minutes
20,000	Fatal after 5 minutes
Butyl acetate	
200	Irritation of upper respiratory tract, nose, eyes, and throat
300	Strong irritation effects
900	Strong irritation, appearance of narcotic effects with sensation of vertigo
1800	Deep narcosis with vertigo and unconsciousness
Ethyl acetate	
0.2	Limit of perception of odor
200	Strong odor perceived
350	Irritation of nose and eyes
700	Narcotic effects without fainting
3800	Distinct narcosis with fainting and significant irritation

Table 2.14 (continued). Response of Humans to Various Concentrations of Some Materials.

Compound and Concentration, ppm	Symptoms
Formaldehyde	
0.05 to 1.0	Threshold of odor
0.08 to 1.6	Slight irritation of eyes and nose
0.25 to 1.6	Threshold of irritation of eyes
0.5	Threshold of irritation of throat
10	Conjunctivitis, rhinitis, and pharyngitis within a few minutes
10 to 15	Dyspnea, cough, pneumonia, bronchitis
over 50	Necrosis of mucous membranes, spasm of larynx, edema of lungs
Hydrogen chloride (HCl)	
1 to 5	Limit of detection by odor
5 to 10	Mild irritation of mucous membranes
35	Irritation of throat on short exposure
50 to 100	Barely tolerable
1000	Danger of lung edema after merely short exposure
Hydrogen cyanide (HCN)	
0.2 to 5.1	Threshold of odor
18 to 36	Slight symptoms, headache, after several hours
45 to 54	Tolerated for 1/2 to 1 hour without difficulty
100	Fatal after 1 hour
110 to 135	Fatal after 1/2 to 1 hour, dangerous to life
135	Fatal after 30 minutes
181	Fatal after 10 minutes
280	Immediately fatal
Hydrogen fluoride (HF)	
3 to 5	Redness of skin, irritation of nose and eyes after one week exposure
32	Irritation of eyes and nose
60	Itching of skin, irritation of respiratory tract from exposure of 1 minute
120	Conjunctival and respiratory irritation just tolerable for 1 minute
50 to 100	Dangerous to life after a few minutes
Hydrogen sulfide (H_2S)	
20 to 30	Conjunctivitis
50	Objection to light after 4 hours, lachrymation
50 to 500	Irritation of respiratory tract
150 to 200	Objection to light, irritation of mucous membranes, headache
200 to 400	Slight symptoms of poisoning after several hours
250 to 600	Pulmonary edema and bronchial pneumonia after prolonged exposure
500 to 1000	Systemic poisoning, painful eye irritation, vomiting
1000	Immediate acute poisoning
1000 to 2000	Lethal after 30 to 60 minutes
over 2000	Acute lethal poisoning
Nitrogen dioxide (NO_2)	
5	Threshold of perception by odor
10 to 20	Mildly irritant to eyes, nose, and upper respiratory tract
25 to 38	Apparently no adverse effects in workers exposed over several years
50	Distinct irritation
80	Tightness of chest after 3 to 5 minutes
90	Pulmonary edema after 30 minutes
100 to 200	Very dangerous within 30 to 60 minutes
250	Death after a few minutes

continued

Table 2.14 (continued). Response of Humans to Various Concentrations of Some Materials.

Compound and Concentration, ppm	Symptoms
Styrene	
60	Threshold of odor, no irritation
100	Strong odor, tolerable
200 to 400	Intolerable odor
216	Unpleasant subjective symptoms
376	Definite signs of neurological impairment
600	Irritation of eyes
800	Immediate irritation of eyes and throat, somnolence, weakness
over 800	Nausea, vomiting, and total weakness
Sulfur dioxide (SO_2)	
3 to 5	Odor threshold
8 to 12	Slight irritation of eyes and throat, resistance of air tracts
20	Coughing and eye irritation
30	Immediate strong irritation, remains very unpleasant
100 to 250	Dangerous to life
600 to 800	Death in a few minutes
Toluene	
190 to 380	No complaints
500 to 1000	Headache, nausea, momentary loss of memory, anorexia, irritation of eyes
1000 to 1500	Palpitation, extreme weakness, loss of coordination, impairment of reaction time
2000 to 2500	Dizziness, nausea, narcosis after 3 hours
10,000	Immediately fatal
Toluene diisocyanate (TDI)	
0.01	No irritation, no odor for 30 minutes
0.018 to 0.02	Odor threshold
0.05	Slight irritation of the eyes
0.1	Tolerable irritation of eyes, nose, and throat
0.5	Heavy irritation of eyes, nose, and throat
1.3	Heavy irritation, coughing, spasms of the bronchi, tracheitis lasting several hours after exposure

Table 2.15. Response of Animals to Various Concentrations of Some Materials.

Compound and Concentration, ppm	Effects
Acetaldehyde	
3000 to 20,000	Edema of lungs and narcotic effects in animals
Acetone	
20,000	Narcosis in guinea pigs after 8 to 9 hours
20, 256	Narcosis in mice after 1 hour
46,000	Narcosis in rats after 2 hours, lethal to mice in 1 hour
50,000	Lethal to guinea pigs in 3 hours
126,000	Lethal to rats in 2 hours
Ammonia (NH_3)	
480 to 570	Irritation in various species after 4 hours
800 to 1070	Strong irritation in various species after 7 hours
820 to 1430	Dyspnea, tracheitis in various species after 5 to 7 hours
2000	No deaths in rats after 4 hours
3420	Minimum lethal concentration for mice after 2 hours
4000	Fatal to rats after 4 hours
4760	LC_{50} for mice in 2 hours
10,930	LC_{50} for rats in 2 hours
20,000 to 30,000	Fatal to various species after a few minutes
Carbonyl fluoride	
360	LC_{50} for rats in 1 hour
Formaldehyde	
250	LC_{50} for rats in 30 minutes
830	LC_{50} for rats in 4 hours
Hydrogen Chloride (HCl)	
50	Tolerable to monkeys for 6 hours daily over 20 days
300	Mild corneal damage in guinea pigs after 6 hours
3200	No mortality in mice after 5 minutes
4300	Lung edema in rabbits, death after 30 minutes
13,745	LC_{50} for mice in 5 minutes
30,000	No mortality in rats after 5 minutes
40,989	LC_{50} for rats in 5 minutes
Hydrogen cyanide (HCN)	
50	Minimum lethal dose for rats in 30 minutes
56	Symptoms in cats after 60 minutes
50 to 90	Symptoms in rats after 30 minutes
100	May be fatal to rats after 30 minutes
110	Fatal to dogs and cats after 30 minutes
120	LC_{50} for rats in 60 minutes
125	Severe symptoms in monkeys after 12 minutes
142	LC_{50} for rats in 30 minutes
200	LC_{50} for rats in 20 minutes
315	Fatal to dogs and cats after 5 minutes
323	LC_{50} for mice in 5 minutes
503	LC_{50} for rats in 5 minutes

continued

Table 2.15 (continued). Response of Animals to Various Concentrations of Some Materials.

Compound and Concentration, ppm	Effects
Hydrogen fluoride (HF)	
7	No effect on various species after daily exposure
40	Guinea pigs and rabbits survived for 41 hours
300	Death in guinea pigs and rabbits after 2 hours
2430	Highest concentration without mortality in mice after 5 minutes
6247	LC_{50} for mice in 5 minutes
10,000	Highest concentration without mortality in rats after 5 minutes
18,200	LC_{50} for rats in 5 minutes
Nitrogen dioxide (NO_2)	
50	No mortality in mice after 4 hours
100	Lethal to mice after 4 hours
260	Highest concentration without mortality in rats and mice after 5 minutes
831 to 832	LC_{50} for rats in 5 minutes
1880	LC_{50} for mice in 5 minutes
Sulfur dioxide (SO_2)	
400	Most species survive after 1 to 5 hours
600 to 800	Lethal to some species
Tetrafluoroethylene	
40,000	LC_{50} for rats in 4 hours
Toluene diisocyanate (TDI)	
9.7	LC_{50} for mice in 4 hours
11.0	LC_{50} for rabbits in 4 hours
12.7	LC_{50} for guinea pigs in 4 hours
13.9	LC_{50} for rats in 4 hours

CHAPTER 3

Fire Response Characteristics

Fire response characteristics are those properties which describe the response of a material when exposed to fire. For both convenience and convention, fire response characteristics are generally considered to include the following:

1. Smolder susceptibility
2. Ignitability
3. Flash-fire propensity
4. Flame spread
5. Heat release
6. Fire endurance
7. Ease of extinguishment
8. Smoke evolution
9. Toxic gas evolution
10. Corrosive gas evolution

SECTION 3.1. TYPES OF FIRE RESPONSE CHARACTERISTICS

Fire response characteristics can be classified according to the types of measurements used to describe them. Five types of measurements are used:

1. Thermophysical: smolder susceptibility, ignitability, flash-fire propensity, flame spread, heat release, fire endurance, ease of extinguishment, and smoke evolution (gravimetric)
2. Optical: smoke evolution
3. Biological: toxic gas evolution
4. Chemical: toxic gas evolution and corrosive gas evolution
5. Electrical: corrosive gas evolution

Thermogravimetric measurements and optical measurements are used to describe smoke evolution. Chemical or biochemical measurements and biological measurements are used to describe toxic gas evolution. Chemical measurements and electrical measurements are used to describe corrosive gas evolution.

57

SECTION 3.2. SMOLDER SUSCEPTIBILITY

Smolder has been defined as to burn without flame, and smoldering has been defined as combusting without flame, usually with incandescence and moderate smoke. Smoldering can continue for relatively long periods of time, during which it may or may not kindle into flame. Smoldering combustion can be considered to be the slow propagation of a combustion wave through porous fuel, characterized by relatively low temperatures and incomplete oxidation controlled by the diffusion rate of oxygen through the porous fuel.

Smoldering combustion generally requires a porous fuel, and charring appears to be a prerequisite for smoldering. Cellulosic materials such as wood, cotton, and rayon often tend to form char and become prone to smoldering. Smoldering has occurred in cellulose fiber insulation and cellulose fiberboard.

Some synthetic materials also tend to form char, either by design or by treatment, and these can become prone to smoldering. Smoldering has occurred in certain phenolic and polyurethane foams.

In flexible polyurethane foams, the porous char formed by smoldering permits the flow of oxygen into the char interior, where it can react with the combustible vapors produced by the primary smolder reaction and, with reduced heat losses, lead to ignition. Once transition to ignition occurs in the char, flame can propagate outward to the surface.

Smolder susceptibility may be defined as the tendency of a material to support smoldering combustion within itself. This characteristic provides a measure of fire hazard in that a material undergoing smoldering combustion may generate increasing amounts of heat and may, under certain conditions, bring itself or an adjacent material to the point of ignition.

Smolder susceptibility test data on some materials are presented in Table 3.1.

SECTION 3.3. IGNITABILITY

Ignitability has been defined as the ease of ignition, especially by a small flame or spark. Ignition has been defined as the initiation of combustion.

The thermal exposure causing ignition is a combination of heat flux and time. The higher the heat flux, the shorter the time before ignition occurs for a given material. Thermal insulation materials permit less heat flux to flow from the exposed surface to the interior of the material, and as a result reach surface ignition more rapidly then less effective insulators.

Ignitability may be defined as the facility with which a material or its pyrolysis products can be ignited under given conditions of temperature, pressure, and oxygen concentration. This characteristic provides a measure of fire hazard in that a material which has an ignition temperature significantly higher than that of another material is less likely to contribute to a fire, all other factors being the same in both cases. For example, if a fire wall could be relied upon to keep the temperature on the side away from the fire below 300°C. for a stated period of time, a material that ignited at 200°C. would be a potential hazard and a material that ignited at 400°C. would not be a potential hazard.

Ignitability test data on some materials are presented in Tables 3.2 and 3.3.

SECTION 3.4. FLASH-FIRE PROPENSITY

Flash fires are a special form of fire hazard which combine the different aspects of ignitability, flammability, heat release, and flame spread,. A flash fire has been defined as a fire (flame front) which propagates through a fuel-air mixture as a result of the energy release from the combustion of that fuel, having required only an ignition source. A flash fire has also been defined as a fire that spreads with extreme rapidity, such as one that races over flammable liquids and through gases. Extreme rapidity of flame propagation seems to be generally considered as characteristic of a flash fire.

It should be emphasized that a flash fire is not the same as flashover. Flashover has been defined as a stage in the development of a contained fire in which all exposed surfaces reach ignition temperature more or less simultaneously and fire spreads throughout the space and flames appear on all surfaces.

Flash fires tend to occur when combustible gases are evolved as a result of thermal exposure, escape burning at the original point of exposure, and accumulate to form fuel-air mixtures which then contact an ignition source.

Flash-fire propensity may be defined as the tendency of a material to produce a fire that spreads with extreme rapidity. This characteristic provides a measure of fire hazard in that a material with high flash-fire propensity is capable of contributing to sudden development of a rapidly spreading fire.

Flash-fire test data on some materials are presented in Table 3.4.

SECTION 3.5. FLAME SPREAD

Flame spread has been defined as the progress of flame over a surface. Because flame spread is a surface phenomenon, it is critically affected by whether combustible gases are evolved at the surface, or are evolved in the interior of the material and escape at the surface.

Flame spread requires that successive sections of surface be brought to the ignition temperature as a result of heat flux from the advancing flame. The ignitability characteristics of the material therefore have a direct bearing of flame spread. Thermal insulation materials tend to reach surface ignition more rapidly, and as a result tend to exhibit more rapid flame spread.

Flame spread may be defined as the rate of travel of a flame front under given conditions of burning. This characteristic provides a measure of fire hazard, in that flame spread can transmit fire to more flammable materials in the vicinity and thus enlarge a fire, even though the transmitting material itself contributes little fuel to the fire.

Flame spread within the interiors of buildings have led to loss of life in fires like the Coconut Grove night club fire in Boston in 1942. Flame spread up the facades of high-rise buildings have led to loss of life in fires like the high-rise building fires in Sao Paulo, Brazil, in the early 1970s.

Flame spread test data on some materials are presented in Table 3.5.

SECTION 3.6. HEAT RELEASE

The amount of heat released from a burning material, and the rate at which that heat is released, influence the temperature of the fire environment and the rate of fire spread.

Heat release may be defined as the heat produced by the combustion of a given weight or volume of material. This characteristic provides a measure of fire hazard in that a material which burns with the evolution of little heat per unit quantity burned will contribute appreciably less to a fire than a material which generates large amounts of heat per unit quantity burned.

Heat release rate has been defined as the quantity of heat liberated during complete combustion, usually expressed per unit time and per unit quantity of material.

Heat release data on some materials are presented in Tables 3.6 and 3.7.

SECTION 3.7. FIRE ENDURANCE

Fire endurance may be defined as the resistance offered by a material to the passage of fire, normal to the exposed surface over which flame spread is measured. This characteristic provides a measure of fire hazard, in that a material which will contain a fire represents more protection than a material which will give way before the same fire, all other factors being the same in both cases.

Fire endurance has been defined as a measure of the elapsed time during which a material or product maintains its design integrity under specified conditions of test and performance. This measure of fire resistance is usually expressed in terms of hours or fractions of hours of time during which a material exposed to a specified fire environment must continue to exhibit a specified level of performance.

Fire endurance data are generally obtained only for complete systems such as floor and wall constructions.

SECTION 3.8. EASE OF EXTINGUISHMENT

The ease with which fire can be extinguished varies with the material which is serving as the fuel for the fire. Some burning materials are more difficult to extinguish than others.

Ease of extinguishment may be defined as the facility with which burning can be extinguished in the case of a specific material. This characteristic provides a measure of fire hazard, in that a material which requires more effort to extinguish is more likely to prolong a fire than one which is easily extinguished, all other factors being the same in both cases.

One measure of ease of extinguishment is the oxygen concentration required to support continued burning, and the oxygen index is useful for this purpose.

Oxygen index test data for some materials are presented in Table 2.6.

SECTION 3.9. SMOKE EVOLUTION

Smoke has been defined as a visible, nonluminous, airborne suspension of particles, originating from combustion or sublimation. Smoke is an important factor in fire because visibility is a factor in the ability of occupants to escape from a burning structure, and in the ability of firefighters to locate and suppress a fire.

Smoke density may be defined as the degree of light or sight obscuration produced by the smoke from the decomposing or burning material under given conditions of decomposition and combustion. This characteristic provides a measure of fire hazard, in that an occupant has a better chance of escaping from a burning structure if he can see the exit, and a firefighter has a better chance of putting out the fire if he can locate it.

Smoke evolution is most often expressed in terms of optical density because light and sight obscuration is the aspect of greatest concern.

Smoke evolution test data on some materials are presented in Table 3.8.

SECTION 3.10. TOXIC GAS EVOLUTION

Toxic has been defined as poisonous, or destructive to body tissues and organs or interfering with body functions. The toxic gases evolved from materials involved in fires may lead to incapacitation, injury, and death, and are the principal threat to life safety when the thermal threat is minor or insignificant.

Toxic gas evolution may be defined as the level of toxicity exhibited by the gases evolved from a material under specified conditions. This characteristic provides a measure of fire hazard in that a material which produces more toxic effects than another material would be expected to be more likely to produce incapacitation and death.

Organic materials when decomposing and burning can evolve a variety of toxic gases, the most common of which is carbon monoxide. In addition, nitrogen-containing materials may evolve hydrogen cyanide and nitrogen oxides, sulfur-containing materials may evolve sulfur oxides and sulfides, and chlorine-containing materials may evolve hydrogen chloride.

The decomposition and combustion products vary with polymer composition, available oxygen, temperature level, rate of temperature rise, endotherms and exotherms, and rate of volatiles evolution. Pyrolysis and thermal oxidation can give very different results. Decomposition products from some materials are presented in Table 3.9.

Because identification and analysis of all toxicants is difficult, and the combined effect of a group of toxicants of varying concentrations and toxicities may not be accurately predicted with the present state of the art, laboratory animals are used to integrate all the toxic effects of a specified mixture of evolved gases. Results can be expressed as time required for a specified effect, such as time to death, or concentration required for a specified effect, such as lethal concentration for 50 per cent of the test animals, or LC50.

Toxic gas evolution test data on some materials are presented in Tables 3.10 and 3.11.

SECTION 3.11. CORROSIVE GAS EVOLUTION

Corrosive has been defined as corroding or causing corrosion, eating into or wearing away gradually, consuming, or destroying, said of the action of chemicals. The gases evolved from materials involved in fires may be corrosive to other materials which survive and are of value.

Corrosive gas evolution may be defined as the level of corrosiveness exhibited by the gases evolved from a material under specified conditions. This characteristic provides a measure of fire hazard, in that a material which produces more corrosive gases than another material would be expected to be more likely to affect electrical and electronic equipment used in firefighting and rescue at the scene of a fire, and, of broader concern, electrical and electronic equipment used in communications necessary for business operations and public safety, and in control of critical installations such as nuclear power plants.

The evolution of corrosive gases from fires may have four kinds of effects on metallic surfaces:

1. Loss of metal, with results such as excessive clearance between mechanical parts.
2. Formation of deposits, with results such as inadequate clearance between mechanical parts.
3. Bridging of conductor circuits, through mechanisms such as formation of conductive deposits, and altering conductivity of materials.
4. Breaking of conductor circuits, through mechanisms such as loss of conductive material, formation of non-conductive deposits, and altering conductivity of materials.

The hot gases from a fire can be corrosive, in that heat increases the rate of oxidation, and therefore corrosion by oxidation, of an exposed surface.

Organic materials when decomposing and burning can evolve a variety of gases which can cause or contribute to corrosion. Water vapor may increase the rate of corrosion by oxidation, and the rate of corrosion by water-soluble gases. Carbon dioxide may act as carbonic acid. Nitrogen-containing compounds may evolve ammonia, and nitrogen oxides which may act as nitrous and nitric acids. Phosphorus-containing compounds may evolved phosphorus oxides, which may act as phosphorous and phosphoric acids. Sulfur-containing compounds may evolve sulfur oxides, which may act as sulfurous and sulfuric acids. Chlorine-containing compounds may evolve hydrogen chloride, which may act as hydrochloric acid.

Identification and analysis of all potentially corrosive gases is difficult, and the combined effect of a group of corrosive gases of varying concentrations and corrosivities on a specific metal cannot be accurately predicted with the present state of the art. Relative corrosivity has been expressed in terms of the pH and conductivity of aqueous solutions of combustion products, the yield of soluble metal ions from corrosion of metal, and change in resistance of a circuit board exposed to corrosive gases.

There are three causes of electrical equipment failure from corrosion:

1. Current leakage resulting from electrolytic corrosion, corrosion driven by an applied potential in the presence of an electrolyte.
2. High contact resistance resulting from galvanic corrosion, corrosion driven by reaction of dissimilar materials.
3. Metal loss resulting from chemical corrosion, corrosion driven by direct attack by a foreign substance.

Corrosive gas evolution test data on some materials are expressed in Table 3.12.

REFERENCES

Bennett, J. G., Kessel, S. L., Rogers, C. E., "Corrosivity Test Methods for Polymeric Materials. Part 4. Cone Corrosimeter Test Method", Journal of Fire Sciences, Vol. 12, No. 2, 175-195 (March/April 1994)

Braun, E., Levin, B.C., "Nylons: A Review of the Literature on Products of Combustion and Toxicity", NBSIR 85-3280, National Bureau of Standards, Gaithersburg, Maryland (February 1986)

Braun, E., Levin, B.C., "Nylons: A Review of the Literature on Products of Combustion and Toxicity", Fire and Materials, Vol. 11, No. 2, 71-88 (June 1987)

Braun, E., Levin, B.C., "Polyesters: A Review of the Literature on Products of Combustion and Toxicity", Fire and Materials, Vol. 10, Nos. 3/4, 107-123 (September/December 1986)

Grand, A. F., "The Use of the Cone Calorimeter to Assess the Effectiveness of Fire Retardant Polymers under Simulated Real Fire Test Conditions", Interflam '96, Cambridge, U.K. (March 1996)

Gurman, J. L., Baier, L., Levin, B. C., "Polystyrenes: A Review of the Literature on the Products of Thermal Decomposition and Toxicity", NBSIR 85-3277, National Bureau of Standards, Gaithersburg, Maryland (March 1986)

Gurman, J. L., Baier, L., Levin, B. C., "Polystyrenes: A Review of the Literature on the Products of Thermal Decomposition and Toxicity", Fire and Materials, Vol. 11, No. 3, 109-130 (September 1987)

Hilado, C. J., "Flammability Handbook for Plastics", 4th Ed., Technomic Publishing Co., Lancaster, Pennsylvania (1990)

Hilado, C. J., "Toxicity of Off-Gases from Food and Plastic Products", Proceedings of the California Conference on Product Toxicity, Vol. 5, 49-55 (1984)

Hilado, C. J., Cumming, H. J., "Flash-Fire Propensity of Materials", Journal of Fire and Flammability, Vol. 8, No. 4, 443-457 (October 1977)

Hilado, C. J., Cumming, H. J., "Screening Materials for Flash-Fire Propensity", Modern Plastics, Vol. 54, No. 11, 56-59 (November 1977)

Hilado, C. J., Huttlinger, P.A., "Review and Update of the Dome Chamber Toxicity Test Method", Proceedings of the International Conference on Fire Safety, Vol. 8, 169-189 (1983)

Hilado, C. J., Huttlinger, P.A., "Screening Materials by the NASA Dome Chamber Toxicity Test", Proceedings of the California Conference on Product Toxicity, Vol. 4, 20-60 (1983)

Hilado, C. J., Huttlinger, P.A., "Toxic Hazards from Common Materials", Fire Technology, Vol. 17, No. 3, 177-182 (August 1981)

Hilado, C. J., Murphy, R. M., "A Simple Laboratory Method for Determining Ignitability of Materials", Journal of Fire and Flammability", Vol. 9, No. 2, 164-175 (April 1978)

Huggett, C., Levin, B. C., "Toxicity of the Pyrolysis and Combustion Products of Poly(Vinyl Chlorides): A Literature Assessment", NBSIR 85-3286, National Bureau of Standards, Gaithersburg, Maryland (April 1986)

Huggett, C., Levin, B. C., "Toxicity of the Pyrolysis and Combustion Products of Poly(Vinyl Chlorides): A Literature Assessment", Fire and Materials, Vol. 11, No. 3, 131-142 (September 1987)

Johnston, P. K., Doyle, E., Orzel, R. A., "Acrylics: A Literature Review of Thermal Decomposition Products and Toxicity", Journal of the American College of Toxicology, Vol 7, No. 2, 139-200 (1988)

Johnston, P. K., Doyle, E., Orzel, R. A., "Phenolics: A Literature Review of Thermal Decomposition Products and Toxicity", Journal of the American College of Toxicology, Vol 7, No. 2, 201-220 (1988)

Levin, B. C., "A Summary of the NBS Literature Reviews on the Chemical Nature and Toxicity of the Pyrolysis and Combustion Products from Seven Plastics: Acrylonitrile-Butadiene-Styrenes (ABS), Nylons, Polyesters, Polyethylenes, Polystyrenes, Poly(Vinyl Chlorides), and Rigid Polyurethane Foams", Fire and Materials, Vol. 11, No. 3, 143-157 (September 1987)

Nelson, G. L., "Carbon Monoxide and Fire Toxicity", Proceedings of the 1996 National Fire Protection Research Foundation Conference, 340-360 (June 1996)

Ohlemiller, T. J., "Smoldering Combustion", NBSIR 85-3294, National Bureau of Standards, Gaithersburg, Maryland (February 1986)

Paabo, M., Levin, B. C., "A Literature Review of the Chemical Nature and Toxicity of the Decomposition Products of Polyethylenes", NBSIR 85-3268, National Bureau of Standards, Gaithersburg, Maryland (January 1986)

Paabo, M., Levin, B. C., "A Literature Review of the Chemical Nature and Toxicity of the Decomposition Products of Polyethylenes", Fire and Materials, Vol. 11, No. 2, 55-70 (June 1987)

Paabo, M., Levin, B. C., "A Review of the Literature on the Gaseous Products and Toxicity Generated from the Pyrolysis and Combustion of Rigid Polyurethane Foams", NBSIR 85-3224, National Bureau of Standards, Gaithersburg, Maryland (December 1985)

Paabo, M., Levin, B. C., "A Review of the Literature on the Gaseous Products and

Toxicity Generated from the Pyrolysis and Combustion of Rigid Polyurethane Foams", Fire and Materials, Vol. 11, No. 1, 1-29 (March 1987)

Purohit, V., Orzel, R. A., "Polypropylene: A Literature Review of the Thermal Decomposition Products and Toxicity", Journal of the American College of Toxicology, Vol 7, No. 2, 221-242 (1988)

Rutkowski, J. V., Levin, B. C., "Acrylonitrile-Butadiene-Styrene Polymers (ABS): Pyrolysis and Combustion Products and Their Toxicity - A Review of the Literature", NBSIR 85-3248, National Bureau of Standards, Gaithersburg, Maryland (December 1985)

Rutkowski, J. V., Levin, B. C., "Acrylonitrile-Butadiene-Styrene Polymers (ABS): Pyrolysis and Combustion Products and Their Toxicity - A Review of the Literature", Fire and Materials, Vol. 10, Nos. 3/4, 93-105 (September/December 1986)

Sandmann, H., Widmer, G., "The Corrosiveness of Fluoride-Containing Gases on Selected Steels", Fire and Materials, Vol. 10, No. 1, 11-19 (March 1986)

Tse, S. D., Fernandez-Pello, A. C., Miyasaka, K., "Controlling Mechanisms in the Transition from Smoldering to Flaming of Flexible Polyurethane Foam", Proceedings of the 26th International Symposium on Combustion, 1505-1513, The Combustion Institute (1996)

Table 3.1. *Smolder Susceptibility of Materials as Measured by the California Bureau of Home Furnishings.*

	Number of Fabrics Tested	Number of Fabrics Igniting	Percent Igniting
Fabric on fiberglass board			
100% Cellulosic fabrics	42	42	100
Cellulosic/thermoplastic fabric blends	28	15	53.6
Miscellaneous fabrics	7	1	14.3
100% Thermoplastic fabrics	11	0	0
Fabric on polyurethane foam (no FR)			
100% Cellulosic fabrics	42	17	40.5
Cellulosic/thermoplastic fabric blends	28	7	25.0
Miscellaneous fabrics	7	1	14.3
100% Thermoplastic fabrics	11	0	0
Fabric on polyurethane foam (FR, 1)			
100% Cellulosic fabrics	42	36	85.7
Cellulosic/thermoplastic fabric blends	28	15	53.6
100% Thermoplastic fabrics	11	0	0
Miscellaneous fabrics	7	0	0
Fabric on polyurethane foam (FR, 2)			
100% Cellulosic fabrics	42	16	38.1
Cellulosic/thermoplastic fabric blends	28	7	25.0
100% Thermoplastic fabrics	11	0	0
Miscellaneous fabrics	7	0	0
Fabric on polyurethane foam (HR)			
100% Cellulosic fabrics	42	35	83.3
Miscellaneous fabrics	7	4	57.1
Cellulosic/thermoplastic fabric blends	28	14	50.0
100% Thermoplastic fabrics	11	0	0
Fabric on cotton batting (no FR)			
100% Cellulosic fabrics	42	42	100
Cellulosic/thermoplastic fabric blends	28	23	82.1
Miscellaneous fabrics	7	5	71.4
100% Thermoplastic fabrics	11	1	9.1
Fabric on cotton batting (FR)			
100% Cellulosic fabrics	42	32	76.2
Cellulosic/thermoplastic fabric blends	28	12	42.9
Miscellaneous blends	7	1	14.3
100% Thermoplastic fabrics	11	0	0
Fabric on non-resinated polyester batting			
100% Cellulosic fabrics	42	14	33.3
Cellulosic/thermoplastic fabric blends	28	2	7.1
100% Thermoplastic fabrics	11	0	0
Miscellaneous fabrics	7	0	0
Fabric on resinated polyester batting			
100% Cellulosic fabrics	42	8	19.0
Cellulosic/thermoplastic fabric blends	28	1	3.6
100% Thermoplastic fabrics	11	0	0
Miscellaneous fabrics	7	0	0

continued

Table 3.1 (continued). Smolder Susceptibility of Materials as Measured by the California Bureau of Home Furnishings.

	Number of Fabrics Tested	Number of Fabrics Igniting	Percent Igniting
Fabric on 70/30 cotton/polyester batting			
100% Cellulosic fabrics	42	33	78.6
Cellulosic/thermoplastic fabric blends	28	9	32.1
Miscellaneous fabrics	7	1	14.3
100% Thermoplastic fabrics	11	0	0
Fabric on neoprene foam			
100% Cellulosic fabrics	42	39	92.9
Cellulosic/thermoplastic fabric blends	28	11	39.3
Miscellaneous fabrics	7	1	14.3
100% Thermoplastic fabrics	11	0	0
Fabric on neoprene foam interliner and cotton batting (no FR)			
100% Cellulosic fabrics	42	8	19.0
Miscellaneous fabrics	7	1	14.3
Cellulosic/thermoplastic fabric blends	28	1	3.6
100% Thermoplastic fabrics	11	0	0

Table 3.2. Ignitability Characteristics of Some Materials as Measured by the USF Ignitability Test.

Material	Heat Flux, W/cm²		
	5.8	8.1	10.5
	Ignition Time, sec		
Aspen poplar, 3/4 in	42	17	7
Beech, 3/4 in	79	20	11
Yellow birch, 3/4 in	*	22	11
Red oak, 3/4 in, 1	42	21	14
Red oak, 3/4 in, 2	55	23	17
Western red cedar, 3/4 in	26	8	6
Douglas fir, 3/4 in	46	23	11
Western hemlock, 3/4 in	44	21	10
Eastern white pine, 3/4 in	30	12	7
Southern yellow pine, 3/4 in	39	16	9
Hemlock, untreated, 3/4 in	43	16	
Hemlock, treated, 3/4 in	74	43	
Cellulose fiberboard, 1/2 in	11	5	3
Fiberboard soundstop, 1/2 in	15	6	4
Fiberboard sheathing, 1/2 in	17	7	5
Medium density hardboard, 1/2 in	34	21	14
Hardboard, 3/8 in	84	26	17
Chipboard, 1/8 in	77	23	14
Hardboard, unfinished, 1/4 in	222	22	12
Lauan plywood, unfinished, 1/4 in	229	17	10
Polymethyl methacrylate, clear, 1/8 in	115	33	31
Polystyrene, white, 1/8 in	108	35	31
Polyisocyanurate rigid foam, 1/2 in	*	*	56
Polyvinyl chloride flooring, 1/8 in	269	95	78
Linoleum, 1/8 in	59	54	12
Wool carpet, foam rubber backing	26	9	6
Polyester carpet, resin backing	79	42	28
Nylon carpet, foam rubber backing	29	17	13
Acrylic carpet, fiber backing	38	15	9
Wool fabric, FR	104	62	15
Wool/nylon fabric, FR	20	16	8
Rayon fabric, FR	12	9	5
Cotton fabric	10	6	3
Nylon fabric	22	26	7

*No ignition.

Table 3.3. Ignitability Characteristics of Some Materials as Measured by the Cone Calorimeter Test.

Material	Heat Flux, kW/m²		
	25	50	75
		Time to Sustained Ignition, sec	
ABS, FR	120	34	17
ABS, non-FR	111	38	17
HIPS, FR	304	106	25
HIPS, non-FR	205	52	24
PC-ABS, FR	267	53	28
PC-ABS, non-FR	189	49	21
UPT, FR	*	159	79
UPT, non-FR	119	42	
XPE, FR	162	63	37
XPE, non-FR	86	37	

*No ignition.

Ref. Grand, A. F., "The Use of the Cone Calorimeter to Assess the Effectiveness of Fire Retardant Polymers under Simulated Real Fire Test Conditions," Interflam '96, Cambridge, U.K. (March 1996).

Table 3.4. Flash-Fire Propensity of Materials as Measured by the USF Flash-Fire Screening Test.

Material	Height of Flash Fire, Inches	Time to Flash Fire, seconds
Hardwoods		
Aspen poplar	20 ± 6	31 ± 5
Beech	15 ± 3	36 ± 12
Yellow birch	22 ± 4	28 ± 4
Red oak	26 ± 0	34 ± 2
Softwoods		
Western red cedar	26 ± 0	29 ± 8
Douglas fir	26 ± 0	39 ± 11
Western hemlock	22 ± 3	25 ± 2
Eastern white pine	23 ± 6	26 ± 5
Southern yellow pine	26 ± 0	33 ± 8
Cellulosic board		
Fiberboard, core board	9 ± 3	35 ± 14
Fiberboard sounds top	10 ± 1	27 ± 5
Fiberboard sheathing, asphalt-imp.	12 ± 4	41 ± 10
Medium density hardboard	8 ± 1	29 ± 1
Hardboard	13 ± 6	29 ± 13
Chipboard	13 ± 3	35 ± 10
Solid plastics		
Polyethylene, 1	26 ± 0	115 ± 20
Polyethylene, 2	26 ± 0	118 ± 7
Polystyrene	8 ± 4	160 ± 37
Polymethyl methacrylate	26 ± 0	61 ± 18
ABS, 1	7 ± 2	127 ± 32
ABS, 2	10 ± 2	189 ± 7
Polycarbonate, 1	6 ± 3	97 ± 13
Polycarbonate, 2	none	none
Nylon 6	26 ± 0	55 ± 7
Nylon 6/6	26 ± 0	95 ± 37
Nylon 6/10	22 ± 8	77 ± 10
Polyvinyl chloride (3 samples)	none	none
Chlorinated polyvinyl chloride	none	none
Polyphenylene oxide, modified	none	none
Polyphenylene sulfide	none	none
Polyether sulfone	none	none
Polyaryl sulfone	none	none
Rigid foam plastics		
Polyester	22 ± 7	91 ± 8
Polyethylene	26 ± 0	141 ± 4
Polyurethane, no FR	3 ± 1	92 ± 4
Polyurethane, 10% FR (Cl)	3 ± 1	86 ± 11
Polyurethane, 7% FR (P)	4 ± 1	98 ± 6
Polyurethane, high density, FR, 1	4 ± 3	79 ± 7
Polyurethane, high density, FR, 2	6 ± 2	48 ± 8
Polyisocyanurate, urethane-mode., 1	2 ± 1	72 ± 5
Polyisocyanurate, urethane-mod., 2	none	none
Polymethacrylimide	8 ± 1	69 ± 4
Polybismaleimide	6 ± 2	50 ± 2

continued

Table 3.4 (continued). Flash-Fire Propensity of Materials as Measured by the USF Flash-Fire Screening Test.

Material	Height of Flash Fire, inches	Time to Flash Fire, seconds
Rigid foam plastics (cont.)		
Polyimide, modified	6 ± 3	44 ± 1
Polystyrene	3 ± 2	203 ± 10
Polyvinyl chloride	none	none
Flexible foam plastics		
Polyurethane (18 samples)	26 ± 1	44 ± 4
Polyurethane, F6	14 ± 7	33 ± 12
Polyurethane, H1	none	none
Polychloroprene (2 samples)	none	none
Upholstery fabrics		
Cotton (4 samples)	20 ± 3	36 ± 16
Rayon (3 samples)	20 ± 7	37 ± 12
Nylon (3 samples)	21 ± 6	83 ± 8
Polypropylene (3 samples)	25 ± 2	106 ± 13
Cushioning materials		
Cotton batting, no FR	14 ± 3	25 ± 3
Cotton batting, 10% boric acid	8 ± 4	29 ± 4
Excelsior	11 ± 3	21 ± 4
Sisal	23 ± 6	23 ± 4
Kapok	13 ± 2	27 ± 5
Loose fill insulation		
Cellulose, untreated	22 ± 3	18 ± 3
Cellulose, 36.3% 5-mol borax	6 ± 2	20 ± 2
Cellulose, 36.7% boric acid	1 ± 0	18 ± 1

Table 3.5. *Flame Spread and Smoke Characteristics of Materials as Measured by the ASTM E 84 Test.*

Material	Flame Spread Classification	Smoke Developed Classification
Red oak, untreated	100	100
Red oak, treated	25 to 50	0 to 45
Oak, treated	35	25
Aspen, treated	30	20
Yellow poplar, treated	25	30
Ash, treated	60	5
Birch, treated	15 to 30	0 to 15
Lauan, treated	15 to 20	5 to 25
Soft maple, treated	20 to 30	0 to 15
Basswood, treated	25 to 35	25 to 45
Cottonwood, treated	25	20 to 60
Virola, treated	15	10
Northern pine, treated	20	20 to 40
Ponderosa pine, treated	25 to 35	30 to 70
Jack pine, treated	25 to 55	15 to 35
Yellow pine, treated	25	25 to 30
Southern yellow pine, treated	10 to 15	0 to 10
Douglas fir, treated	15 to 25	0 to 15
White fir, treated	20	0
Hemlock, treated	15 to 25	0 to 10
Western hemlock, treated	20	0
Redwood, treated	20	0
Western red cedar, treated	45	25 to 40
Molded plastic	10 to over 200	65 to over 450
Reinforced plastic	15 to over 200	140 to over 450

Table 3.6. Heat Release Characteristics of Materials as Measured by the OSU Release Rate Test.

Material	Orientation, V-Vertical H-Horizontal	Applied Heat Flux, W/cm²	Maximum Heat Release Rate, Btu/ft²min	Total Heat Release, Btu/ft²	
				3 min	10 min
Oak, 1 in	V	1.0	300	250	750
		2.0	420	800	3,500
		2.5	550	1,000	4,000
Pine, 1 in	V	1.5	600	900	6,000
		2.5	800	1,200	8,000
Red oak, flooring	H	1.0	300	500	800
		2.0	450	900	3,500
		2.75	600	1,200	4,500
Hardboard, 0.25 in	V	1.0	800	400	4,500
		2.0	1,200	800	5,500
Particle board, 0.5 in	V	1.0	400	70	1,800
		2.5	800	1,800	5,000
Particle board, FR, 0.5 in	V	1.5	20	0	50
		2.5	350	200	1,200
Fiberboard, low density, 0.625 in	V	1.0	550	1,000	1,500
		2.0	700	1,500	1,600
Exterior plywood, 0.5 in	V	1.0	400	700	1,200
		2.0	600	1,000	3,000
Polypropylene, sheet, 0.125 in	H	1.0	2,200	250	7,000
Polyvinyl chloride, rigid, pipe	V	1.0	20	0	50
		2.6	300	100	1,500
Polyvinyl chloride, flexible, sheet, 90 mil	V	1.0	300	500	2,000
		2.0	500	1,000	2,000
Polystyrene, light diffuser	H	0	500	75	2,400
		1.0	750	250	2,500
ABS, sheet, 0.125 in	H	0	600	75	800
		1.0	1,800	600	1,800
		2.5	1,800	1,650	2,400
Polymethyl, methacrylate, light diffuser	H	0	500	175	3,000
		1.0	800	300	3,500
Polysulfone, sheet, 95 mil	H	1.0	0	0	0
		2.5	450	50	1,500
Polyurethane rigid foam, 1 in	V	0	300	500	500
		1.0	500	500	500
Polyurethane rigid foam, FR, 1 in	V	0.5	0	0	0
		1.0	300	350	350
		2.5	400	650	650
Polyisocyanurate rigid foam, 1 in	V	1.0	0	0	0
		2.75	400	100	150

Table 3.7. **Heat Release Characteristics of Some Materials as Measured by the Cone Calorimeter Test.**

Material	Heat Flux, kW/m²	Peak HRR, kW/m²	Avg HRR, kW/m²	Eff Hc, MJ/kg
ABS, FR	25	439.2	211.0	10.3
ABS, FR	50	413.5	247.2	10.0
ABS, FR	75	493.5	289.3	10.0
ABS, non-FR	20	670.8	382.8	29.0
ABS, non-FR	50	1005.4	538.4	28.3
ABS, non-FR	75	1215.4	691.3	29.4
HIPS, FR	25	303.7	232.2	11.0
HIPS, FR	50	252.0	210.5	10.2
HIPS, FR	75	301.2	220.0	9.8
HIPS, non-FR	20	834.2	331.7	29.8
HIPS, non-FR	50	1038.9	473.6	28.2
HIPS, non-FR	75	1217.7	615.2	26.4
PC-ABS, FR	25	351.1	226.4	17.8
PC-ABS, FR	50	320.7	259.8	18.4
PC-ABS, FR	75	453.1	313.3	17.0
PC-ABS, non-FR	20	436.1	316.9	22.4
PC-ABS, non-FR	50	468.8	339.6	22.4
PC-ABS, non-FR	75	590.3	416.4	22.2
UPT, FR	50	110.7	73.9	15.8
UPT, FR	75	125.1	102.8	14.3
UPT, non-FR	25	568.4	367.8	23.2
UPT, non-FR	50	941.6	600.8	23.8
XPE, FR	25	227.1	77.8	20.0
XPE, FR	50	293.1	221.7	22.0
XPE, FR	75	493.8	324.1	20.7
XPE, non-FR	20	931.0	322.2	36.1
XPE, non-FR	50	1517.0	891.1	39.9

Ref. Grand, A. F., "The Use of the Cone Calorimeter to Assess the Effectiveness of Fire Retardant Polymers under Simulated Real Fire Test Conditions," Interflam '96, Cambridge, U.K. (March 1996).

Table 3.8. Smoke Density from Various Materials (National Bureau of Standards Test).

	Thickness, mils	Maximum Specific Optical Density, Dm		Time to Ds = 16, minutes	
		Nonflaming	Flaming	Nonflaming	Flaming
Polyethylene					
UCC DXM-100	250	526		4.52	
UCC DFDA-6311	125	719	387	2.74	0.91
UCC DHDA-1811	125	739	375	3.76	1.09
UCC DMDA-7075	125	357	280	3.32	1.09
unidentified		468	150	5.5	4
unidentified	125		68		3.2
Polypropylene					
UCC JMD-8500	250	780	119	3.00	4.18
40% glass	100	691	428	2.13	1.57
rug	180	456	110	2.3	1.7
rug, burlap back	220	621	292		
Polytetrafluoroethylene		0	53	NR	11
Tetrafluoroethylene-vinylidene					
fluoride copolymer	71	75	109	2.5	1.2
Polyvinyl fluoride	2	1	4	NR	NR
Polyvinyl chloride					
UCC QCA-2460	15	11	98	NR	0.45
	20	23	153	2.66	0.28
	40	139	326	1.16	0.40
UCC QYTQ	250	315	780	3.25	0.49
rigid, filled	250	490	530	1.6	0.5
rigid, unfilled	125	270	525	2.1	0.5
	250	470	535	2.1	0.6
unidentified	10		100		
	15		210		
	30	120	430		
unidentified	250	300	660	3.9	0.8
fabric	26	261	198	1.4	0.3
Polyvinyl butyral					
UCC XYHL	250	33	31	4.71	6.50
Polystyrene					
UCC SMD-3500	250	395	780	4.00	0.63
unidentified		460	468	4.0	1.2
unidentified	125		660		0.6
unidentified	250	372	660	7.3	1.3
Styrene-acrylonitrile					
UCC C-11	250	389	249	4.13	1.11
40% glass	100	687	684	2.28	0.47

	Thickness, mils	Maximum Specific Optical Density, Dm		Time to Ds = 16, minutes	
		Nonflaming	Flaming	Nonflaming	Flaming
Arylonitrile-butadiene-styrene (ABS)					
Cycolac	250	780	780	2.98	0.57
25/10/65, sheet	45	76	660	4.3	0.6
	80	167	660	3.5	0.7
unidentified	22	188			
	32	220	450		
unidentified	46	71	660	4.8	0.6
unidentified, sheet	70	152	660	1.7	0.4
Polymethylmethacrylate					
FR UV-Abs	250	380	480	3.8	1.8
HR UV-Abs	250	195	90	6.3	2.2
HR UV-Trans	250	190	140	6.5	2.3
unidentified, sheet	180	203	383	6.0	1.9
unidentified	219	156	107	9.2	2.6
unidentified, sheet	230	304	660	4.5	1.4
unidentified, sheet	500	328		9.0	
unidentified	250	142	81		
Acrylic					
rug, Acrilan	300	319	159	1.5	0.6
carpet, 82 oz/sq. yd	375	470	220		
Modacrylic					
fabric	24	54	21	0.5	4.0
fabric, 5.8 oz/sq. yd	13	41	39	0.8	0.4
fabric, 5.9 oz/sq. yd	13	41	39	0.6	0.6
fabric, 8.0 oz/sq. yd	28	48	76	0.8	0.4
fabric, 9.3 oz/sq. yd	30	52	66	0.9	0.5
fabric, 9.6 oz/sq. yd	30	50	72	0.7	0.5
fabric, 10 oz/sq. yd	15	66	50	1.2	1.2
fabric, 10 oz/sq. yd	35	18	90	4.6	0.6
rug, 56 oz/sq. yd	25	324	464	1.5	0.5
Polyacetal, Celcon	125		6		NR
Nylon					
40% glass	100	487	41	3.63	9.20
unidentified		300	95	7.2	4.8
fabric, 13 oz/sq. yd	50	6	16	NR	15.0
rug	300	320	269	2.8	1.8
carpet, foam backing, 86 oz	390	310	270		
Polyamide					
aromatic, sheet, 17 oz/sq. yd	20	3	6	NR	NR
aromatic, sheet, 64 oz/sq. yd	63	7	14	NR	NR
aromatic, fabric, 4.4 oz/sq. yd	15	0	10	NR	NR
aromatic, fabric, 6.1 oz/sq. yd	15	5	8	NR	NR
aromatic, fabric, 9.9 oz/sq. yd	35	6	8	NR	NR
aromatic, fabric, 11 oz/sq. yd	35	30	32	3.4	2.4
aromatic, rug, 45 oz/sq. yd	300	65	51	4.8	3.2
aromatic, carpet, 49 oz/sq. yd	220	175	105		

continued

	Thickness, mils	Maximum Specific Optical Density, Dm		Time to Ds = 16, minutes	
		Nonflaming	Flaming	Nonflaming	Flaming
Polysulfone					
sheet, 28 oz/sq. yd	31	1	28	NR	5.9
sheet, 54 oz/sq. yd	60	4	40	NR	2.6
P-1700	250	111	370	12.61	1.89
3M-360	125		55		8.3
Polycarbonate					
Lexan 100	125		226		2.0
Lexan 113-T	250	48	324	10.85	1.95
Lexan 140	125		162		1.7
Lexan 2014	125		544		1.8
Lexan F6000	125	20	214		
Lexan MR4000	125	53	196		
Lexan F2000	125	38	334		
Lexan	62	3	275		
Lexan	62	1	165		
unidentified	125	12	174	NR	2.1
unidentified		21	324	27	2.0
unidentified, SE		44	660	14	1.6
Phenolic					
Genal 4000	125	110	50	5	5.8
Genal 4200	125	137	55	5	5.5
Genal 4300	125	35	71	8	3.1
Genal 4301	125	20	41	20	5.6
Polyester					
Paraplex P-43	125	780	780	2.66	0.59
Hetron 92	125	595		3.20	
glass reinforced	110	420	720		
Urethane rigid foam, polyether, sucrose/HT/PMDI					
2.5 pcf	250	49	45	0.42	0.24
	500	70	80	0.46	0.20
	1000	111	71	0.44	0.24
	2000	221	113	0.35	0.22
3.2 pcf	250	62	47	0.52	0.60
	500	103	95	0.44	0.20
	1000	195	124	0.44	0.22
	2000	272	119	0.45	0.23
7.5 pcf	250	158	90	0.46	0.31
	500	272	112	0.53	0.25
	1000	446	134	0.52	0.24
	2000	499	183	0.57	0.33
16 pcf	250	363	161	0.72	0.38
	500	580	220	0.75	0.24
	1000	652	311	0.72	0.33
	2000	647	270	0.65	0.35

Table 3.8 (continued). Smoke Density from Various Materials (National Bureau of Standards Test).

	Thickness, mils	Maximum Specific Optical Density, Dm		Time to Ds = 16, minutes	
		Nonflaming	Flaming	Nonflaming	Flaming
Urethane rigid foam, polyether, sucrose/HT/PMDI (cont.)					
35 pcf	250	718	486	1.67	0.52
	500	658	710	1.61	0.52
	1000	652	529	1.46	0.41
	2000	710	789	1.32	0.66
62 pcf	250	789	664	2.22	0.69
	500	789	333	2.19	0.72
	1000	661	737	2.60	0.78
	2000	626	449	2.55	0.85
sucrose/MDI, FR, 2 pcf		170	500	0.30	0.13
TDI, FR, 1.8 pcf		290	285	0.44	0.17
TDI, 13 pcf		515	319	0.52	0.80
PAPI, FR, 2 pcf, CO_2 blown		119	196	0.36	0.21
PAPI, FR, 2 pcf		112	252	0.40	0.14
Urethane rigid foam, polyester					
TDI, 2.5 pcf		161	70	0.43	0.20
TDI, 25 pcf		431	534	1.0	0.5
PAPI, 4 pcf		454	525	0.39	0.18
Urethane rubber					
TDI polyether		57	210	4.3	2.1
TDI polyester		131	230	4.0	1.5
Urethane foam, unidentified	500	156	20	0.5	0.5
Urethane foam, polyether	625	156	35	0.5	1.4
Urethane foam, polyester, 9.2 oz	210	77	58	0.6	1.9
Urethane foam, polyether, 35 oz	600	164	229	0.4	0.2
Urethane foam, polyether, FR	1000	318	30	0.5	2.5
Urethane foam, polyether, FR	1000	286	262	0.7	0.2
Urethane foam, polyether	1000	300	41	0.7	0.6
Red oak	250	395	76	4.1	8.0
Red oak	250	552			
Red oak	500	372	118	3.8	11.2
Red oak, flooring	750	505	300		
Red oak	780	660	117	7.1	7.8
White oak	250	420	107	3.5	6.6
White oak	500	409	56	4.0	11.8
Elm	50	150	35		
Elm	125	270			
Elm	185	390	65		
Elm	240	510			
Douglas fir	250	380	156	2.1	4.6
Douglas fir	500	438	110	3.0	6.8
Clear spruce	125	147	45	5.1	5.5
Clear spruce	250	275	115	4.3	4.2
Clear spruce	500	378	145	4.6	4.3

continued

Table 3.8 (continued). Smoke Density from Various Materials (National Bureau of Standards Test).

	Thickness, mils	Maximum Specific Optical Density, Dm		Time to Ds = 16, minutes	
		Nonflaming	Flaming	Nonflaming	Flaming
Clear spruce	750	421	310	4.8	4.8
Sitka spruce	500	263	130	4.2	7.7
White pine	250	325	155	2.3	2.7
Southern pine	500	431	156	3.0	7.8
Yellow birch	500	419	79	4.2	7.0
Black walnut	250	460	91	3.4	7.5
Western larch	500	323	111	3.8	7.4
Red lauan (Philippine mahogany)	500	374	96	2.5	5.4
Mahogany	500	320	109	3.5	5.2
Sugar maple	500	448	118	4.7	10.4
Eastern red cedar	500	372	76	2.9	4.6
Redwood	250	260	133	2.7	2.5
Redwood (Sequoia)	500	390	85	3.2	4.7
Marine plywood	250	285	62	2.7	5.4
Douglas fir interior plywood	250	350	96	3.4	5.3
Douglas fir exterior plywood					
uncoated	250	287	112	4.6	4.7
latex paint	254	258	108	3.8	5.6
alkyd paint	253	236	84	4.6	4.8
varnish	254	262	71	5.0	4.3
wallpaper, 5 coats	285	285	132	3.3	4.4
wall cloth	261	79	56	5.1	4.7
enameled wall covering	305	184	79	4.1	4.3
vinyl film	254	268	75	3.7	1.7
vinyl counter top	320	355	355	2.7	0.7
Douglas fir exterior plywood					
uncoated	375	404	14	1.9	4.8
primer, acrylic emulsion paint		336	4	2.0	NR
alkyd resin flat paint		195	21	3.1	7.0
phenolic spar varnish		422	14	1.6	6.3
shellac		488	12	1.9	3.8
alkyd enamel undercoat, alkyd resin titanium base flat paint		225	19	4.2	5.9
alkyd enamel undercoat, semigloss alkyd enamel		305	20	2.9	16.8
alkyd enamel undercoat, gloss alkyd enamel		263	12	2.0	6.2
alkyd enamel undercoat, latex paint		255	23	2.1	15.1

Table 3.8 (continued). Smoke Density from Various Materials (National Bureau of Standards Test).

Thickness, mils	Maximum Specific Optical Density, Dm		Time to Ds = 16, minutes	
	Nonflaming	Flaming	Nonflaming	Flaming
Douglas fir exterior plywood (cont.)				
oil base primer, linseed oil base paint	240	5	2.0	NR
linseed oil base paint	339	17	2.5	10.2
alkyd enamel undercoat, fire retardant latex paint	80	55	1.7	1.4
modified synthetic resin base fire retardant paint	44	69	1.7	2.4
modified epoxy fire retardant paint	195	137	0.9	0.3
sealer, modified polyurethane fire retardant varnish	361	117	0.8	0.2
Paperboard, homogeneous	359	68	1.3	4.2
Paperboard, laminated	169	60	2.0	10.9

81

Table 3.9. Decomposition and Combustion Products from Some Polymers.

Polyethylene: pyrolysis (major products)
Pentene
Hexene-1
n-Hexane
Heptene-1
n-Heptane
Octene-1
n-Octane
Nonene-1
Decene-1
Polyethylene: thermo-oxidation (major products)
Propanal
Pentene
n-Pentane
Butyraldehyde
Valeraldehyde
Polyethylene: combustion (major products)
Pentene
Butyraldehyde
Hexene-1
n-Hexane
Benzene
Valeraldehyde
Polypropylene: pyrolysis (major products)
Propylene
Isobutylene
Methylbutene
Pentane
2-pentene
2-methyl-1-pentene
Cyclohexane
2,4-dimethyl-1-heptene
2,4,6-trimethyl-8-nonene
Polypropylene: thermal oxidation (major products)
Propylene
Formaldehyde
Acetaldehyde
Butene
Acetone
Cyclohexane
Polyvinyl chloride: pyrolysis (major products)
Chloromethane
Benzene
Toluene
Dioxane
Xylene
Indene
Naphthalene
Chlorobenzene
Divinylbenzene
Methyl ethyl cyclopentane
Hydrogen chloride

Table 3.9 (continued). Decomposition and Combustion Products from Some Polymers.

Polyvinyl chloride: thermo-oxidation (major products)
  Benzene
  Toluene
  Chlorobenzene
  Divinylbenzene
  Hydrogen chloride
Polystyrene: combustion (products found)
  Acetaldehyde
  Acetophenone
  Acetylene
  Acrolein
  Acryldehyde
  Allylbenzene
  Benzaldehyde
  Benzene
  Benzoic acid
  Benzyl alcohol
  Butane
  1-Butene
  Carbon dioxide
  Carbon monoxide
  Cinnamaldehyde
  Cumene
  1,3-diphenylpropane
  1,3-diphenylpropene
  Ethane
  Ethylbenzene
  Ethylene
  Ethylmethylbenzene
  Formaldehyde
  Formic acid
  Methane
  Methanol
  Methyl phenol
  alpha-methylstyrene
  beta-methylstyrene
  Phenol
  1-phenylethanol
  Propane
  Isopropylbenzene
  n-propylbenzene
  Propylene
  alpha-styrene
  Styrene monomer
  Styrene dimer
  Styrene trimer
  Styrene oxide
  Toluene

continued

Table 3.9 (continued). Decomposition and Combustion Products from Some Polymers.

Acrylonitrile-butadiene-styrene (ABS): pyrolysis and
  oxidative combustion (products found)
    Acetophenone
    Acids (combustion only)
    Acrolein
    Acrylonitrile
    Aldehydes (combustion only)
    Benzaldehyde
    Carbon dioxide (combustion only)
    Carbon monoxide (combustion only)
    Cresol
    Dimethylbenzene
    Ethanal
    Ethylbenzene
    Ethylmethylbenzene
    Hydrogen cyanide
    Isopropylbenzene
    alpha-methylstyrene
    beta-methylstyrene
    Nitric oxide (combustion only)
    Nitrogen dioxide (combustion only)
    Phenol
    Phenyl cyclohexane
    2-phenyl-2-propanol
    3-phenyl-1-propene
    n-propylbenzene
    Styrene
    4-vinyl-1-cyclohexane
Polymethyl methacrylate (PMMA): pyrolysis (major products)
    Methyl methacrylate
Poly (n-butyl methacrylate): pyrolysis (major products)
    n-butyl methacrylate
Aliphatic polyesters: pyrolysis (major products)
    Acetaldehyde
    Butadiene
    Carbon dioxide
    Carbon monoxide
    Cyclopentanone
    Propanaldehyde
    Water
Semi-aromatic polyesters: pyrolysis (major products)
    Carbon dioxide
    Carbon monoxide
    Cyclohexene
    1,5-hexadiene
Aromatic polyesters: pyrolysis (major products)
    Benzene
    Carbon dioxide
    Quinone
    Water

***Table 3.9 (continued). Decomposition and Combustion
Products from Some Polymers.***

Nylon 6: pyrolysis (major products)
   Caprolactam
   Benzene
   Acetonitrile
   Hydrocarbons with five carbons or less
Nylon 6/6: pyrolysis (major products)
   Hydrocarbons with five carbons or less
   Cyclopentanone
Polyurethane foams: pyrolysis (major products found)
   Hydrogen cyanide
   Acetonitrile
   Acrylonitrile
   Benzene
   Benzonitrile
   Naphthalene
   Pyridine
   Toluene
   Propene
   Acetone
Phenol-formaldehyde: pyrolysis (major products)
   Propanols
   Methylphenols
   Dimethylphenols

Table 3.10. Time to Death for Various Materials in Dome Chamber Toxicity Test (Condition 1).

Material	Number of Samples	Time to Death, min.
Feathers/down 75/25	1	7.3
Wool fabric	4	7.6 ± 1.3
Silk fabric	2	9.2 ± 0.4
Polyester fabric	3	10.7 ± 2.2
Mashed potatoes	1	10.9
Polyether sulfone	3	12.3 ± 2.1
Cotton fabric, including upholstery fabric	10	13.1 ± 2.1
Polyphenylene sulfide	4	13.2 ± 3.8
Polyaryl sulfone	2	13.5 ± 3.2
Polyimide flexible foam	2	13.7 ± 1.4
Sugar, soft drink mix	2	14.0 ± 1.0
Wood	12	14.0 ± 1.5
Cotton fabric, FR	6	14.2 ± 3.6
Viscose fiber, including FR	4	14.2 ± 0.4
Polyamide (nylon)	3	14.4 ± 1.7
Rayon fabric, including upholstery fabric	10	15.4 ± 2.4
Polyphenyl sulfone	1	15.5
Polyurethane rigid foam	7	15.5 ± 4.1
Polymethyl methacrylate (PMMA)	1	15.6
Polyvinylidene fluoride	1	15.9
Saltine crackers, vanilla wafers	2	16.0 ± 1.2
Cellulosic board	8	16.6 ± 3.5
Polyvinyl chloride (PVC), unplasticized	2	16.6 ± 0.3
Polypropylene fabric, including upholstery fabric	4	16.6 ± 2.8
Nylon fabric, including upholstery fabric	9	16.8 ± 3.5
Acrylonitrile-butadiene-styrene (ABS)	3	17.1 ± 2.4
Phenolic	1	17.2
Polyethylene, including foam	5	17.3 ± 3.7
Polyetherimide	1	17.4
Margarine, cooking oil	2	17.8 ± 0.6
Chocolate flavoring, freeze-dried coffee	2	18.7 ± 0.2
Acrylonitrile rubber (NBR)	3	19.1 ± 2.9
Rice, flour, spaghetti	3	19.8 ± 0.4
Polyphenylene oxide, modified	1	20.0
Polyurethane flexible foam	14	20.0 ± 1.7
Bisphenol A polycarbonate	3	20.4 ± 3.8
Polyvinyl fluoride	1	20.5
Ethylene-propylene-diene (EPDM)	2	20.7 ± 0.1
Chlorosulfonated polyethylene	2	20.9 ± 2.1
Polypropylene, pellet and fiber	2	20.9 ± 0.8
Dry milk, cocoa mix	2	21.0 ± 0.1
Polyisocyanurate rigid foam	2	21.7 ± 1.4
Polyisoprene (natural rubber)	1	22.1
Chlorinated polyvinyl chloride (CPVC)	2	22.2 ± 0.7
Polybutylene	2	22.5 ± 1.2
Polystyrene	2	23.1 ± 4.3
Polyphosphazene flexible foam	2	24.0 ± 0.1
Styrene-butadiene rubber (SBR)	1	24.1
Silicone foam	1	25.0
Polychloroprene flexible foam	6	25.9 ± 1.2
Chlorinated polyethylene	2	26.1 ± 1.8

Table 3.11. LC_{50} for Various Materials in Dome Chamber Toxicity Test (Condition 1).

Material	LC_{50}, mg/litre
Wool/nylon 90/10 fabric	8.8
Acrylonitrile-butadiene-styrene (ABS)	9.5, 10.4
Nylon 6	9.5, 10.8
Polyethylene	11.8, 11.9, 14.3
Polyurethane flexible foam	13.2, 16.6
Polycarbonate	15.0
Nylon, treated	15.2
Polyaryl sulfone	16.6, 21.8, 21.9
Acrylonitrile-butadiene-styrene (ABS)	21.4
Polyether sulfone	22.1, 22.4, 25.0, 30.1
Douglas fir	22.4, 22.4
Wool, treated	22.5
Phenolic C/7781 glass composite, 25% resin	24.1
Wool fabric	25.0, 27.6
Polychloroprene flexible foam	25.5
Polyphenylene sulfide	26.4, 31.0, 32.1, 33.1
Furane	29.0
Polycarbonate	32.4
Polyphenylene oxide, modified	34.0
Chlorinated polyvinyl chloride (CPVC)	34.2, 35.9
Fluorene polycarbonate, uncured	34.4
Aromatic polyamide fabric	35.0, 43.6
Polyvinyl fluoride film/acrylic adhesive	41.3
Aromatic phenolic fabric	42.0
Bisphenol A polycarbonate (reference material)	52.4
Polyether sulfone/glass fabric	64.0
Polyvinyl fluoride film	65.2
Epoxy B/120 glass composite, 47% resin	71.4
Red oak	73.6, 73.8
Polyquinoxaline/polyamide composite	77.5
Epoxy A/7781 glass composite, 42% resin	78.0
Polychloroprene flexible foam	84.5
Phenolic F/120 glass composite, 48.7% resin	87.3
Polyvinylidene fluoride	98.6
Phenolic B/120 glass composite, 47.4% resin	119.2
Phenolic D/120 glass composite, 41.7% resin	119.3, 133.0
Phenolic E/7781 glass composite, 28.3% resin	228.6
Polychloroprene flexible foam	232.0
Fluorene polycarbonate, cured	232.5

Table 3.12. Corrosivity Data on Some Materials Using the Cone Corrosimeter Method at 50 kW/m².

	Material	24 hr Metal Loss Angstroms/45,000 Angstrom Target, Average
1	XL olefin elastomer, metal hydrate filler	600
2	Blend of HDPE and chlorinated PE elastomer	4718
3	Chlorinated PE, fillers	2148
4	EVA polyolefin, ATH filler	398
5	Blend of polyphenylene oxide and polystyrene	480
6	Polyetherimide	203
7	Polyetherimide/siloxane copolymer	540
8	Polypropylene, intumescent	540
9	Nylon, mineral filler	
10	Polyolefin copolymer, mineral filler	615
11	XL polyolefin copolymer, mineral filler	622
12	XL polyolefin copolymer, ATH filler	75
13	XL polyolefin copolymer, ATH filler	645
14	EVA polyolefin, mineral filler	495
15	Polyolefin, mineral filler	563
16	XL polyethylene copolymer, chlorinated additive	1965
17	Polyvinylidene fluoride material	2453
18	Polytetrafluoroethylene material	
19	PVC material	5205
20	PVC building wire compound	
21	Polyethylene homopolymer	
22	Douglas fir	555
23	EVA polyolefin copolymer	
24	Nylon 6/6	1710
25	XL polyethylene copolymer, brominated additive	435

Ref. Bennett, J. G., Kessel, S. L., Rogers, C. E., "Corrosivity Test Methods for Polymeric Materials. Part 4. Cone Corrosimeter Test Method," Journal of Fire Sciences, Vol. 12, No. 2, 175–195 (March/April 1994).

CHAPTER 4

Flammability Tests

Because expressing information as numbers facilitates comparison, the behavior of plastic materials when exposed to fire is generally expressed as numbers obtained through flammability tests. It is hoped that these tests will provide some correlation with actual fire exposure.

A warning issued by the American Society for Testing and Materials with its standard flammability test methods is appropriate for all flammability tests:

> This standard should be used to measure and describe the response of materials, products, or assemblies to heat and flame under controlled conditions and should not be used to describe or appraise the fire-hazard or fire-risk of materials, products, or assemblies under actual fire conditions. However, results of the test may be used as elements of a fire-hazard assessment or a fire-risk assessment which takes into account all of the factors which are pertinent to an assessment of the fire hazard or fire risk of a particular end use.

Most tests are relatively small in scale for three reasons:

1. Experimental materials are often available in relatively small quantities.
2. It is prohibitively costly to burn complete structures repeatedly.
3. Small-scale tests are more easily and more economically replicated than large-scale fires.

Many tests have an inherent deficiency in that they fail to reproduce the massive effect of heat present in a large-scale fire, and therefore give results that can be misleading if applied in the wrong context.

The market for plastics is an international market. The most prominent international standards organization is the International Standardization Organization (ISO), P.O. Box 56, CH-1211 Geneva 20, Switzerland. Some ISO standards relevant to fire safety and plastics are shown in Table 4.1.

Most of the flammability tests discussed in this chapter are tests used in the United States of America.

89

There are at least five standards organizations in the United States which issue standards relevant to fire safety and plastics:

1. American National Standards Institute (ANSI), 11 West 42nd Street, New York, New York 10036, Tel 212-642-8908, Fax 212-398-0023
2. American Society for Testing and Materials (ASTM), 100 Barr Harbor Drive, West Conshohocken, Pennsylvania 19428-2959, USA, Tel 610-832-9500, Fax 610-832-9555. Some ASTM standards relevant to fire safety and plastics are shown in Table 4.2.
3. Underwriters Laboratories (UL), 300 Pfingsten Road, Northbrook, Illinois 60062-2096, USA, Tel 847-272-8800. Some UL standards relevant to fire safety and plastics are shown in Table 4.3.
4. National Fire Protection Association (NFPA), 1 Batterymarch Park, P.O. Box 9101, Quincy, Massachusetts 02269-9101, USA, Tel 617-770-3000. Some NFPA standards relevant to fire safety and plastics are shown in Table 4.4.
5. International Conference of Building Officials (ICBO), 5360 South Workman Mill Road, Whittier, California 90601, USA, Tel 562-699-0541. Some Uniform Building Code (UBC) standards relevant to fire safety and plastics are shown in Table 4.5.

The standards of other predominantly English-speaking countries are similar to American standards in varying degrees.

There are at least three standards organizations in Canada which issue standards relevant to fire safety and plastics:

1. Standards Council of Canada (SCC)
2. Canadian Standards Association (CSA), 178 Rexdale Boulevard, Etobicoke, Ontario M9W 1R3, Tel 416-747-4033, Fax 416-747-2475. Some CSA standards relevant to fire safety and plastics are shown in Table 4.6.
3. Underwriters Laboratories of Canada (ULC), 7 Crouse Road, Scarborough, Ontario M1R 3A9, Tel 416-757-3611, Fax 416-757-9540. Some ULC standards relevant to fire safety and plastics are shown in Table 4.7.

The standards organization in Australia is Standards Association of Australia (SAA). Some Australian standards (AS) standards relevant to fire safety and plastics are shown in Table 4.8.

The standards organization in New Zealand is Standards New Zealand (SNZ).

The standards organization in the United Kingdom is British Standards Institution (BSI), 2 Park Street, London W1A 2BS, United Kingdom. Some British standards (BS) relevant to fire safety and plastics are shown in Table 4.9.

Some standards of other countries, relevant to fire safety and plastics, are listed in Section 6.9 (The International Market).

Many flammability tests exist, have existed, and will exist in multiple versions which differ, in apparatus, specimen, and procedure, to varying degrees with time and with place. In some cases, test standards differ because of rounding-off during conversions between English and metric measurements, with 4 inches, for example, becoming 100 to 102 mm after conversion. Some tests undergo sudden periods of rapid change. Some test standards have been withdrawn or discontinued by their issuing organizations but continue to be cited and used. For these and other reasons, only those tests and test details which the author believes are and may remain reasonably constant will be described in this chapter, and the details of apparatus, specimen, and procedure may vary from those described.

SECTION 4.1. TYPES OF FLAMMABILITY TESTS

Flammability tests can be classified according to the fire response characteristics which they are designed to measure:

1. Tests for smolder susceptibility, such as the tests developed by the California Bureau of Home Furnishings, National Bureau of Standards (NBS), and Upholstered Furniture Action Council (UFAC)
2. Tests for ignitability, such as the ASTM D 1929 test for ignition temperatures
3. Tests for flash-fire propensity, such as the NBS and Douglas flash-fire tests
4. Tests for flame spread, such as the ASTM E 84 or UL 25-foot tunnel test and the ASTM E 162 or NBS radiant panel test
5. Tests for heat release, such as the ASTM E 906 or Ohio State University (OSU) release rate test or the ASTM E or NBS cone calorimeter test
6. Tests for fire endurance, such as the ASTM E 119 and UL 181 tests
7. Tests for ease of extinguishment, such as the ASTM D 2863 test for oxygen index.
8. Tests for smoke evolution, such as the ASTM E 662 and ASTM D 2843 tests
9. Tests for toxic gas evolution, such as the University of Pittsburgh test, the German DIN 53436 test, and the Japanese JIS A1321 test
10. Tests for corrosive gas evolution, such as the British CEGB test and the French CNET test

Tests can be divided into two general groups: those which are primarily research tests, and those which are primarily acceptance tests. The first group includes those test methods which are standard primarily because they are recognized as desirable methods for evaluating the fire response characteristics of materials on the basis of technical soundness and scientific value. The second group includes those test methods which are standard primarily because they are a requirement for doing business. Most of the acceptance tests have, at some time or other, been used as research tests. Hopefully, the best research tests will eventually become acceptance tests.

Flammability tests can be divided into three groups on the basis of severity:

1. Low-severity tests, which provide exposure to relatively small amounts of heat for relatively short periods of time, simulating small accidental fires

2. Medium-severity tests, which provide exposure to significant heat for longer periods of time, simulating developing fires.
3. High-severity tests, which simulate the large-scale fire with massive heat exposure.

The characteristics of some ignition sources define their severity. Tests which employ sparks, electric arcs, hot surfaces, and open flames, even apparently large burner flames, for not more than 30 seconds can not be expected to deliver more than 100 KJ, and are low-severity tests. Burning crumpled or folded paper can deliver 200 to 4,000 KJ in 1 to 8 minutes, and may provide medium-severity exposure. Burning bedding can deliver 130,000 KJ in 20 minutes and may provide high-severity exposure.

Flammability tests can be divided into those which are intended for individual materials, and those designed for complete systems. Most tests for fire endurance and many tests for surface flame spread are designed for complete systems. Most tests for ease of ignition and many tests for surface flame spread and smoke evolution are intended for individual materials.

Flammability tests for plastic materials can be divided into four groups on the basis of origin:

1. Tests originally designed for plastics. These tend to be low-severity tests because the initial uses for plastics were in small articles and in relatively small parts for electrical appliances.
2. Tests originally designed for fabrics and textiles, and pertinent to plastics because of their use in home furnishings and vehicle interiors.
3. Tests originally designed for materials of construction. These tend to be high-severity tests, and are pertinent to plastics because of their use in construction.
4. Tests originally designed for measuring fire response characteristics in a particular application or under a specific type of exposure, such as hot-bolt and arc-ignition tests, and tests for electrical wire and cable, for floor coverings, for furniture and furnishings, and for pipe insulation.

SECTION 4.2. TESTS FOR SMOLDER SUSCEPTIBILITY

Smolder susceptibility may be defined as the tendency of a material to support smoldering combustion within itself. This characteristic provides a measure of fire hazard in that a material undergoing smoldering combustion may generate increasing amounts of heat and may, under certain conditions, bring itself or an adjacent material to the point of ignition.

The simplest test methods for smolder susceptibility involve placing lighted cigarettes at specified locations on the actual products to be tested. This is the basis for smolder susceptibility tests for mattresses described in Federal flammability standard FF 4-72 (16 CFR 1632) and California Technical Bulletin 106, and smolder susceptibility tests for upholstered furniture described in California Technical Bulletin 116.

Test methods for smolder susceptibility have been developed by the National Bureau of Standards (NBS), now the National Institute of Standards and Technology (NIST), and by the California Bureau of Home Furnishings, now the California Bureau of Home Furnishings and Thermal Insulation. Both laboratories used a small-scale mock-up or mini-mock-up 203 mm (8 in) wide primarily for comparing furniture components such as fabrics and foams, and a larger scale prototype or mock-up 450 by 550 mm (18 by 22 in) simulating complete assemblies. The ignition source was a king-size non-filter cigarette covered with one layer of laundered 100% cotton bed sheeting material. Some smolder susceptibility test data from the California Bureau are presented in Table 3.1.

A review of the data showed that smolder susceptibility was not a function of the fabric alone, or of the substrate alone, but of the specific fabric/substrate combination.

The 100% cellulosic fabrics exhibited the greatest susceptibility to smoldering, but this susceptibility was greatly reduced when the substrate was resinated polyester batting or neoprene foam interliner. This smolder susceptibility was also reduced when the substrate was non-resinated polyester batting or certain polyurethane flexible foams. The 100% thermoplastic fabrics exhibited no smolder susceptibility except when untreated cotton batting was the substrate.

Smoldering combustion generally requires a porous fuel, and charring appears to be a prerequisite for smoldering. Cellulosic fabrics such as cotton and rayon are, like wood, prone to char and smolder. The char formation promoted by some fire retardants also may encourage smoldering. Thermoplastic fabrics such as nylon and polyolefin inhibit smoldering, by not forming char and by melting down to seal off areas of char to air diffusion and spread of smoldering. Wool and wool blend fabrics, which both melt and char, tend to exhibit enough melting to inhibit smoldering combustion.

The marked differences between these types of fabrics were brought out in smolder susceptibility test data using the USF smolder susceptibility test method, using a California-type mini-mock-up and polyurethane flexible foam substrates of the appropriate densities for seat back and bottom cushions, fire-retarded to meet both FAA and California flame resistance requirements. These data are presented in Table 4.10.

The mini-mock-up and cigarette ignition are used in tests described in California Technical Bulletin 117, Section D Part II for resilient cellular materials in upholstered furniture, in the UFAC voluntary standard for upholstered furniture, and in ASTM E 1353 and NFPA 260 for furniture components. The mock-up and cigarette ignition are used in ASTM E 1352 and NFPA 261 for furniture and in California Technical Bulletin 116 for upholstered furniture.

SECTION 4.3. TESTS FOR IGNITABILITY

Ignitability may be defined as the facility with which a material or its pyrolysis products can be ignited under given conditions of temperature, pressure, and oxygen concentration. This characteristic provides a measure of fire hazard in that a material which has an ignition temperature significantly higher than that of another material is less likely to contribute to a fire, all other factors being the same in both cases. For example, if a fire wall could be relied upon to keep the temperature on the side away from the fire below 300°C. for a stated period of time, a material that ignited at 200°C. would be a potential hazard and a material that ignited at 400°C. would not be a potential hazard. Some measures of ease of ignition are autoignition temperature, flash ignition temperature, and ignition sensitivity.

Many tests provide a measure of both ease of ignition and surface flammability. Some tests have, in essence, become measures of surface flammability for easily ignited materials, and measures of ease of ignition for materials that are difficult to ignite. For convenience, all tests which provide some measure of surface flammability will be discussed in Section 4.5; in general, this group includes all tests which provide a measurable distance available for relatively free flame travel.

Almost any material can be made to ignite, given enough heat, enough oxygen, and enough time. Ease of ignition can therefore be measured by the amount of heat required under fixed conditions of oxygen and time, by the amount of oxygen required under fixed conditions of heat and time, or by the amount of time required under fixed conditions of heat and oxygen. The most simple tests provide fixed conditions of heat, oxygen, and time, and the specimen either ignites or does not ignite under those fixed conditions.

There are four possible sources of heat energy which can cause ignition:

1. Chemical heat energy, such as heat of combustion
2. Electrical heat energy, such as heat from resistance elements and from arcing
3. Mechanical heat energy, such as heat from friction
4. Nuclear heat energy, such as heat of fission

In general, only two sources of energy are used in tests for ease of ignition:

1. Chemical heat energy, in the form of a direct flame or heated object
2. Electrical heat energy, in the form of resistance elements and arcs

Section 4.3.1. Heated Air and Radiant Heat

Electrical heat energy is used in the ignitability tests which are the most precisely controlled, providing either air heated to specified temperatures, or radiant heat flux at specified levels. Air heated to specified temperatures is used in the ASTM E 136 and ASTM D 1929 tests. Radiant heat flux at specified levels is used in several tests.

Both the ASTM E 136 and ASTM D 1929 tests employ a vertical furnace tube 254 mm (10 in) long with 102 mm (4 in) inside diameter, heated by electric heating coils, and an inner refractory tube 154 mm (10 in) long with 76 mm (3 in) inside diameter, inside which the specimen is placed. Air is admitted at a controlled rate, and its temperature is measured by means of thermocouples.

In the ASTM E 136 test, a specimen 51 mm (2 in) long, 38 mm (1.5 in) wide, and 38 mm (1.5 in) thick is placed in a stream of air at 750°C. (1382°F.) moving at 3 m (10 ft) per minute. A material passes this test if specimen temperatures do not increase more than 30°C. (54°F.) and there is no flaming after the first 30 seconds. In earlier versions of the test, a material which passed the test was considered noncombustible.

The ASTM E 136 test is similar to those described in the following standards: UBC 2-1.

In the ASTM D 1929 test, a specimen is exposed to air at successively higher temperatures until ignition is observed. Flash-ignition temperature is defined as the lowest initial temperature of air passing around the specimen at which a sufficient amount of combustible gas is evolved to be ignited by a small external pilot flame. Self-ignition temperature is defined as the lowest initial temperature of air passing around the specimen at which, in the absence of an ignition source, the self-heating properties of the specimen lead to ignition or ignition occurs of itself, as indicated by an explosion, flame, or sustained glow. This test is also known as the Setchkin ignition test. Data for various materials are presented in Table 2.5.

The ASTM D 1929 test is similar to those described in the following standards: UBC 26-6.

The classic approach to the study of ignition behavior of materials involves the measurement of time to ignition at different heat flux levels. The University of San Francisco (USF) ignitability test employs a 50/50 platinum/rhodium wire electric heater to provide heat flux up to 13 W/cm^2. A specimen 76.2 mm (3 in) square is supported vertically in a frame such that an area of front surface 65.1 mm (2-9/16 in)(square in exposed to a specified level of heat flux. Time to ignition is recorded. Some data obtained using this test are presented in Table 3.2.

Time to ignition generally decreases with increasing heat flux. Because heat flux is a function of distance from the heat source and is reduced by intervening materials, design and placement of the material in the system can often determine whether or not ignition will occur in a particular fire situation.

Several flammability tests which employ electric heaters to provide more precisely controlled radiant heat flux levels were originally designed to measure other fire response characteristics such as heat release, but are capable of measuring time to ignition at various radiant heat flux levels.

Time to ignition can be measured at various radiant heat flux levels, or the radiant heat flux level needed for ignition can be determined, in several tests, or modifications of tests, which include the following:

ASTM D 5485 (cone corrosimeter)
ASTM E 648 (flooring, radiant panel, gas-fired)
ASTM E 662 (smoke, NBS chamber)
ASTM E 906 (heat release, OSU apparatus)
ASTM E 1354 (heat release, NBS cone calorimeter)
BS 476 Part 13 or ISO 5657

In the ASTM D 5485 and ASTM E 1354 tests, a radiant heat source, using an electrical heater rod, rated at 5000 W and 240 V, tightly wound into the shape of a truncated cone, is used to generate heat flux up to 100 kW/m². Some ignitability data obtained using the ASTM D 5485 test are presented in Table 4.11. Some ignitability data obtained using the ASTM E 1354 test are presented in Table 3.3.

Section 4.3.2. Hot Surface, Hot Wire, and Arc

Electrical heat energy is also used in ignitability tests which are less precisely controlled but which realistically simulate possible ignition in electrical applications. These include hot-surface ignition, hot-wire ignition, and arc ignition.

Hot-surface ignition (HSI) resistance is expressed as the number of seconds required to ignite a specimen by an electrically heated surface operating at a specified temperature.

The specimen is heated by electrically-heated surfaces in the following tests:

ASTM D 757, Globar at 950°C. (1742°F.)
UL 746A, Sec. 33, glow wire

Hot-wire ignition (HWI) resistance is expressed as the number of seconds needed to ignite standard specimens that are wrapped with resistance wire that dissipates a specified level of electrical energy.

The specimen is heated by electrically heated wire coils in several tests which include the following:

ASTM D 229
ASTM D 3874
UL 746A, Sec. 30. (hot wire ignition, similar to ASTM D 3874)
Federal Test Method Standard No. 406, Method 2023

High-current arc ignition (HCAI) resistance is expressed as the number of arc-rupture exposures (standardized as to electrode type and shape and electrical circuit) that are necessary to ignite a material when they are applied at a standard rate on the surface of the material.

High-voltage arc ignition (HVAI) resistance is expressed as the number of seconds needed to ignite a material subjected to repeated high-voltage, low-current arcing on its surface under specified conditions.

The specimen is heated by electric arcs in several tests which include the following:

UL 746A, Sec. 31. (high-current arc ignition)
UL 746A, Sec. 32. (high-voltage arc ignition)
Federal Test Method Standard No. 406, Method 4011

Section 4.3.3. Burner Flames

Direct flame contact and flame radiation differ in various ways as ignition sources from radiant heat from electric heaters, from heated air, from hot wires, from electric arcs, and from hot surfaces, and do not necessarily give the same relative results or rankings of materials.

Flammability tests using burner flames generally require longer or larger specimens, because the igniting flame covers part of the specimen. The longer dimensions permit measurement of flame travel distance if ignition occurs. Flammability tests using burner flames therefore tend to be ignitability tests for materials which are difficult to ignite, and flame spread tests for materials which ignite.

The specimen is mounted horizontally and ignited at one end by a flame in several tests which include the following:

ASTM D 470
ASTM D 635
ASTM F 776
UL 44
UL 82
UL 94, Sec. 7. (horizontal test, HB)
UBC 52-4
MVSS 302 (49 CFR 571.302)
Federal Test Method Standard No. 191, Method 5906
Federal Test Method Standard No. 406, Method 2021

The specimen is mounted at a 45° angle and ignited at the bottom by a flame in several tests which include the following:

ASTM D 1230
ASTM D 1433
16 CFR 1610 (clothing textiles)
16 CFR 1611 (vinyl plastic film)
Federal Test Method Standard No. 191, Method 5908

The specimen is mounted vertically and ignited at the bottom by a flame in several tests which include the following:

ASTM D 568
ASTM D 2633
ASTM D 3014
ASTM D 3801
ASTM F 501
UL 44
UL 62
UL 83
UL 94, Sec. 8. (vertical test, V-0, V-1, or V-2)
UL 94, Sec. 9. (vertical bar, 5VA, horizontal plaque, 5VB)
UL 94, Sec. 11. (vertical test, VTM-0, VTM-1, or VTM-2)
UL 214
UL 224
UL 651
Federal standard FF 3-71 (16 CFR 1615) (children's sleepwear sizes 0 through 6X)
Federal standard FF 5-74 (16 CFR 1616) (children's sleepwear sizes 7 through 14)
California Technical Bulletin 117, Sec. A. Part I.
Federal Test Method Standard No. 191, Method 5903
Federal Test Method Standard No. 191, Method 5904
Federal Test Method Standard No. 406, Method 2022

Section 4.3.4. Liquid Fuels

Few flammability tests use liquid fuels as ignition sources. Liquid fuels offer the advantages of convenience and precise measurement, but they present special hazards in storage and use, and are not realistic for many fire scenarios. Some tests of this type are rarely used or discontinued.

Liquid fuels are used as ignition sources in several tests which include the following:

ASTM D 1360 (5 ml ethyl alcohol)
ASTM D 1361 (4 ml ethyl alcohol)
Federal Test Method Standard No. 191, Method 5900 (0.3 ml ethyl alcohol)

Section 4.3.5. Solid Combustibles

Solid combustibles are relatively safe and convenient ignition sources.

The best-known simple solid combustible ignition source is the Eli Lilly No. 1588 methenamine tablet, used in several tests which include the following:

ASTM D 2859 (floor coverings)
FF 1-70 (16 CFR 1630) (carpets and rugs)
FF 2-70 (16 CFR 1631) (small carpets and rugs)
California Technical Bulletin 117, Section A, Part III (expanded polystyrene
    beads)
CAN4-S117.1 (textile floor covering materials)
BS 6307 (textile floor coverings)

Cigarettes are used as ignition sources in several tests which include the following:

ASTM E 1352 (furniture mock-up)
ASTM E 1353 (furniture components)
NFPA 260 (furniture components)
NFPA 261 (furniture mock-up)
Federal standard FF 4-72 (16 CFR 1632) (mattresses and mattress pads)
California Technical Bulletin 106 (mattresses and mattress pads)
California Technical Bulletin 116 (upholstered furniture)
California Technical Bulletin 117, Section D, Part II (resilient cellular materials)
BS 5852 Part 1 (upholstered furniture)

Known amounts of paper of known characteristics are ignition sources which are
more convenient, less time-consuming, less expensive, and more realistic for some fire
scenarios than elaborately constructed wood cribs or brands. On the other hand, wood
cribs and brands can provide more severe ignition sources, and have history and
acceptance among fire protection officials.

Known amounts of paper of known characteristics are used as ignition sources in
several tests which include the following:

California Technical Bulletin 121 (mattresses for high-risk occupancies)
California Technical Bulletin 133 (seating furniture for public occupancies)

Wood cribs or brands are used as ignition sources in several tests which include
the following:

ASTM E 108 (roof coverings)
UL 790 (roof coverings, similar to ASTM E 108)
NFPA 256 (roof coverings, similar to ASTM E 108, UL 790)
BS 5852 Part 2 (upholstered furniture)

SECTION 4.4. TESTS FOR FLASH-FIRE PROPENSITY

Flash-fire propensity may be defined as the tendency of a material to produce a
fire that spreads with extreme rapidity. This characteristic provides a measure of fire
hazard in that a material with high flash-fire propensity is capable of contributing to
sudden development of a rapidly spreading fire.

The basic approach to laboratory evaluation of flash-fire propensity involves propagation through fuel-air mixtures in a tube. Most laboratory methods use upward propagation. The flammable limits are wider for upward propagation than for horizontal or downward propagation, and upward propagation is favored for safety considerations.

The National Bureau of Standards (NBS) flash-fire test employs a tube 50 mm in diameter and 500 mm long, into which pyrolysis gases are introduced and ignited by an electrical spark. A sample of known weight is introduced into the furnace at 500°C. to produce the gas mixture which is recirculated to improve uniformity. Some NBS flash-fire data are presented in Table 4.12.

The Douglas Aircraft Company flash-fire test uses a combustion tube of similar design, with an improved pyrolysis method and without recirculation. A 0.50 g sample is pyrolyzed at a heating rate of approximately 500°C./min to produce the pyrolysis gases. Some Douglas flash-fire data are presented in Table 4.13.

The University of San Francisco (USF) flash-fire test uses a vertical combustion tube 45mm in diameter and 660 mm high, into which pyrolysis gases from a horizontal tube furnace are introduced, permitted to mix with air, and ignited by the hot pyrolyzing surfaces. A 0.10 g sample is introduced into the furnace at 800°C. to produce the pyrolysis gases. The height of the flash fire and the time to flash fire are recorded. Some USF flash-fire data are presented in Table 3.4.

Polyurethane flexible foams were among the materials exhibiting the greatest flash-fire propensity in all three tests, and polychloroprene flexible foams were among the materials exhibiting the least flash-fire propensity in the Douglas and USF tests. The materials which appeared to be the least prone to flash fires included polyvinyl chloride, polyphenylene oxide, polyphenylene sulfide, polyether sulfone, polyaryl sulfone, and polychloroprene.

Flash fires developed most rapidly with cellulosic fabrics and cushioning materials, wood, and polyurethane flexible foam. The largest flash fires were observed to occur with wood, polyolefin plastics and fabrics, nylon plastics and fabrics, and polyurethane flexible foam. For these materials which are large-volume products and not easily modified, careful design of the application and limitation of the amount exposed seem advisable.

Dust explosions are similar to flash fires in that they can occur only within certain ranges of combustible concentrations and are characterized by extremely rapid movement of a flame front. Dust explosions are influenced by additional factors such as particle size and dispersion.

The dust explosibility apparatus which has been most extensively used is the Hartmann apparatus developed by the U.S. Bureau of Mines. The basic Hartmann apparatus is a vertical cylinder of polymethyl methacrylate (PMMA), with a height of 12 in (304 mm) and a volume of 75 in³ (0.00123 m³), mounted on a precision-machined base which serves as the sample holder. Several test procedures have been developed. The ignition sensitivity is calculated from the minimum ignition temperature, the minimum ignition energy, and the minimum explosive concentration. The explosion severity is calculated from the maximum explosion pressure and the maximum rate of pressure rise. The explosibility index is the product of the ignition sensitivity and the explosion severity. Some dust explosibility data are presented in Table 4.14.

This test method, using a steel cylinder, is described in ASTM E 789.

SECTION 4.5. TESTS FOR FLAME SPREAD

Flame spread may be defined as the rate of travel of a flame front under given conditions of burning. This characteristic provides a measure of fire hazard, in that flame spread can transmit fire to more flammable materials in the vicinity and thus enlarge a fire, even though the transmitting material itself contributes little fuel to the fire. Some measures of flame spread are burning rate, combustion rate, burning extent, distance of flame travel, flame spread factor, and flame height.

Tests for flame spread can be classified and numerically described in Cartesian coordinates by the angle formed by the exposed surface with the horizontal. This angle formed by the exposed surface with the horizontal is defined as the surface angle Θ. The surface angle Θ is $0°$ for a horizontal material burning on its upper surface with the flame front moving away from the origin, as in ASTM E 648. Rotating this surface through the first quadrant until Θ becomes $45°$ gives a burning upper surface such that the flame travels upward at an angle of $45°$ from the horizontal, as in ASTM D 1230. Further rotation through the first quadrant until Θ becomes $90°$ gives a burning vertical surface on which the flame front travels upward, as in ASTM D 3014. Further rotation through the second quadrant until Θ becomes $180°$ gives a horizontal material burning on its lower surface, as in ASTM E 84. Further rotation through the third quadrant until Θ becomes $240°$ gives a burning lower surface such that the flame travels downward at an angle of $30°$ from the vertical, as in ASTM E 162.

The surface angle Θ determines the extent to which hot combustion gases will preheat the surface ahead of the advancing flame front. Because these gases travel directly upward unless obstructed, the degree to which preheating by this mechanism occurs is a function of sin Θ, which is the preheat factor. Where forced air flow is controlling, rather than natural air flow, then the preheat factor becomes a function of the direction of forced air flow rather than surface angle Θ, with forced air flow in the direction of flame front travel giving a preheat factor of essentially 1.0. A preheat factor of 1.0 represents the most severe conditions from the viewpoint of preheat, and a preheat factor of -1.0 represents the least severe conditions.

The surface angle Θ also determines the extent to which hot combustion gases will promote penetration into the material by exerting pressure perpendicular to the exposed surface, and by the extent to which heavy gases will move perpendicular to the exposed surface. The degree to which these phenomena occur is a function of -cos Θ, which is the buoyancy factor. Where the combustion gases are restrained by confinement in a limited space, the buoyancy factor must be adjusted to a more realistic value.

The extent to which the heat of the exposure fire and the heat of combustion are concentrated in proximity to the exposed surface is a function of the dimensions of the enclosure relative to the dimensions of the exposed surface. The effective enclosure volume per unit of surface area is the concentration factor, a measure of the severity of the test. Severity of exposure, however, does not increase indefinitely with increasing concentration factor. Where enclosure dimensions approach specimen dimensions, the combustion gases tend to smother the flame by reducing access of oxygen.

In this discussion, tests for flame spread are organized in seven groups:

1. Burning along the top of a horizontal surface
2. Burning upward along the top of a surface at an angle of 0 to 45° from the horizontal
3. Burning upward along a vertical surface
4. Burning upward along the bottom of a surface at an angle of 6° to 30° from the horizontal
5. Burning along the bottom of a horizontal surface
6. Burning along the side of a vertical surface
7. Burning downward along the bottom of a surface at an angle of 60° from the horizontal
8. Burning within an enclosure

Section 4.5.1. Along Top of Horizontal Surface

The most simple test for flame spread along the top of a horizontal surface is the ASTM D 2859 test for floor coverings. It employs a test chamber measuring 305 by 305 by 305 mm (12 by 12 by 12 in) and a specimen measuring 230 by 230 mm (9 by 9 in) placed on the floor of the chamber. The ignition source is an Eli Lilly No. 1588 methenamine tablet. With varying differences, this test method is described in the following standards:

> FF 1-70 (16 CFR 1630) (carpets and rugs)
> FF 2-70 (16 CFR 1631) (small carpets and rugs)
> CAN4-S117.1 (textile floor covering materials)
> BS 6307 (textile floor coverings)

The most common small-scale horizontal test for plastics employs a specimen measuring 125 by 12.5 mm (5 by 0.5 in) by the supplied thickness, held in a horizontal position and ignited at one end by the flame from a burner. With varying differences, this test method is described in the following standards:

> ASTM D 470

ASTM D 635, with specimen 125 by 12.5 mm (5 by 0.5 in)
ASTM D 757, with specimen 121 by 12.7 by 3.17 mm (5 by 1/2 by 1/8 in) and
    Globar instead of burner
ASTM F 776
UL 44
UL 82
UL 94, Sec. 7. (HB)
UBC 52-4
MVSS 302 (49 CFR 571.302), with specimen 14 by 4 in
Federal Test Method Standard No. 191, Method 5906, with specimen 12.5 by 4.5
    in
Federal Test Method Standard No. 406, Method 2021

A similar test for foam plastics employs a specimen measuring 150 by 50 mm (6 by 2 in) by up to 13 mm (0.5 in) thick, held in a horizontal position and ignited at one end by the flame from a burner. With varying differences, this test method is described in the following standards:

ASTM D 1692
ASTM D 4986
UL 94, Sec. 12. (HBF, HF-1, HF-2)

Horizontal tests on a larger scale permit observation of flame travel for longer distances, evaluation under more severe fire exposure conditions, evaluation of larger and more representative samples of products, and more realistic simulation of actual fire scenarios. They require larger facilities, including more building floor space, and a large enclosure over the specimen or forced air flow or both in order to control the direction of flame travel.

Larger scale horizontal tests which provide a large enclosure over the specimen include the ASTM E 648 radiant panel test for floor covering systems and the ASTM E 970 radiant panel test for attic floor insulation.

The ASTM E 648 radiant panel test for floor covering systems employs a gas-fueled porous refractory radiant panel, with a radiation surface of 305 by 457 mm (12 by 18 in), inclined at 30° to and directed downward at a horizontally mounted specimen 250 by 1050 mm (9.9 by 41.4 in), with an enclosure 1400 by 500 by 710 mm (55 by 19.5 by 28 in) above the test specimen. The radiant panel is maintained at a temperature of 500°C. (932°F.), and the radiant energy flux is plotted as a function of distance along the specimen plane to produce a flux profile curve. The critical radiant flux at flame-out or the point of farthest advance of the flame front is calculated from the flux profile curve. Some test data are presented in Table 4.15.

The ASTM E 648 radiant panel test for floor covering systems is similar to those described in the following standard: NFPA 253.

The ASTM E 970 radiant panel test for attic floor insulation employs a gas-fueled porous refractory radiant panel, with a radiation surface of 305 by 457 mm (12 by 18 in), inclined at 30° to and directed downward at a horizontally mounted specimen tray 250 by 1000 mm (10 by 40 in) by 50 mm (2 in) deep, with an enclosure 1400 by 500 by 710 mm (55 by 19.5 by 28 in) above the test specimen. The radiant panel is maintained at a temperature of 485 ± 25°C. (839 ± 45°F.), and the radiant energy flux is plotted as a function of distance along the specimen plane to produce a flux profile curve. The critical radiant flux at flame-out or the point of farthest advance of the flame front is calculated from the flux profile curve.

The ASTM E 970 radiant panel test for attic floor insulation is similar to those described in the following standards.

The LIFT apparatus used in the ASTM E 1321 test can be used in the horizontal orientation for flame spread along the top of a horizontal surface.

Larger scale horizontal tests which provide forced air flow include the UL 910 test for cables in air-handling spaces.

The UL 910 test for cables in air-handling spaces employs a horizontal tunnel 7.62 by 0.445 by 0.305 m (300 by 17.5 by 12 in) in which cable trays 305 mm (12 in) wide and 7.62 m (25 ft) long are mounted. The fire source, two gas burners, is adjusted to deliver 300,000 Btu/hr (87.9 kW). The average air velocity is 73.2 ± 1.5 m/min (240 ± 5 ft/min).

The UL 910 test for cables in air-handling spaces is similar to those described in the following standards: NFPA 262.

Section 4.5.2. Upward along Surface at 0 to 45°

The test for flame spread upward at 45° employs a specimen measuring 152 by 51 mm (6 by 2 in), ignited at the bottom end with a flame from a hypodermic needle. With varying differences, this test method is described in the following standards:

ASTM D 1230, with specimen 152 by 51 mm (6 by 2 in)
ASTM D 1433, with specimen 228 by 76 mm (9 by 3 in)
16 CFR 1610, with specimen 6 by 2 in
16 CFR 1611, with specimen 9 by 3 in
Federal Test Method Standard No. 191, Method 5908, with specimen 6 by 2 in

The ASTM E 108 test for roof coverings involves flame spread upward at an angle adjustable between 0 and 45°, and includes three test procedures.

In the intermittent flame test, a specimen measuring 3 ft 4 in by 4 ft 4 in (1.0 by 1.3 m), mounted at a specified slope, is exposed to flame at 760°C. (1400°F.) for Class

A and Class B, or 704°C. (1300°F.) for Class C, and a 12 mph air current, intermittently according to a specified sequence.

In the flame spread test, a specimen measuring 3 ft 4 in by 13 ft (1.0 by 4.0 m), mounted at a specified slope, is exposed to flame at 760°C. (1400°F.) for 10 min for Class A and Class B, or 704°C. (1300°F.) for 4 min for Class C, and a 12 mph air current.

In the burning brand test, a specimen measuring 3 ft 4 in by 4 ft 4 in (1.0 by 1.3 m), mounted at a specified slope, is exposed to a 12 mph air current and a burning brand, which weighs 2000 g of Douglas fir for Class A, 500 g of Douglas fir for Class B, or 9.25 g of white pine for Class C.

The ASTM E 108 test for roof coverings is similar to those described in the following standards:

UL 790
NFPA 256
UBC 32-7
CAN/ULC-S107

Section 4.5.3. Upward on Vertical Surface

The common small-scale vertical test with upward burning employs a specimen mounted vertically and ignited at its bottom end with a flame from a burner. The vertical test requires a longer specimen than the horizontal test for a given size of igniting flame, because the igniting flame usually envelops a certain length of the specimen, and because flame generally travels more rapidly upward along a vertical surface. With varying differences, this test method is described in the following standards:

ASTM D 568, with specimen 450 by 25 mm (18 by 1 in) and 25mm (1 in) flame
ASTM D 2633
ASTM D 3014, with specimen 254 by 19 by 19 mm (10 by 0.75 by 0.75 in)
    enclosed in chimney
ASTM D 3801
ASTM E 69, with specimen 9.5 by 19 by 1016 mm (3/8 by 3/4 by 40 in) enclosed
    in tube
ASTM F 501
UL 44
UL 62
UL 83, with wire specimen 18 in long
UL 94, Sec. 8, with specimen 127 by 12.7 mm (5 by 0.5 in) and 19 mm (3/4 in)
    blue flame
UL 94, Sec. 9, Method A, with specimen 127 by 12.7 mm (5 by 0.5 in) and 127
    mm (5 in) flame with 38 mm (1.5 in) blue inner cone

UL 94, Sec. 11, with specimen 200 by 50 mm (8 by 2 in) and 19 mm (3/4 in) blue flame

UL 214, small-flame test, with specimen 10 by 2.75 in

UL 224

UL 651, with conduit specimen 18 in long and 5 in flame with 1.5 in blue inner cone

Federal Standard FF 3-71 (16 CFR 1615), with specimen 25.4 by 8.9 cm (10 by 3.5 in)

Federal Standard FF 5-74 (16 CFR 1616), with specimen 25.4 by 8.9 cm (10 by 3.5 in)

California Technical Bulletin 117, Sec. A., Part I, with specimen 12 by 3 in and 1.5 in flame

Federal Test Method Standard No. 191, Method 5903, with specimen 12 by 2.75 in

Federal Test Method Standard No. 191, Method 5904, with specimen 5 by 2 in (candle flame)

Federal Test Method Standard No. 406, Method 2022

Vertical tests on a larger scale permit observation of flame travel for longer distances, evaluation under more severe fire exposure conditions, evaluation of larger and more representative samples of products, and more realistic simulation of actual fire scenarios. They require larger facilities, including more building height, and a large enclosure over the sample or forced air flow or both in order to control the direction of flame travel.

The NASA 8060.1B upward flame spread test exposes a sample area 5 cm by 30 cm at its bottom edge to a flaming ignition source, an electrically-initiated chemical igniter, hexamethylenetetramine, at 1100°C. flame temperature, 6.4 cm peak flame height, 25 sec duration, and 750 cal. or 3140 joules total energy release.

A NIST modification of the NASA upward flame spread test employs a radiant panel with twelve 500 watt linear quartz halogen lamps, providing incident heat flux on the sample surface of 18 to 32 kW/m^2.

The UL 1581 test for wires and cables in vertical trays employs a vertical tray 8 ft high and 10 in wide, on which wires and cables occupy a width of 6 in.

The UL 1581 test for wires and cables in vertical trays is similar to those described in the following standard: IEEE 383.

The UL 1666 test for wires and cables in vertical shafts or risers employs a vertical shaft 20 ft high and 8 ft square, in which wires and cables are exposed to fire horizontally at the bottom and vertically for 12 ft.

The UL 1666 test for wires and cables in vertical shafts or risers is similar to those described in the following standards.

Facade fire tests employ large vertical facade specimens to simulate the facades of high-rise buildings.

The Canadian CAN/ULC-S134 facade test employs a facade specimen 5 m wide by about 10.3 m high, ignited at the bottom by propane-fueled flames.

The German facade test employs a facade specimen 3 m wide by 6 m high, ignited at the bottom by flames from a 25 kg wood crib.

The Swedish SP 105 facade test employs a facade specimen 4.18 m wide by 6.0 m high, ignited at the bottom by flames from a pan containing 60 L of heptane.

Section 4.5.4. Upward Along Bottom of Surface at 6° to 30°

Burning upward along the bottom of a surface at an angle from the horizontal approaches burning along the bottom of a horizontal surface, such as along the ceiling of a corridor, and uses natural convection from combustion to replace forced air flow. The most prominent examples are the ASTM E 286 or FPL 8-foot tunnel test, and the ASTM D 3806 or Monsanto 2-foot tunnel test.

The ASTM E 286 test employs a specimen 8 ft (2.44) long and 13.75 in (349 mm) wide, mounted horizontally so as to form the roof of a tunnel and slope at a 6° angle from end to end. The heat supply rate for initial trials is 3,400 Btu/min.

The ASTM D 3806 test employs a specimen measuring 24 by 4 in, inclined 28° from the horizontal so as to form the roof of a tunnel.

The Pittsburgh-Corning 30-30 tunnel employs a specimen measuring 30 by 3-7/8 in inclined 30° from the horizontal so as to form the roof of a tunnel. The flame source is a burner supplied with 2,600 cc/min of natural gas, giving a 6-in flame and delivering about 90 Btu/min.

Section 4.5.5. Along Bottom of Horizontal Surface

Burning along the bottom of a horizontal surface represents the most realistic simulation of certain important fire scenarios, especially flame spread along the ceiling of a compartment such as a room or corridor.

Tests for flame spread along the bottom of a horizontal surface require forced air flow in order to control the direction of flame travel, and as a result have longer igniting flames and require larger specimens. Such horizontal tests on a larger scale permit observation of flame travel for longer distances, evaluation under more severe fire exposure conditions, evaluation of larger and more representative samples of products, and more realistic simulation of actual fire scenarios. They require larger facilities, including more building floor space.

The most established test for flame spread along the bottom of a horizontal surface is the ASTM E 84 or Underwriters Laboratories (UL) 25-foot tunnel test. The ASTM E 84 test requires a specimen 7.62 by 0.496 m (300 by 19.5 in), mounted face down so as to form the roof of a tunnel 7.62 by 0.445 by 0.305 m (300 by 17.5 by 12 in). The fire source, two gas burners 305 mm (12 in) from the fire end of the sample and 190 mm (7.5 in) below the surface of the sample, is initially adjusted to deliver 5,000 Btu/min (5.3 MJ/min or 87.9 kW), and is finally adjusted so that a test sample of select-grade red oak flooring would spread flame 5.94 m (19.5 ft) from the end of the igniting fire in 5.5 min ± 15 sec. The end of the igniting flame is considered as being 1.37 m (4.5 ft) from the burners, this flame length being due to an average air velocity of 73.2 ± 1.5 m/min (240 ± 5 ft/min). Materials are rated on a flame spread classification with red oak as 100. Some test data are presented in Table 3.5.

The ASTM E 84 test is similar to those described in the following standards:

UL 723
NFPA 255
UBC 8-1
CAN/ULC-S102

The UL 25-foot tunnel test is so widely used that smaller-scale tunnel tests were developed to predict performance in the 25-foot tunnel test using smaller specimens.

The Union Carbide 4-foot tunnel test employs a specimen measuring 47.5 in long and 7.5 in wide, mounted horizontally so as to form the roof of a tunnel 6.75 in deep. The heat supply rate is 325 Btu/min, compared to an initial trial rate of 3,400 Btu/min for ASTM E 286 and an initial rate of 5,000 Btu/min for ASTM E 84.

Different variations of 4-foot tunnel tests have been used by BASF, Mobay (now Bayer), and other companies.

The different variations of 4-foot tunnel tests have been useful in predicting performance in the 25-foot tunnel test with smaller specimens, and the companies which found them useful had no interest in developing a standard 4-foot tunnel test.

Some tests developed to predict performance in the 25-foot tunnel with smaller specimens used flame spread, not along the bottom of a horizontal surface, but along the bottom of a surface inclined at an angle from the horizontal, and used natural convection from combustion to replace forced air flow. The angle is 6° in the FPL 8-foot tunnel, 28° in the Monsanto 2-foot tunnel, and 30° in the Pittsburgh Corning 30-30 tunnel.

Section 4.5.6. Along Side of Vertical Surface

Burning along the side of a vertical surface represents the most realistic simulation of certain important fire scenarios, especially flame spread along the wall of a compartment such as a room or corridor.

Tests for flame spread along the side of a vertical surface do not require forced air flow in order to control the direction of flame travel. Such tests on a larger scale permit observation of flame travel for longer distances, evaluation under more severe fire exposure conditions, evaluation of larger and more representative samples of products, and more realistic simulation of actual fire scenarios. They require larger facilities, including more building floor space.

An apparatus known as the lateral ignition and flame spread test apparatus (LIFT), developed from the International Maritime Organization (IMO) test A.653(16), became the apparatus for the ASTM E 1317 and ASTM E 1321 tests.

The ASTM E 1317 test exposes vertically mounted specimens to the heat from a vertical air-gas fueled porous refractory radiant-heat energy source, with heated surface dimensions of 280 by 483 mm, inclined at 15° to the specimen. The specimen is 155 mm wide and 800 mm long.

Means are provided for observing the times to ignition, spread, and extinguishment of flame along the length of the specimen, and measuring the temperature of the stack gases during burning. Results are reported in terms of heat for ignition, heat for sustained burning, critical flux at extinguishment, and heat release of the specimen during burning.

The ASTM E 1317 test is similar to ISO 5658-2.

The ASTM E 1321 test exposes vertically mounted specimens to the heat from a vertical air-gas fueled porous refractory radiant-heat energy source, with heated surface dimensions of 280 by 483 mm, inclined at 15° to the specimen.

For the ignition test, a series of 155 by 155 mm specimens are exposed to a nearly uniform heat flux, and the time to flame attachment, using piloted ignition, is determined.

For the flame spread test, a 155 by 800 mm specimen is exposed to a graduated heat flux that is approximately 5 kW/m² higher at the hot end than the minimum heat flux necessary for ignition. The specimen is preheated to thermal equilibrium, the preheat time being derived from the ignition test. After using piloted ignition, the pyrolyzing flame-front progression along the horizontal length of the specimen as a function of time is tracked.

The ASTM E 1321 test is similar to ISO 5658-3.

The LIFT apparatus can be used in the horizontal orientation for flame spread along the top of a horizontal surface.

Section 4.5.7. Downward Along Bottom of Surface at 60°

The most established test for flame spread downward along the bottom of a surface at an angle of 60° from the horizontal is the ASTM E 162 or NBS radiant panel test. The ASTM E 162 test employs a radiant heat source consisting of a 305 by 457 mm (12 by 18 in) vertically mounted porous refractory panel maintained at 670 ± 4°C. (1238 ± 7°F.). A specimen measuring 152 by 457 mm (6 by 18 in) is supported in front of it with the 457 mm (18 in) dimension inclined 30°C. from the vertical. A pilot burner ignited the top of the specimen, 121 mm (4.75 in) away from the radiant panel, so that the flame front progresses downward along the underside exposed to the radiant panel. Some test data are presented in Tables 4.16 and 4.17.

The ASTM E 162 test is similar to those described in the following standards:

ASTM D 3675
Federal Test Method Standard No. 501a, Method 6421

Section 4.5.8. Within an Enclosure

Burning of a surface within an enclosure represents greater severity of fire exposure than burning of the same surface without an enclosure, because of the effect of reradiation from the surfaces of the enclosure and confinement of hot combustion gases. Burning of multiple surfaces within an enclosure represents even greater severity of fire exposure, because the burning surfaces can radiate energy to each other and accelerate combustion and flame spread. Tests within an enclosure provide this effect.

The best example of two burning surfaces radiating energy to each other is provided by the Schlyter test. This test is more than 35 years old and has appeared in several versions. The Schlyter test employs two specimens, each 12 in wide and 31 in high, held in a vertical position with their faces 2 in apart. The bottom of the assembly is subjected for 3 min to flames from burners delivering either 37 Btu/min or 291 Btu/min.

Tests which require entire rooms as enclosures can be classified as room tests or room fire experiments. The rooms usually have 8 ft ceilings, and vary considerably in size. Three sizes appear to be the most frequently used:

1. 8 by 12 ft (2.4 by 3.7 m)
2. 10 by 10 ft (3.0 by 3.0 m)
3. 12 by 12 ft (3.7 by 3.7 m)

The ignition sources most frequently used are:

1. Matches
2. Polyolefin waste containers, 1.75 gal (6.6 liter), 7 gal (26.5 liter), and 32 gal (121 liter), with 12, 24, and 72 milk cartons, respectively

3. Wood cribs, 5 lb (2.3 kg), 20 lb (9.1 kg), 30 lb (13.6 kg), and 50 lb (22.7 kg)

The ignition source is placed in one of three locations:

1. At the center of the test area
2. At one wall
3. At one corner

Room tests are described in the following standards:

ASTM E 603 (room fire experiments)
UBC 42-2 (room fire growth contribution of textile wall covering, room 8 by 12 ft by 8 ft high).

SECTION 4.6. TESTS FOR HEAT RELEASE

Heat release may be defined as the heat produced by the combustion of a given weight or volume of material. This characteristic provides a measure of fire hazard in that a material which burns with the evolution of little heat per unit quantity burned will contribute appreciably less to a fire than a material which generates large amounts of heat per unit quantity burned. Some measures of heat release are heat evolution factor and fuel contribution index.

Tests for heat release provide a measure of the amount of heat produced by the material in the process of burning. Such tests generally provide information, not on the absolute amount of heat produced, but on the relative amount of heat produced, or on the equivalent amount of heat required to produce similar results, for purposes of comparison with other materials.

The heat release information produced by some tests may be used to calculate other flammability characteristics and may be used in fire hazard assessment and fire models.

The ASTM E 906 or Ohio State University (OSU) release rate test employs a chamber 890 by 410 by 200 mm (35 by 16 by 8 in) with a pyramidal top section 395mm (15.5 in) high connecting to the outlet. A radiant heat source, using four silicon carbide Globar Type LL elements, is used to generate heat flux up to 100 kW/m². Specimens 160 by 150 mm (6 by 6 in) are tested in vertical orientation, and specimens 110 by 150 mm (4.5 by 6 in) are tested in horizontal orientation. A radiation reflector is used for horizontally mounted specimens. The total air flow of 0.04 m³/sec (84 ft³/min) leaves the apparatus through a rectangular exhaust stack 133 by 70 mm (5.25 by 2.75 in) in cross section and 254 mm (10 in) high. The temperature difference between the air entering and the air leaving the apparatus is measured by a thermopile having 3 hot junctions spaced across the top of the exhaust stack and 3 cold junctions located in the pan at the bottom. Some OSU heat release data are presented in Table 3.6 (pre-ASTM E 906), Table 4.18 (ASTM E 906), and Table 4.19 (FAR modification).

This test method, with various modifications, is described in the following standards:

ASTM E 906
NFPA 263
Federal Aviation Regulations (FAR) Parts 25 and 121

The ASTM E 1354 or National Bureau of Standards (NBS) cone calorimeter is an apparatus 1680 by 1625 by 686 mm. A radiant heat source, using an electrical heater rod, rated at 5000 W and 240 V, tightly wound into the shape of a truncated cone, is used to generate heat flux up to 100 kW/m^2. Specimens 100 by 100 mm are tested in either horizontal or vertical orientation. A paramagnetic oxygen analyzer with a range from 0 to 25% oxygen is used to measure the oxygen concentration in the exhaust gas. The rate of heat release is measured by the principle of oxygen consumption. The amount of heat released is calculated from the amount of oxygen consumed. Smoke evolution is measured. Some heat-release data are presented in Table 4.20.

This test method, with various modifications, is described in the following standards:

ISO 5660-1
ASTM E 1354
ASTM E 1474 (furniture components)
ASTM E 1740 (wallcovering composites)
ASTM F 1550 (correctional facility furnishings)
NFPA 264
BS 476 Part 15

The intermediate scale calorimeter (ICAL) is an apparatus consisting of a vertically mounted radiant panel assembly facing a vertically mounted specimen, at a distance adjusted by means of a trolley. The radiant panel assembly, with three rows of ceramic-faced natural gas burners, is used to generate heat flux up to 50 kW/m^2. Specimens 1000 by 1000 mm are tested in vertical orientation. An oxygen analyzer with a range from 0 to 25% oxygen is used to measure the oxygen concentration in the exhaust gas. The rate of heat release is measured by the principle of oxygen consumption. The amount of heat released is calculated from the amount of oxygen consumed. Smoke evolution and carbon monoxide evolution are measured.

This test method is described in ASTM E 1623.

The Factory Mutual fire propagation apparatus (FPA) is an apparatus consisting of four infrared heaters and four types of specimen holders. The four infrared heaters, each containing six tungsten filament tubular quartz lamps, are used to generate heat flux up to 50 kW/m^2. The four types of specimens are horizontal square 0.10 by 0.10 m (4 by 4 in), horizontal circular 0.097 m (3.8 in) diameter, vertical 0.305 by 0.076 m (19 by 5.2 in), and vertical cable 0.81 m (32.5 in) long and up to 51 mm (2 in) diameter.

This test method is described in ASTM E5 Z6880Z.

The ASTM E 1537 or California Technical Bulletin 133 (TB 133) test is perhaps the most widely recognized large-scale test for heat release. The test specimen is a full-size manufactured item of upholstered furniture, a representative prototype of the upholstered furniture, or a mock-up of the upholstered furniture. The specimen is ignited with a propane gas burner, used at a flow rate of 13 ± 0.25 L/min for 80 seconds (equivalent to 19.3 kW). It approximates the ignition propensity of five crumpled sheets of newspaper located on the seating cushion. An oxygen analyzer with a range from 0 to 21 % oxygen is used to measure the oxygen concentration in the exhaust gas. The rate of heat release is measured by the principle of oxygen consumption. The amount of heat released is calculated from the amount of oxygen consumed. Smoke evolution and carbon monoxide evolution are measured.

One of three test configurations is used in the test method:

1. A test room 3.66 by 2.44 by 2.44 m high
2. A test room 3.66 by 3.05 by 2.44 m high
3. An open calorimeter (or furniture calorimeter)

Test room configurations are described for items of furniture less than 1 m across, such as a chair, and for items of furniture between 1 and 2.44 m across, such as a sofa.

This test method, with various modifications, is described in the following standards:

ASTM E 1537
California Technical Bulletin 133
UL 1056
NFPA 266
Boston BFD IX-10

The ASTM E 1590 or California Technical Bulletin 129 (TB 129) test is a large-scale test for heat release used for mattresses. The test specimen is a full-size manufactured mattress or a prototype thereof. The specimen is ignited with a propane gas burner, used at a flow rate of 12 ± 0.25 L/min for 180 seconds (equivalent to 17.8 kW). An oxygen analyzer with a range from 0 to 21 % oxygen is used to measure the oxygen concentration in the exhaust gas. The rate of heat release is measured by the principle of oxygen consumption. The amount of heat released is calculated from the amount of oxygen consumed. Smoke evolution and carbon monoxide evolution are measured.

One of three test configurations is used in the test method:

1. A test room 3.66 by 2.44 by 2.44 m high

2. A test room 3.66 by 3.05 by 2.44 m high
3. An open calorimeter (or furniture calorimeter)

This test method, with various modifications, is described in the following standards:

ASTM E 1590
California Technical Bulletin 129
UL 1895
NFPA 267

The ASTM E 1822 test is a large-scale test for heat release used for stacking chairs. The test specimen is a stack of five identical stacking chairs or prototype thereof. The specimen is ignited with a propane gas burner, used at a flow rate of 12 ± 0.25 L/min for 180 seconds (equivalent to 17.8 kW). An oxygen analyzer with a range from 0 to 21 % oxygen is used to measure the oxygen concentration in the exhaust gas. The rate of heat release is measured by the principle of oxygen consumption. The amount of heat released is calculated from the amount of oxygen consumed. Smoke evolution and carbon monoxide evolution are measured.

One of three test configurations is used in the test method:

1. A test room 3.66 by 2.44 by 2.44 m high
2. A test room 3.66 by 3.05 by 2.44 m high
3. An open calorimeter (or furniture calorimeter)

SECTION 4.7. TESTS FOR FIRE ENDURANCE

Fire endurance may be defined as the resistance offered by a material to the passage of fire, normal to the exposed surface over which flame spread is measured. This characteristic provides a measure of fire hazard, in that a material which will contain a fire represents more protection than a material which will give way before the same fire, all other factors being the same in both cases. Some measures of fire endurance are penetration time and resistance rating.

Tests for fire endurance are generally more concerned with complete systems than with individual materials, because it is widely recognized that the performance of the individual materials comprising a system is not necessarily indicative of the performance of the system as a whole.

The ASTM E 119 test standard for building construction and materials provides for exposure of various structural components to a standard fire, the character of which is determined by a standard time-temperature curve. The main points on the curve are:

538°C. (1000°F.) at 5 minutes
704°C. (1300°F.) at 10 minutes

843°C. (1550°F.) at 30 minutes
927°C. (1700°F.) at 1 hour
1010°C. (1850°F.) at 2 hours
1093°C. (2000°F.) at 4 hours
1260°C. (2300°F.) at 8 hours or over

The performance is defined as the period of resistance to standard exposure elapsing before the first critical point in behavior is observed, and is expressed in the time ratings, such as 2-hour and 4-hour.

For bearing walls and partitions, the area exposed to fire shall not be less than 100 ft^2 (9 m^2), with neither dimension less than 9 ft (2.7 m).

For nonbearing walls and partitions, the area exposed to fire shall not be less than 100 ft^2 (9 m^2), with neither dimension less than 9 ft (2.7 m).

For columns, the length of the column shall not be less than 9 ft (2.7 m).

For protected structural steel columns, the length of the column shall not be less than 8 ft (2.4 m).

For floors and roofs, the area exposed to fire shall not be less than 180 ft^2 (16 m^2), with neither dimension less than 12 ft (3.7 m).

For loaded restrained beams, the length of beam exposed to fire shall not be less than 12 ft (3.7 m). A section of representative floor or roof construction not more than 7 ft (2.1 m) wide may be included with the test specimen.

For protected solid structural steel beams and girders, the length of beam or girder exposed to fire shall not be less than 12 ft (3.7 m). A section of representative floor construction not less than 5 ft (1.5 m) wide shall be included in the test assembly.

The ASTM E 119 test is similar to those described in the following standards:

UL 263
NFPA 251
UBC 7-1

The standard time-temperature curve described in ASTM E 119 is used to define the exposure fire for the following tests:

ASTM E 152 (door assemblies)
ASTM E 163 (window assemblies)
ASTM E 814 (through-penetration fire stops)
UL 9 (window assemblies, similar to ASTM E 163)
UL 10B (door assemblies, similar to ASTM E 152)

NFPA 252 (door assemblies, similar to ASTM E 152, UL 10B)
NFPA 257 (window assemblies, similar to ASTM E 163, UL 9)
UBC 7-2 (door assemblies, similar to ASTM E 152, UL 10B)
UBC 7-4 (window assemblies, similar to ASTM E 163, UL 9)

Similar severity of exposure is intended in the Union Carbide fire endurance tests for thermal insulation materials on pipes and vessels. In these tests, the flow of gasoline fuel to the pool fire is controlled so that the fire exposure approximates that of the ASTM E 119 standard time-temperature curve.

The UL 181 test for air ducts employs a horizontal specimen 19 by 19 in forming the roof of the test furnace. Gas burners below the specimen are the fire source.

In the hydrocarbon processing industry (HPI), large, free-burning (that is, outdoors), fluid-hydrocarbon-fueled pool fires can occur. One of the most distinguishing features of the pool fire is the rapid development of high temperatures and heat fluxes that can subject exposed structural members and assemblies to a thermal shock much greater than that associated with the ASTM E 119 standard time-temperature curve.

The ASTM E 1529 test standard specifies a standard fire exposure that simulates total continuous engulfment in the luminous flame (fire plume) area of a large free-burning-fluid-hydrocarbon pool fire. The average heat flux is 50,000 Btu/ft^2.h ± 2500 Btu/ft^2.h (158 kW/m^2 ± 8 kW/m^2). The temperature of the environment that generates this heat flux shall be:

1. At least 1500°F. (815°C.) after the first 3 minutes
2. Between 1850°F. (1010°C.) and 2150°F (1180°C.) at all times after the first 5 minutes

The ASTM E 1529 test standard describes three methods:

1. Method A. Column Tests: The length of the column assembly subjected to the fire exposure shall not be less than 9 ft for loaded specimens and 8 ft for unloaded steel specimens.
2. Method B. Beam Tests: The length of the beam assembly subjected to the fire exposure shall not be less than 12 ft (3.7 m).
3. Method C. Tests of Fire-Containment Capability of Walls: The test specimen shall have a fire-exposed surface of not less than 50 ft^2 (4.65 m^2) and a height of not less than 8 ft (2.44 m).

Sample size may be specified by a particular regulation. For example, 46 CFR 164.007 (merchant vessels) requires the sample to be 40 by 60 in.

SECTION 4.8. TESTS FOR EASE OF EXTINGUISHMENT

Ease of extinguishment may be defined as the facility with which burning can be extinguished in the case of a specific material. This characteristic provides a measure of fire hazard, in that a material which requires more effort to extinguish is more likely to prolong a fire than one which is easily extinguished, all other factors being the same in both cases.

The ASTM D 2863 oxygen index test employs a vertical heat-resistant glass tube at least 75 mm in diameter and at least 450 mm high, in which a specimen 70 to 150 mm long, 6.5 mm wide, and 3.0 mm thick is held vertically by a clamp at its bottom end. A mixture of oxygen and nitrogen of known composition is metered into the bottom of the tube, passing through a bed of glass beads 3 to 5 mm in diameter and 80 to 100 mm deep at the bottom to smooth the flow of the gas. The gas flow rate in the column is 4 ± 1 cm/sec. The specimen is ignited at its upper end with an igniting flame which is then withdrawn, and the atmosphere which just permits steady burning down the specimen is determined. The oxygen index is the minimum concentration of oxygen in an oxygen-nitrogen mixture which will just permit the sample to burn, that is, burn 3 minutes or 50 mm. Data for various materials are presented in Table 2.6.

The ASTM D 2863 test is similar to those described in several standards which include the following:

ISO 4589-2
AS 2122, Part 2 (Australia)
BS 2782, Part 1, Method 141 (United Kingdom)
JIS K 7201 (Japan)
NF T 51-071 (France)
NT FIRE 013 (Nordic countries)

SECTION 4.9. TESTS FOR SMOKE EVOLUTION

Smoke density may be defined as the degree of light or sight obscuration produced by the smoke from the decomposing or burning material under given conditions of decomposition and combustion. This characteristic provides a measure of fire hazard, in that an occupant has a better chance of escaping from a burning structure if he can see the exit, and a firefighter has a better chance of putting out the fire if he can locate it. Some measures of smoke density are degree of light absorption, specific optical density, and smoke development factor.

Tests for smoke evolution generally involve the measurement of the fraction of light absorbed or obstructed by smoke evolved from a decomposing or burning material. The degree of obscuration is a function of the number and size of particles, refractive index, light scattering, rate of movement, extent of ventilation, and distance through which light must travel. The practical effects of smoke can be studied without detailed consideration of particle size and distribution, and the smoke production of various materials can be compared under equal conditions of test.

Tests for smoke-producing characteristics employ one of two measurement techniques: optical and gravimetric.

Optical methods involve measurement of the fraction of light absorbed or obstructed by the smoke evolved. The most widely used optical methods are the ASTM E 662 or National Bureau of Standards (NBS) smoke chamber test, the ASTM D 2843 or Rohm and Haas smoke chamber test, the ASTM E 84 or Underwriters Laboratories (UL) 25-foot tunnel test, the ASTM E 906 or Ohio State University (OSU) release rate test, and the ASTM E 1354 or NBS cone calorimeter test.

The tests which measure the density of smoke accumulated in an enclosure are:

ASTM E 662
ASTM D 2843

The tests which measure the density of smoke flowing past a specific location are:

ASTM E 84
ASTM E 906
ASTM E 1354

The ASTM E 662 smoke chamber is a completely closed cabinet, 914 by 610 by 914 mm (36 by 24 by 36 in), in which a specimen 76.2 mm (3 in) square is supported vertically in a frame such that an area 65.1 mm (2-9/16 in) square is exposed to heat under either piloted (flaming) or nonpiloted (smoldering) conditions. The heat source is an electric furnace, adjusted with the help of a circular foil radiometer to give a heat flux of 2.5 W/cm^2 (2.2 Btu/sec-ft^2) at the specimen surface. A vertical photometer path for measuring light absorption is employed to minimize measurement differences due to smoke stratification which could occur with a horizontal photometer path at a fixed height; the full 914 mm (3 ft) height of the chamber is used to provide an overall average for the entire chamber. Some test data are presented in Table 3.8.

The ASTM E 662 test is similar to those described in the following standard: FPA 258.

The ASTM D 2843 or Rohm and Haas test employs a chamber 300 by 300 by 790 mm (12 by 12 by 31 in), completely sealed except for 25 by 230 mm (1 by 9 in) openings in the four sides of the bottom of the chamber. A specimen 25.4 by 25.4 by 6.2 mm (1 by 1 by 1/4 in) is exposed to the propane-air flame from a burner with a 0.13 mm (0.005 in) orifice operating at a pressure of 2.8 kgf/cm^2 (40 psi). A horizontal photometer path, 480 mm (19-3/4 in) above the base of the chamber, is used for measuring light absorption. Some data obtained this test are shown in Table 4.21.

The ASTM D 2843 test is similar to those described in the following standard: UBC 26-5.

The ASTM E 84 or UL 25-foot tunnel test requires a specimen 7.62 by 0.496 m (300 by 19.5 in), mounted face down so as to form the roof of a tunnel 7.62 by 0.445 by 0.305 m (300 by 17.5 by 12 in). The fire source, two gas burners 305 mm (12 in) from the fire end of the sample and 190 mm (7.5 in) below the surface of the sample, is initially adjusted to deliver 5,000 Btu/min (5.3 MJ/min or 87.9 kW), and is finally adjusted so that a test sample of select-grade red oak flooring would spread flame 5.94 m (19.5 ft) from the end of the igniting fire in 5.5 min ± 15 sec. The end of the igniting flame is considered as being 1.37 m (4.5 ft) from the burners, this flame length being due to an average air velocity of 73.2 ± 1.5 m/min (240 ± 5 ft/min). A vertical photometer path at the vent pipe 406 mm (16 in) in diameter is used for measuring light absorption. Some ASTM E 84 smoke data are presented in Table 3.5.

The ASTM E 84 test is similar to those described in the following standards:

UL 723
NFPA 255
UBC 8-1
CAN/ULC-S102

The ASTM E 906 or Ohio State University (OSU) release rate test employs a chamber 890 by 410 by 200 mm (35 by 16 by 8 in) with a pyramidal top section 395mm (15.5 in) high connecting to the outlet. A radiant heat source, using four silicon carbide Globar Type LL elements, is used to generate heat flux up to 100 kW/m^2. Specimens 160 by 150 mm (6 by 6 in) are tested in vertical orientation, and specimens 110 by 150 mm (4.5 by 6 in) are tested in horizontal orientation. A radiation reflector is used for horizontally mounted specimens. The total air flow of 0.04 m^3/sec (84 ft^3/min) leaves the apparatus through a rectangular exhaust stack 133 by 70 mm (5.25 by 2.75 in) in cross section and 254 mm (10 in) high. A horizontal photometer path above the outlet is used for measuring light absorption. Some OSU smoke release data are presented in Table 4.22 (pre-ASTM E 906), Table 4.18 (ASTM E 906) and Table 4.19 (FAR modification).

This test method, with various modifications, is described in the following standards:

ASTM E 906
NFPA 263
Federal Aviation Regulations (FAR) Parts 25 and 121

The ASTM E 1354 or National Bureau of Standards (NBS) cone calorimeter is an apparatus 1680 by 1625 by 686 mm. A radiant heat source, using an electrical heater rod, rated at 5000 W and 240 V, tightly wound into the shape of a truncated cone, is used to generate heat flux up to 100 kW/m^2. Specimens 100 by 100 mm are tested in either horizontal or vertical orientation. A horizontal photometer path in the exhaust system is used for measuring light absorption.

This test method, with various modifications, is described in the following standards:

ASTM E 1354

        ASTM E 1474
        ASTM E 1740
        ASTM F 1550
        NFPA 264

Gravimetric methods involve determining the weight of smoke particles deposited on a filter under specified conditions. The best known gravimetric method is the ASTM D 4100 or Arapahoe smoke test.

The ASTM D 4100 or Arapahoe test employs a vertical cylindrical combustion chamber 125 mm (5 in) in diameter and 175 mm (7 in) high, a cylindrical chamber stack 75 mm (3 in) in diameter and 450 mm (18 in) high, and a filter assembly at the top of the stack. A propane burner mounted at an angle of 10° from the horizontal is mounted in the base of the combustion chamber, and is fed with approximately 90 cm^3/min of propane to produce a well-defined blue flame about 25 mm (1 in) long. A specimen is exposed to the burner flame, and the smoke particles are collected on the surface of the glass fiber filter paper at the top of the stack by drawing air through the filter under controlled vacuum. Some data obtained this test are presented in Table 4.23.

SECTION 4.10. TESTS FOR TOXIC GAS EVOLUTION

Toxic gas evolution may be defined as the level of toxicity exhibited by the gases evolved from a material under specified conditions. This characteristic provides a measure of fire hazard in that a material which produces more toxic effects than another material would be expected to be more likely to produce incapacitation and death.

Tests for toxic gas evolution generally fall into two types:

1. Tests concerned with the identification and analysis of the chemical compounds in the gaseous combustion products
2. Tests concerned with studying the effects of gaseous combustion products on laboratory animals

Both types of tests suffer from common problems: interference from other compounds or other factors, and loss of materials or change in materials between the point of fire and the point of test. The latter problem, which can present serious difficulty in accounting for all the materials in the system, presents an interesting question of validity. If some of the gaseous products have characteristics in such that they are condensed on or absorbed by various materials on the way from the combustion chamber to the point of sampling or test, eliminating all possibility of condensation or absorption in a laboratory investigation would produce results which would have little relation to the fire gases reaching persons at or near an actual fire. If high temperature is required to keep such wayward compounds in the gas stream, the high gas temperatures involved would probably be above the upper temperature limit for survival, and toxicity would be irrelevant to a person dead from or fatally injured by heat.

Section 4.10.1. Chemical Methods

The chemical methods of identification and analysis which may be useful for studying fire gases include:

1. Infrared analysis
2. Gas chromatography
3. Mass spectroscopy
4. Chemiluminescence
5. Polarographic methods
6. Paramagnetic methods
7. Ion-selective electrodes
8. Titrimetric methods
9. Colorimetric methods

Tests for identification and analysis of gases evolved during pyrolysis and combustion are described in several standards which include the following:

ASTM E 800 (measurement of gases present or generated during fires)
ASTM E 1678 (smoke toxicity)
JIS K 7217 (analytical method for gases from burning plastics, Japan)
NF C 20-454 (tube furnace method, analysis of gases from pyrolysis or combustion of materials in electrotechnics, France)
NF X 70-100 (tube furnace method, analysis of pyrolysis and combustion gas, France)
NF X 70-101 (test chamber method, analysis of gases from combustion or pyrolysis, France)
CSN 64 0758 (hydrogen cyanide in combustion product of plastics, Czechoslovakia)

The ASTM E 1678 test employs a transparent polycarbonate or polymethylmethacrylate chamber with a nominal volume of 0.2 m^3 (200 L) and inside dimensions of 1220 by 370 by 450 mm (48 by 14-1/2 by 17-3/4 in). A test specimen no larger than 76 by 127 mm (3 by 5 in) and no thicker than 50 mm (2 in) is placed in a combustion cell, a horizontal quartz tube with a 127 mm (5 in) inside diameter and approximately 320 mm (12.5 in) long, and exposed to a radiant heat flux of 50 kW/m^2 for 15 min, using four quartz infrared lamps (with tungsten filaments), rated at 2000 W at 240 V. A stainless steel chimney approximately 30 by 300 mm (1.25 by 11.75 in) inside dimensions, and 300 mm (11.75 in) wide, connects the combustion cell to the animal exposure chamber. Without the exposure of test animals, the gases evolved are analyzed for CO, O_2, and CO_2, and, if present, HCN, HCl, and HBr.

Section 4.10.2. Biological Methods

Toxicological studies on laboratory animals generally involve exposure of small animals, usually rats or mice, to the gaseous products of decomposition and combustion under carefully controlled conditions, with observations of the number and nature of deaths, incapacitations, and injuries, followed by hematology (blood) and pathology (tissues) to determine the cause of death, if fatalities occur, and the cause and extent of incapacitation and injuries.

The laboratory animals most often used in tests for toxic gas evolution are rats and mice. Each of these two species has its proponents and critics, and each strain of these two species has its proponents and critics. Other animal species such as guinea pigs, rabbits, dogs, and monkeys may provide additional information, but have the disadvantages of greater cost and substantial opposition from animal rights groups.

The principal biological endpoints in bioassay procedures are lethality (death) and incapacitation.

The determination of death is simple, sensitive, reliable, and has good reproducibility. Lethality under specified conditions may be reported in one or more of the following ways:

1. Per cent lethality, between 0 and 100°
2. Time to death
3. Dose lethal for 50% of the test animals, or LD_{50}
4. Concentration lethal for 50% of the test animals, or LC_{50}
5. Time lethal for 50% of the test animals, or LT_{50}

The determination of incapacitation has used a variety of methods and endpoints, each with its proponents and critics.

The test conditions under which a material is decomposed and burned generally have more effect on test results than do the bioassay procedures for determining lethality and incapacitation. The tests for toxic gas evolution will therefore be arranged according to the method of generating the toxic gases.

1. Tests that employ tube furnaces
2. Tests that employ crucible furnaces
3. Tests that employ radiant furnaces

The tests for toxic gas evolution which employ tube furnaces include:

1. The DIN 53436 test, developed by Deutsches Institut fur Normung (DIN), Germany
2. The FAA-CAMI or FAA or CAMI test, developed by the Civil Aeromedical Institute of the Federal Aviation Administration
3. The USF-NASA or USF or PSC or dome chamber test, developed at the University of San Francisco with the support of the National Aeronautics and Space Administration, and continued by Product Safety Corporation
4. The University of Pittsburgh or UPitt test, developed at the University of Pittsburgh

The tests for toxic gas evolution which employ crucible furnaces include:

1. The crucible furnace test developed by the National Bureau of Standards
2. The Polish Fire Safety Centre test

The tests for toxic gas evolution which employ radiant furnaces include:

1. The JIS A 1321 Announcement No. 1231 test, developed by the Japanese Standards Association (JSA)
2. The FAA-CAMI or FAA or CAMI 265-liter test, developed by the Civil Aeromedical Institute of the Federal Aviation Administration
3. The UPitt II test, developed at the University of Pittsburgh
4. ASTM E 1678 test, developed by the National Institute of Building Sciences

The DIN 53436 method employs a horizontal tube 1000 ± 300 mm long and 40 ± 1 mm in diameter. An annular electric oven tightly enclosing the tube is moved along the tube, in a direction opposite to the direction of the air stream flowing over the test sample. The speed of the annular oven determines the burning rate of the sample, and, together with the length of the test sample, determines the time of test duration. The test samples are heated at different constant furnace temperatures between 200°C. and 600°C. The combustion gases from the tube are diluted and cooled with secondary air. The animal exposure chamber varies with the laboratory. From 5 to 20 rats are used in each test, with either head-only or whole-body exposure.

The FAA-CAMI test employs a horizontal tube heated by two 425-watt semicylindrical heating units. Three rats in individual compartments in a rotating cage are used in each test, with whole-body exposure. The assembled system has a total volume of 12.6 liters. This test was used in studies for the Federal Aviation Administration (FAA) and Urban Mass Transportation Administration (UMTA).

The USF or dome chamber test employs a horizontal tube inside a tube furnace. The animal exposure chamber is hemispherical and has a volume of 4.2 liters. Four mice are used in each test, with whole-body exposure. Four appears to be the optimum number of mice in each test to avoid excessive oxygen consumption and increase in temperature and humidity during the test period. The test animals are observed for time to responses such as staggering, convulsions, collapse, and death. Lethality is determined either as time to death or as LC_{50}. The rising temperature at 40°C./min from 200°C. to 800°C. has been used to evaluate over 300 materials under conditions which are intended to simulate the pre-ignition and pre-flashover stages of a fire. Some data obtained using this test method are presented in Table 3.10 (time to death) and Table 3.11 (LC_{50}).

The Polish Fire Safety Centre test employs an animal exposure chamber unit consisting of two sections, a combustion chamber and an exposure chamber, each with a volume of 4.5 liters, made of quality steel. A transparent polymethyl methacrylate door allows continuous observation during the exposure. Four mice are used in each test. Four appears to be the optimum number of mice in each test to avoid excessive oxygen consumption and increase in temperature and humidity during the test period. A sample not exceeding 2.0 g is placed in a quartz pyrolysis crucible and is heated to a temperature determined individually for each material tested. Duration of exposure is 20 min. Some data obtained using this test method are presented in Table 4.24 (LC_{50}).

The University of Pittsburgh (UPitt) test employs a vertical tube inside a tube furnace. The sample is subjected to rising temperature at 20°C./min. When a weight loss of 1% is detected, the combustion products from the sample are carried from the furnace at a flow rate of 20 l/min, diluted with cool air, and carried into the animal exposure chamber. The animal exposure chamber has a volume of 2.3 liters. Four mice are used in each test, with head-only exposure.

One variation of the UPitt test is intended for use with electrical insulating materials. A sample weighing 1 to 10 g is heated in a furnace capable of heating at 20°C ± 1°C per min with linearity to 1000°C. The sample is subjected to rising temperature at 20°C./min. When a weight loss of 1% is detected, the combustion products from the sample are carried from the furnace at a flow rate of 20 l/min, diluted with cool air, and carried into the animal exposure chamber. Four mice are used in each test. This variation of the UPitt test is described in ASTM Z6520Z.

The UPitt II test employs a radiant heat source. The radiant heat source, using an electrical heater rod, rated at 5000 W and 240 V, tightly wound into the shape of a truncated cone, can be used to generate heat flux up to 100 kW/m^2. A specimen 100 by 100 mm is tested in the horizontal orientation. The apparatus was placed in a cone hood with a volume of 120 liters. The animal exposure chamber consists of two sections, each 110 mm in diameter and 270 mm long and fitted with eight smaller cylinders 29 mm in diameter and 90 mm long to hold the mice.

The UPitt test has changed with time and exists in slightly different versions. One version of this test is the combustion toxicity test required for building products in New York State, in accordance with Article 15, Part 1120, of the New York State Fire Prevention and Building Code. Some test data are presented in Table 4.25.

The NBS crucible furnace test employs a 1000 ml quartz beaker 92 mm in diameter and 150 mm high inside a 9-in cubical furnace. The animal exposure chamber has a nominal volume of 200 liters and interior dimensions of 122 by 36 by 46 cm. Six rats are used in each test, with head-only exposure.

The JIS A 1321/1231 test exposes an area of specimen measuring 18 by 18 cm to both gas flame and electric heater. The combustion gases enter a dilution chamber measuring 0.5 by 0.5 by 0.5 m for adjustment to suitable temperature and concentration, and then an exposure chamber measuring 0.5 by 0.5 by 0.5 m containing 8 mice in revolving cages.

The FAA-CAMI 265-liter test employs four 2000-watt quartz lamps as its radiant heat source. The animal exposure chamber has an internal volume of 203 liters, with a

42-liter plenum for cooling and diluting the hot combustion gases before entry into the animal exposure chamber. Four rats in individual rotating cages are used in each test, with whole-body exposure. The assembled system has a total volume of 265 liters. This test was used in studies for the Federal Aviation Administration.

The ASTM E 1678 test employs a transparent polycarbonate or polymethylmethacrylate chamber with a nominal volume of 0.2 m³ (200 L) and inside dimensions of 1220 by 370 by 450 mm (48 by 14-1/2 by 17-3/4 in). A test specimen no larger than 76 by 127 mm (3 by 5 in) and no thicker than 50 mm (2 in) is placed in a combustion cell, a horizontal quartz tube with a 127 mm (5 in) inside diameter and approximately 320 mm (12.5 in) long, and exposed to a radiant heat flux of 50 kW/m² for 15 min, using four quartz infrared lamps (with tungsten filaments), rated at 2000 W at 240 V. A stainless steel chimney approximately 30 by 300 mm (1.25 by 11.75 in) inside dimensions, and 300 mm (11.75 in) wide, connects the combustion cell to the animal exposure chamber. Six rats are used in each test, with head-only exposure.

The NBS crucible furnace test and the ASTM E 1678 test use essentially the same animal exposure chamber, and differ in apparatus for decomposing and burning materials, and for introducing the gases into the animal exposure chamber.

SECTION 4.11. TESTS FOR CORROSIVE GAS EVOLUTION

Corrosive gas evolution may be defined as the level of corrosiveness exhibited by the gases evolved from a material under specified conditions. This characteristic provides a measure of fire hazard, in that a material which produces more corrosive gases than another material would be expected to be more likely to affect electrical and electronic equipment used in firefighting and rescue at the scene of a fire, and, of broader concern, electrical and electronic equipment used in communications necessary for business operations and public safety, and in control of critical installations such as nuclear power plants.

Tests for corrosive gases generally fall into three types:

1. Tests concerned with measuring the pH and ionic conductivity of aqueous solutions of combustion products, on the assumption that pH and ionic conductivity are major factors in corrosion.
2. Tests concerned with measuring the yield of soluble metal ions from corrosion of metal, on the assumption that loss of metal is a major effect of corrosion.
3. Tests concerned with measuring the change in resistance of a circuit exposed to combustion products, on the assumption that damage to electrical and electronic equipment is a major concern related to corrosion.

There are several tests which measure the pH of aqueous solutions of combustion products, including:

IEC 754-1 (23 to 800°C for 40 min ramp, 20 min isothermal)
BS 602 Part 1
JCS C No. 53/397
CSA 22-2 No. 0.3M (800°C for six 5 min steps, 10 min isothermal)
SAA AS 1660.5.3 (800°C for 10 min)

There are several tests which measure the pH and ionic conductivity of aqueous solutions of combustion products, including:

IEC 754-2 (935°C for 30 min)
BS 602 Part 2
DIN 57472 Part 813
CEGB E/TSS/EX5/8056 Part 3

There are several tests which measure the loss of metal from corrosion, including:

ISO 11907-2 (800°C for 2.5 min)
ISO 11907-3 (600°C for 20 min)
ASTM D 5485 (typically 50 kW/m^2 with 70% mass loss)
ASTM E5 Z3314Z (typically 50 kW/m^2 for 15 min)

There are several tests which measure the change in resistance of a circuit exposed to combustion products, including CNET DEC-0611/C.

The modified NIBS test employs a transparent polycarbonate or polymethylmethacrylate chamber with a nominal volume of 0.2 m^3 (200 L) and inside dimensions of 1220 by 370 by 450 mm (48 by 14-1/2 by 17-3/4 in). A test specimen no larger than 76 by 127 mm (3 by 5 in) and no thicker than 50 mm (2 in) is placed in a combustion cell, a horizontal quartz tube with a 127 mm (5 in) inside diameter and approximately 320 mm (12.5 in) long, and exposed to a radiant heat flux of 50 kW/m^2 for 15 min, using four quartz infrared lamps (with tungsten filaments), rated at 2000 W at 240 V. A stainless steel chimney approximately 30 by 300 mm (1.25 by 11.75 in) inside dimensions, and 300 mm (11.75 in) wide, connects the combustion cell to the 200 L chamber. The combustion gases from a 2 by 2 in specimen are condensed on a copper corrosion probe. Corrosion is expressed as average metal loss in angstroms (1 angstrom = 39.37 x 10^{-9} in.) as determined by the resistance increase of the probe. Some data are presented in Table 4.26.

The ASTM D 5485 is a modification of the ASTM E 1354 test. A specimen measuring 100 by 100 mm is subjected to radiant heat, and a spark igniter is used to ignite the combustible vapors. A radiant heat source, using an electrical heater rod, rated at 5000 W and 240 V, tightly wound into the shape of a truncated cone, is used to generate heat flux up to 100 kW/m^2. The cone heater is similar to that described in ASTM E 1354. A portion of the products of decomposition or combustion flows continuously through an 11.2-liter exposure chamber which holds the corrosion targets until the specimen has lost an average of 70% of the total combustible mass or for a period of 60 min, whichever is less. The corrosion of the target is determined by exposure of the target to combustion products for 1 hour, followed by 24-hour exposure of the target to an environment of $23 \pm 2°C$. and $75 \pm 5\%$ relative humidity in a separate chamber. The increase in electrical resistance of each target is monitored, and the reduction in thickness of the metal on the target is calculated from the increase in electrical resistance. This reduction in thickness is referred to as corrosion-by-metal-loss. Some data are presented in Table 3.12.

The DIN 57472 Part 813 method employs a horizontal tube 1000 ± 300 mm long and 40 ± 1 mm in diameter. An annular electric oven tightly enclosing the tube is moved along the tube, in a direction opposite to the direction of the air stream flowing over the test sample. The speed of the annular oven determines the burning rate of the sample, and, together with the length of the test sample, determines the time of test duration. The test samples are heated at different constant furnace temperatures between 200°C. and 600°C. The combustion gases from the tube are absorbed in an aqueous solution, and the aqueous solution is measured for pH and conductivity. Some data are presented in Table 4.27.

The CNET test method was developed at Centre National d'Etudes et de Telecommunications employs a test chamber which is a 20-liter Pyrex of polymethyl methacrylate cylinder 11.75 in. inside diameter and 17.75 in. long. The ignition system consists of an Inconel electrical resistance coil which is calibrated to 800°C. The sample, a mixture of 600 mg of the material to be tested and 100 mg of a standard polyethylene, is placed on top of the heater coil in a quartz crucible. The combustion gases condense out on a corrosion target, a 30 by 60 mm printed circuit board with copper tracks 130,000 angstroms thick and a resistance of approximately 8 ohms. Some data are presented in Table 4.28.

SECTION 4.12. MULTICAPABILITY TEST METHODS

Fire hazard is the summation of all the fire response characteristics of a material or product in a specific environment. The desirability of studying as many fire response characteristics as possible with a single apparatus has produced a class of fire test methods which obtain information on two or more fire response characteristics in order to provide more information for evaluating fire hazard. These are the multicapability fire test methods.

Twelve tests will be reviewed to show the development of multicapability fire test methods over 68 years, from 1929 to 1997:

1. The Forest Products Laboratory (FPL) fire-tube apparatus, first published in 1929, which became ASTM E 69
2. The Underwriters Laboratories (UL) 25-foot tunnel, first published in 1944, which became ASTM E 84
3. The National Bureau of Standards (NBS) radiant panel, first published in 1956, which became ASTM E 162

4. The Forest Products Laboratory (FPL) 8-foot tunnel, first published in 1957, which became ASTM E 286
5. The National Bureau of Standards (NBS) smoke chamber, first published in 1967, which became ASTM E 662
6. The Ohio State University (OSU) release rate apparatus, first published in 1972, which became ASTM E 906
7. The LIFT apparatus, first published in 1979, which became ASTM E 1317 and ASTM E 1321
8. The National Bureau of Standards (NBS) cone calorimeter, first published in 1982, which became ASTM E 1354
9. The intermediate scale calorimeter (ICAL), which became ASTM E 1623
10. The National Institute of Building Sciences (NIBS) toxicity test, which became ASTM E 1678
11. The Factory Mutual (FM) fire propagation apparatus, which is proposed as an ASTM standard (ASTM E5 Z6880Z)
12. The California Technical Bulletin 133 test, first published in 1984, which became ASTM E 1537

The fire-tube apparatus, which became ASTM E 69, was developed at the USDA Forest Products Laboratory (FPL) in Madison, Wisconsin, by T. R. Truax et al. and first published in 1929.

The ASTM E 69 fire-tube apparatus is a vertically mounted tube 4.75 in. in diameter and 38 in. long, in which a specimen 3/8 by 3/4 in (9.5 by 19 mm) and 40 in (1016 mm) is mounted vertically. The specimen is ignited at the bottom by a burner flame 11 in. (279 mm) in height, with a tall distinct inner cone. The weight loss of the specimen during burning provides a measure of combustibility, but not a measure of a particular fire response characteristic.

The FPL fire-tube apparatus is not easily adapted to obtain information on two or more fire response characteristics and become a multicapability fire test method.

A thermocouple is not easily installed to measure heat release, so that the area under the time-temperature curve can be used as a measure of heat release.

A light beam is not easily installed to measure smoke density, so that the area under the time-absorption curve can be used as a measure of smoke evolution.

The first prominent multicapability fire test method appeared 15 years later.

The 25-foot tunnel, which became ASTM E 84, was developed at Underwriters Laboratories (UL) by Albert J. Steiner and first published in 1944.

The ASTM E 84 test requires a specimen 7.62 by 0.496 m (300 by 19.5 in), mounted face down so as to form the roof of a tunnel 7.62 by 0.445 by 0.305 m (300 by 17.5 by 12 in). The fire source, two gas burners 305 mm (12 in) from the fire end of the sample and 190 mm (7.5 in) below the surface of the sample, is initially adjusted to deliver 5,000 Btu/min (5.3 MJ/min or 87.9 kW), and is finally adjusted so that a test sample of select-grade red oak flooring would spread flame 5.94 m (19.5 ft) from the end of the igniting fire in 5.5 min ± 15 sec. The end of the igniting flame is considered as being 1.37 m (4.5 ft) from the burners, this flame length being due to an average air velocity of 73.2 ± 1.5 m/min (240 ± 5 ft/min). Test duration is 10 min.

Materials are rated for flame spread on a scale on which asbestos-cement board is zero and select-grade red oak flooring is 100.

A thermocouple located 25.4 mm (1 in) from the exposed surface of the test sample and within 305 mm (1 ft) of the vent end of the sample is used to measure heat release. The area under the time-temperature curve for the 10 min test is used as a measure of heat release.

A light source mounted on a horizontal section of the vent pipe with the light beam directed downward is used to measure smoke density. The area under the time-absorption curve is used as a measure of smoke evolution.

Gas samples from the tunnel can be used to measure toxic gas evolution using analytical techniques. Analysis of the combustion products is not required by the test method.

The ASTM E 84 test is similar to those described in the following standards:

UL 723
NFPA 255
UBC 8-1
CAN/ULC-S102

The UL 25-foot tunnel was the first prominent test to obtain information on two or more fire response characteristics and was the first prominent multicapability fire test method.

This test method provides measurements of flame spread, heat release, and smoke evolution, and may be used to measure ignitability and toxic gas evolution.

The radiant panel, which became ASTM E 162, was developed at National Bureau of Standards (NBS) by Alex F. Robertson et al. and first published in 1956.

The ASTM E 162 test employs a radiant heat source consisting of a 305 by 457 mm (12 by 18 in) vertically mounted porous refractory panel maintained at 670 ± 4°C. (1238 ± 7°F.). A specimen measuring 152 by 457 mm (6 by 18 in) is supported in front of it with the 457 mm (18 in) dimension inclined 30°C. from the vertical. A pilot burner ignited the top of the specimen, 121 mm (4.75 in) away from the radiant panel, so that the flame front progresses downward along the underside exposed to the radiant panel.

The temperature rise recorded by stack thermocouples, above their base level of 180 to 230°C. (356 to 446°F.) is used as a measure of heat release.

The ASTM E 162 test is similar to those described in the following standards:

ASTM D 3675
Federal Test Method Standard No. 501a, Method 6421

In the original test, a smoke sampling device which collected smoke particles on glass fiber filter paper was used to measure smoke evolution.

The NBS radiant panel was the second prominent test to obtain information on two or more fire response characteristics and was a prominent multicapability fire test method.

This test method provided measurements of flame spread, heat release, and smoke evolution, and could be used to measure ignitability.

The 8-foot tunnel, which became ASTM E 286, was developed at the USDA Forest Products Laboratory (FPL) in Madison, Wisconsin, by H. D. Bruce et al. and first published in 1957.

The ASTM E 286 test employs a specimen 8 ft (2.44) long and 13.75 in (349 mm) wide, mounted horizontally so as to form the roof of a tunnel and slope at a 6° angle from end to end. The heat supply rate for initial trials is 3,400 Btu/min.

The FPL 8-foot tunnel provides measurements of ignitability, flame spread, heat release, and smoke evolution, and is a multicapability fire test method.

The smoke chamber, which became ASTM E 662, was developed at National Bureau of Standards (NBS) by Daniel Gross et al. and first published in 1967.

The ASTM E 662 smoke chamber is a completely closed cabinet, 914 by 610 by 914 mm (36 by 24 by 36 in), in which a specimen 76.2 mm (3 in) square is supported vertically in a frame such that an area 65.1 mm (2-9/16 in) square is exposed to heat under either piloted (flaming) or nonpiloted (smoldering) conditions. The heat source is an electric furnace, adjusted with the help of a circular foil radiometer to give a heat flux of 2.5 W/cm^2 (2.2 Btu/sec-ft^2) at the specimen surface. A vertical photometer path for measuring light absorption is employed to minimize measurement differences due to smoke stratification which could occur with a horizontal photometer path at a fixed height; the full 914 mm (3 ft) height of the chamber is used to provide an overall average for the entire chamber.

The ASTM E 662 test is similar to those described in the following standard: NFPA 258.

Visual observations have been used to measure ignitability, and gas samples have been used to measure toxic gas evolution.

The NBS smoke chamber was the second NBS test to obtain information on two or more fire response characteristics and was a multicapability fire test method. This test method can provide measurements of ignitability, smoke evolution, and toxic gas evolution.

The release rate apparatus, which became ASTM E 906, was developed at the Ohio State University (OSU) by Edwin E. Smith and first published in 1972.

The ASTM E 906 release rate test employs a chamber 890 by 410 by 200 mm (35 by 16 by 8 in) with a pyramidal top section 395mm (15.5 in) high connecting to the outlet. A radiant heat source, using four silicon carbide Globar Type LL elements, is used to generate heat flux up to 100 kW/m^2. Specimens 160 by 150 mm (6 by 6 in) are tested in vertical orientation, and specimens 110 by 150 mm (4.5 by 6 in) are tested in horizontal orientation. A radiation reflector is used for horizontally mounted specimens. The total air flow of 0.04 m^3/sec (84 ft^3/min) leaves the apparatus through a rectangular exhaust stack 133 by 70 mm (5.25 by 2.75 in) in cross section and 254 mm (10 in) high. The temperature difference between the air entering and the air leaving the apparatus is measured by a thermopile having 3 hot junctions spaced across the top of the exhaust stack and 3 cold junctions located in the pan at the bottom. A horizontal photometer path above the outlet is used for measuring light absorption.

This test method, with various modifications, is described in the following standards:

ASTM E 906
NFPA 263
Federal Aviation Regulations (FAR) Parts 25 and 121

This test method provides measurements of heat release and smoke evolution, and may be used to measure ignitability, flame spread, and toxic gas evolution.

The apparatus which is used in ASTM E 1317 and ASTM E 1321, known as the lateral ignition and flame spread test apparatus (LIFT), was developed from International Maritime Organization (IMO) test A.653(16) at the National Bureau of Standards (NBS) by Alex F. Robertson and first published in 1979.

The test method which became ASTM E 1317 was developed for marine surface finishes.

The ASTM E 1317 test exposes vertically mounted specimens to the heat from a vertical air-gas fueled porous refractory radiant-heat energy source, with heated surface dimensions of 280 by 483 mm, inclined at 15° to the specimen. The specimen is 155 mm wide and 800 mm long.

Means are provided for observing the times to ignition, spread, and extinguishment of flame along the length of the specimen, and measuring the temperature of the stack gases during burning. Results are reported in terms of heat for ignition, heat for sustained burning, critical flux at extinguishment, and heat release of the specimen during burning.

The ASTM E 1317 test provides information on ignitability, flame spread, heat release, and ease of extinguishment, and is a multicapability fire test method.

The test method which became ASTM E 1321 was developed at the National Bureau of Standards (NBS) by James Quintiere and Margaret Harkelroad and first published in 1984.

The ASTM E 1321 test exposes vertically mounted specimens to the heat from a vertical air-gas fueled porous refractory radiant-heat energy source, with heated surface dimensions of 280 by 483 mm, inclined at 15° to the specimen.

For the ignition test, a series of 155 by 155 mm specimens are exposed to a nearly uniform heat flux, and the time to flame attachment, using piloted ignition, is determined.

For the flame spread test, a 155 by 800 mm specimen is exposed to a graduated heat flux that is approximately 5 kW/m^2 higher at the hot end than the minimum heat flux necessary for ignition. The specimen is preheated to thermal equilibrium, the preheat time being derived from the ignition test. After using piloted ignition, the pyrolyzing flame-front progression along the horizontal length of the specimen as a function of time is tracked.

The ASTM E 1321 test provides information on ignitability and flame spread, and is a multicapability fire test method.

The cone calorimeter, which became ASTM E 1354, was developed at the National Bureau of Standards (NBS) by Vytenis Babrauskas et al. and first published in 1982.

The ASTM E 1354 cone calorimeter is an apparatus 1680 by 1625 by 686 mm. A radiant heat source, using an electrical heater rod, rated at 5000 W and 240 V, tightly wound into the shape of a truncated cone, is used to generate heat flux up to 100 kW/m^2. Specimens 100 by 100 mm are tested in either horizontal or vertical orientation. A paramagnetic oxygen analyzer with a range from 0 to 25 % oxygen is used to measure the oxygen concentration in the exhaust gas. The rate of heat release is measured by the principle of oxygen consumption. The amount of heat released is calculated from the amount of oxygen consumed. Smoke evolution is measured.

This test method, with various modifications, is described in the following standards:

ISO 5660-1
ASTM E 1354
ASTM E 1474 (furniture components)
ASTM E 1740 (wallcovering composites)
ASTM F 1550 (correctional facility furnishings)
NFPA 264
BS 476 Part 15

The ASTM E 1354 test provides measurements of ignitability, heat release, and smoke evolution, and is a multicapability fire test method. The combustion gases can be evaluated for toxic gas evolution and corrosive gas evolution.

The cone heater is used in the following standard: ASTM D 5485 (cone corrosimeter).

The intermediate scale calorimeter, which became ASTM E 1623, was developed at Weyerhaeuser Company.

The intermediate scale calorimeter (ICAL) is an apparatus consisting of a vertically mounted radiant panel assembly facing a vertically mounted specimen, at a distance adjusted by means of a trolley. The radiant panel assembly, with three rows of ceramic-faced natural gas burners, is used to generate heat flux up to 50 kW/m². Specimens 1000 by 1000 mm are tested in vertical orientation. An oxygen analyzer with a range from 0 to 25 % oxygen is used to measure the oxygen concentration in the exhaust gas. The rate of heat release is measured by the principle of oxygen consumption. The amount of heat released is calculated from the amount of oxygen consumed. Smoke evolution and carbon monoxide evolution are measured.

This test method is described in ASTM E 1623.

The ASTM E 1623 test provides measurements of ignitability, flame spread, heat release, smoke evolution, and toxic gas evolution, and is a multicapability fire test method.

The combustion toxicity test, which became ASTM E 1678, was developed by the National Institute of Building Sciences (NIBS) and first published in 1990.

The animal exposure chamber was developed at the National Bureau of Standards and first published in 1982. The radiant heat source was developed at Weyerhaeuser Company and first published in 1984.

The ASTM E 1678 test employs a transparent polycarbonate or polymethylmethacrylate chamber with a nominal volume of 0.2 m³ (200 L) and inside dimensions of 1220 by 370 by 450 mm (48 by 14-1/2 by 17-3/4 in). A test specimen no larger than 76 by 127 mm (3 by 5 in) and no thicker than 50 mm (2 in) is placed in a combustion cell, a horizontal quartz tube with a 127 mm (5 in) inside diameter and approximately 320 mm (12.5 in) long, and exposed to a radiant heat flux of 50 kW/m^2 for 15 min, using four quartz infrared lamps (with tungsten filaments), rated at 2000 W at 240 V. A stainless steel chimney approximately 30 by 300 mm (1.25 by 11.75 in) inside dimensions, and 300 mm (11.75 in) wide, connects the combustion cell to the animal exposure chamber. Without the exposure of test animals, the gases evolved are analyzed for CO, O$_2$, and CO$_2$, and, if present, HCN, HCl, and HBr.

This test provides measurements of ignitability and toxic gas evolution, and is a multicapability fire test method. It can not provide measurements of flame spread, heat release, and smoke evolution.

The fire propagation apparatus (FPA), which is proposed as an ASTM method, was developed at Factory Mutual Research Corporation by Archibald Tewarson and first published in 1988.

The Factory Mutual fire propagation apparatus is an apparatus consisting of four infrared heaters and four types of specimen holders. The four infrared heaters, each containing six tungsten filament tubular quartz lamps, are used to generate heat flux up to 50 kW/m^2. The four types of specimens are horizontal square 0.10 by 0.10 m (4 by 4 in), horizontal circular 0.097 m (3.8 in) diameter, vertical 0.305 by 0.076 m (19 by 5.2 in), and vertical cable 0.81 m (32.5 in) long and up to 51 mm (2 in) diameter.

This test method is described in ASTM E5 Z6880Z.

This test provides measurements of ignitability, flame spread, heat release, and smoke evolution, and is a multicapability fire test method.

The California Technical Bulletin 133 test, which became ASTM E 1537, was developed at the California Bureau of Home Furnishings by Gordon H. Damant et al. and first published in 1984.

The ASTM E 1537 or California Technical Bulletin 133 (TB 133) test is perhaps the most widely recognized large-scale test for heat release. The test specimen is a full-size manufactured item of upholstered furniture, a representative prototype of the upholstered furniture, or a mock-up of the upholstered furniture. The specimen is ignited with a propane gas burner, used at a flow rate of 13 ± 0.25 L/min for 80 seconds (equivalent to 19.3 kW). It approximates the ignition propensity of five crumpled sheets of newspaper located on the seating cushion. An oxygen analyzer with a range from 0 to 21 % oxygen is used to measure the oxygen concentration in the exhaust gas. The rate of heat release is measured by the principle of oxygen consumption. The amount of heat released is calculated from the amount of oxygen consumed. Smoke evolution and carbon monoxide evolution are measured.

One of three test configurations is used in the test method:

1. A test room 3.66 by 2.44 by 2.44 m high

2. A test room 3.66 by 3.05 by 2.44 m high
3. An open calorimeter (or furniture calorimeter)

Test room configurations are described for items of furniture less than 1 m across, such as a chair, and for items of furniture between 1 and 2.44 m across, such as a sofa.

This test method, with various modifications, is described in the following standards:

ASTM E 1537
California Technical Bulletin 133
UL 1056
NFPA 266
Boston BFD IX-10

This test provides measurements of heat release, smoke evolution, and toxic gas evolution, and is a multicapability fire test method. It can provide measurements of ignitability and flame spread over various surface orientations.

REFERENCES

Alexeeff, G. V., Packham, S. C., "Use of a Radiant Furnace Fire Model to Evaluate Acute Toxicity of Smoke", Journal of Fire Sciences, Vol. 2, No. 4, 306-320 (July/August 1984)

Babrauskas, V., "Development of the Cone Calorimeter: A Bench-Scale Heat Release Rate Apparatus Based on Oxygen Consumption", NBSIR 82-2611, National Bureau of Standards, Gaithersburg, Maryland (November 1982)

Babrauskas, V., "Development of the Cone Calorimeter: A Bench-Scale Heat Release Rate Apparatus Based on Oxygen Consumption", Fire and Materials, Vol. 8, No. 2, 81-95 (June 1984)

Babrauskas, V., Parker, W. J., "Ignitability Measurements with the Cone Calorimeter", NBSIR 86-3445, National Bureau of Standards, Gaithersburg, Maryland (September 1986)

Babrauskas, V., Parker, W. J., "Ignitability Measurements with the Cone Calorimeter", Fire and Materials, Vol. 11, No. 1, 31-43 (March 1987)

Babrauskas, V., Harris, R. H., Gann, R. G., Levin, B. C., Lee, B. T., Peacock, R. D., Paabo, M., Twilley, V., Yoklavich, M. F., Clark, H. M., "Fire Hazard Comparison of Fire-Retarded and Non-Fire Retarded Products", NBS Special Publication 749, National Bureau of Standards, Gaithersburg, Maryland (July 1988)

Babrauskas, V., "Smoke and Gas Evolution Rate Measurements on Fire-Retarded Plastics with the Cone Calorimeter", Fire Safety Journal, Vol. 14, 135-142 (1989)

Babrauskas, V., "Ten Years of Heat Release Research with the Cone Calorimeter", III-1 to III-8, in "Heat Release and Fire Hazard", Vol. 1, Y. Hasemi, ed., Building Research Institute, Tokyo, Japan (1993)

Babrauskas, V., Wetterlund, I., "The Role of Flame Flux in Opposed-Flow Flame Spread", Fire and Materials, Vol. 19, 275-281 (1995)

Babrauskas, V., "Specimen Heat Fluxes for Bench-Scale Heat Release Rate Testing", Fire and Materials, Vol. 19, 243-252 (1995)

Babrauskas, V., Wetterlund, I., "The CBUF Cone Calorimeter Test Protocol", SP Report 1996:12, Swedish National Testing and Research Institute, Boras, Sweden (1996)

Babrauskas, V., Facade Fire Tests: Towards an International Test Standard", Proceedings of the International Conference on Fire Safety, Vol. 21, 100-109 (January 1996)

Babrauskas, V., "Sandwich Panel Performance in Full-Scale and Bench-Scale Fire Tests", Fire and Materials, Vol. 21, 53-65 (1997)

Babrauskas, V., White, J. A., Urbas, J., "Testing for Surface Spread of Flame: New Tests to Come into Use", Building Standards, 13-17 (March/April 1997)

Barth, E., Muller, B., Prager, F. H., Wittbecker, F.-W., "Corrosive Effects of Smoke: Decomposition with the DIN Tube According to DIN 53436", Journal of Fire Sciences, Vol. 10, No. 5, 432-454 (September/October 1992)

Beeson, H. D., Hshieh, F.-Y., Hirsch, D. B., "Ignitability of Advanced Composites in Liquid and Gaseous Oxygen", ASTM STP 1319 (1997)

Bennett, J. G., Kessel, S. L., Rogers, C. E., "Corrosivity Test Methods for Polymeric Materials. Part 3. Modified DIN Test Method", Journal of Fire Sciences, Vol. 12, No. 2, 155-174 (March/April 1994)

Bennett, J. G., Kessel, S. L., Rogers, C. E., "Corrosivity Test Methods for Polymeric Materials. Part 4. Cone Corrosimeter Test Method", Journal of Fire Sciences, Vol. 12, No. 2, 175-195 (March/April 1994)

Bottin, M.-F., "The ISO Static Test Method for Measuring Smoke Corrosivity", Journal of Fire Sciences, Vol. 10, No. 2, 160-168 (March/April 1992)

Braun, E., Shields, J. R., Harris, R. H., "Flammability Characteristics of Electrical Cables Using the Cone Calorimeter", NISTIR 88-4003, National Institute of Standards and Technology, Gaithersburg, Maryland (January 1989)

Briber, A. A., "Fire Tests for Surface Flammability", Proceedings of the International Conference on Fire Safety, Vol. 24, 238-248 (July 1997)

Bridgman, A. L., Nelson, G. L., "Heat Release Rate Calorimetry of Engineering Plastics", Journal of Fire and Flammability, Vol. 13, No. 2, 114-134 (April 1982)

Bridgman, A. L., Nelson, G. L., "Radiant Panel Tests on Engineering Plastics", Proceedings of the International Conference on Fire Safety, Vol. 8, 191-226 (January 1983)

Briggs, P. J., "Fire Hazard Analysis of New Insulation Materials", Proceedings of the International Conference on Fire Safety, Vol. 13, 357-366 (January 1988)

Brown, J. E., Braun, E., Twilley, W. H., "Cone Calorimeter Evaluation of the Flammability of Composite Materials", NBSIR 88-3733, National Bureau of Standards, Gaithersburg, Maryland (March 1988)

Bruce, H. D., Miniutti, V. P., "Small Tunnel Furnace Test for Measuring Surface Flammability", USDA Forest Products Laboratory Report No. 2097, U.S. Department of Agriculture, Madison, Wisconsin (November 1957)

Caldwell, D. J., Alarie, Y. C., "A Method to Determine the Potential Toxicity of Smoke from Burning Polymers: I. Experiments with Douglas Fir", Journal of Fire Sciences, Vol. 8, No. 1, 23-62 (January/February 1990)

Caldwell, D. J., Alarie, Y. C., "A Method to Determine the Potential Toxicity of Smoke from Burning Polymers: II. The Toxicity of Smoke from Douglas Fir", Journal of Fire Sciences, Vol. 8, No. 4, 275-309 (July/August 1990)

Caldwell, D. J., Alarie, Y. C., "A Method to Determine the Potential Toxicity of Smoke from Burning Polymers: III. Comparison of Synthetic Polymers to Douglas Fir Using the UPitt II Flaming Combustion/Toxicity of Smoke Apparatus", Journal of Fire Sciences, Vol. 9, No. 6, 470-518 (November/December 1991)

Caudill, L. M., Chapin, J. T., Comizzoli, R. B., Gandhi, P., Peins, G. A., Sinclair, J. D., "Current State of Fire Corrosivity Testing: Preliminary Electrical Leakage Current Measurements", Proceedings of the 1995 International Wire and Cable Symposium, 432-437 (1995)

Chapin, J. T., Caudill, L. M., Gandhi, P., Backstrom, R., "Leakage Current Smoke Corrosivity Testing: Comparison of Cable and Material Data", Proceedings of the 1996 International Wire and Cable Symposium, 184-193 (1996)

Christy, M. R., Petrella, R. V., Penkala, J. J., "Controlled-Atmosphere Cone Calorimeter", in "Fire and Polymers II", Ed. G. L. Nelson, ACS Symposium Series 599, 498-517 (1995)

Coaker, A. W., Hirschler, M. M., "Fire Characteristics of Standard and Advanced PVC Wire and Cable Compounds", Proceedings of the International Conference on Fire Safety, Vol. 13, 397-416 (January 1988)

Crane, C. R., Sanders, D. C., Endecott, B. R., Abbott, J. K., Smith, P. W., "Inhalation Toxicology: I. Design of a Small-Animal Test System. II. Determination of the Relative Toxic Hazards of 75 Aircraft Cabin Materials". FAA-AM-77-9, Civil Aeromedical Institute, Federal Aviation Administration, Oklahoma City, Oklahoma (March 1977)

Crane, C. R., Sanders, D. C., Endecott, B. R., Abbott, J. K., "Electrical Insulation Fire Characteristics. Volume II. Toxicity", UMTA-MA-06-0025-79-2,II, Civil Aeromedical Institute, Federal Aviation Administration, Oklahoma City, Oklahoma (March 1979)

Crane, C. R., Sanders, D. C., Endecott, B. R., Abbott, J. K., "Inhalation Toxicology: III. Evaluation of the Thermal Degradation Products from Aircraft and Automobile Engine Oils, Aircraft Hydraulic Fluid, and Mineral Oil", FAA-AM-83-12, Civil Aeromedical Institute, Federal Aviation Administration, Oklahoma City, Oklahoma (April 1983)

Crane, C. R., Sanders, D. C., Endecott, B. R., Abbott, J. K., "Inhalation Toxicology: VI. Evaluation of the Relative Toxicity of Thermal Decomposition Products from Nine Aircraft Panel Materials", DOT/FAA/AM-86/3, Civil Aeromedical Institute, Federal Aviation Administration, Oklahoma City, Oklahoma (February 1986)

Crane, C. R., Sanders, D. C., Endecott, B. R., Abbott, J. K., "Combustibility of Electrical Wire and Cable for Rail Rapid Transit Systems. Volume II. Toxicity", UMTA-MA-06-0025-83-7, Urban Mass Transit Administration, Washington, D. C. (May 1983)

Czerczak, S., Stetkiewicz, J., "Toxicity Classification of Thermal Degradation Products of Chemical Materials Used in Construction", Journal of Fire Sciences, Vol. 14, 367-378 (September/October 1996)

Damant, G. H., McCormack, J. A., Mikami, J. F., Wortman, P. S., Hilado, C. J., "The California Technical Bulletin 133 Test: Some Background and Experience", Proceedings of the International Conference on Fire Safety, Vol. 14, 1-12 (January 1989)

Damant, G. H., Nurbakhsh, S., Hilado, C. J., "Flammability of Seating Furniture: California Technical Bulletin 133 Test History and Development", Proceedings of the International Conference on Fire Safety, Vol. 16, 7-25 (January 1991)

Damant, G. H., Nurbakhsh, S., "Developing a 'Code of Practice' for Technical Bulletin 133 Testing", Proceedings of the International Conference on Fire Safety, Vol. 18, 48-71 (January 1993)

Dietenberger, M. A., "Forest Product Laboratory's New Experimental and Analytical Protocol for LIFT Apparatus", Proceedings of the International Conference on Fire Safety, Vol. 19, 272-276 (January 1994)

Dietenberger, M. A., "Ignitability Analysis of Siding Materials Using a Modified Protocol for LIFT Apparatus", Proceedings of the International Conference on Fire Safety, Vol. 20, 297-306 (January 1995)

Dietenberger, M. A., "Protocol for Ignitability, Lateral Flame Spread, and Heat Release Rate Using LIFT Apparatus", in "Fire and Polymers II", Ed. G. L. Nelson, ACS Symposium Series 599, 435-449 (1995)

Dietenberger, M. A., "Ignitability Analysis Using the Cone Calorimeter and the LIFT Apparatus", Proceedings of the International Conference on Fire Safety, Vol. 22, 189-197 (July 1996)

Dorsett, H. G., Jacobson, M., Nagy, J., Williams, R. P., "Laboratory Equipment and Test Procedures for Evaluating Explosibility of Dusts", U.S. Bureau of Mines Report of Invest. 5624 (1960)

Eickner, H. W., "Surface Flammability Measurements for Building Materials and Related Products", in "Treatise on Analytical Chemistry", Part 3, Vol. 4, John Wiley and Sons (1977)

Finley, G. F., "Factors Affecting the Performance of Wall Linings in the ASTM E84 Test", Proceedings of the International Conference on Fire Safety, Vol. 17, 314-335 (January 1992)

Fritz, T. W., Hunsberger, P. L., "Testing Floor Coverings at an Elevated Exposure Flux in the ASTM E648 Apparatus", Proceedings of the International Conference on Fire Safety, Vol. 21, 179-184 (January 1996)

Gallagher, J. A., Smiecinski, T. M., Grace, O. M., "The OSU Heat Release Unit as a Screening Tool for California TB 133", Proceedings of the International Conference on Fire Safety, Vol. 16, 73-81 (January 1991)

Gardner, W. D., Thomson, C. R., "Flame Spread Properties of Forest Products: Comparison and Validation of Prescribed Australian and North American Flame Spread Test Methods", Fire and Materials, Vol. 12, No. 2, 71-85 (June 1988)

Grand, A. F., "Evaluation of the Corrosivity of Smoke from Fire Retarded Products", Journal of Fire Science, Vol. 9, No. 1, 44-58 (January/February 1991)

Grand, A. F., "Evaluation of the Corrosivity of Smoke Using a Laboratory Radiant Combustion/Exposure Apparatus", Journal of Fire Science, Vol. 10, No. 1, 72-93 (January/February 1992)

Grand, A. F., "The Use of the Cone Calorimeter to Assess the Effectiveness of Fire Retardant Polymers under Simulated Real Fire Test Conditions", Interflam '96, Cambridge, U.K. (March 1996)

Grayson, S. J., Hirschler, M. M., "National and International Developments in Standards for Buildings and Contents", Proceedings of the International Conference on Fire Safety, Vol. 19, 75-88 (January 1994)

Grenier, A. T., Janssens, M. L., "An Improved Method for Analyzing Ignition Data of Composites", Proceedings of the International Conference on Fire Safety, Vol. 23, 253-264 (January 1997)

Gross, D., Loftus, J. J., Robertson, A. F., "Method for Measuring Smoke from Burning Materials", ASTM Spec. Tech. Publ. No. 422, 166-204, American Society for Testing and Materials, Philadelphia (1967)

Harkleroad, M. F., "Fire Properties Database for Textile Wall Coverings", NISTIR 89-4065, National Institute of Standards and Technology, Gaithersburg, Maryland (March 1989)

Hilado, C. J., Way, D. H., "Fire Performance of Spray-Applied Rigid Urethane Foam as Vessel Insulation", Journal of Fire and Flammability, Vol. 1, No. 1, 30-35 (January 1970)

Hilado, C. J., Burgess, P. E., "A Four-Foot Tunnel Test Apparatus for Measuring Surface Flame Spread", Journal of Fire and Flammability, Vol. 3, No. 2, 154-163 (April 1972)

Hilado, C. J., "The Multicapability Fire Test Method", Proceedings of the International Conference on Fire Safety, Vol. 1, 201-212 (1976)

Hilado, C. J., "The Multicapability Fire Test Method", Journal of Fire and Flammability, Vol. 7, No. 2, 248-256 (April 1976)

Hilado, C. J., Cumming, H. J., "Studies with the Arapahoe Smoke Chamber", Journal of Fire and Flammability, Vol. 8, No. 3, 300-308 (July 1977)

Hilado, C. J., Cumming, H. J., "Flash Fire Propensity of Materials", Journal of Fire and Flammability, Vol. 8, No. 4, 443-457 (October 1977)

Hilado, C. J., Cumming, H. J., "Screening Materials for Flash Fire Propensity", Modern Plastics, Vol. 54, No. 11, 56-59 (November 1977)

Hilado, C. J., Cumming, H. J., "Relative Toxicity of Pyrolysis Gases from Materials: Effects of Chemical Composition and Test Conditions", Fire and Materials, Vol. 2, No. 2, 68-79 (April 1978)

Hilado, C. J., Murphy, R. M., "A Simple Laboratory Method for Determining Ignitability of Materials", Journal of Fire and Flammability, Vol. 9, No. 2, 164-175 (April 1978)

Hilado, C. J., Machado, A. M., "Smoke Studies with the Arapahoe Chamber", Journal of Fire and Flammability, Vol. 9, No. 2, 240-244 (April 1978)

Hilado, C. J., Brandt, D. L., Damant, G. H., "Smoldering Tests of Furniture and Aircraft Seat Fabrics", Journal of Consumer Product Flammability, Vol. 5, No. 3, 121-125 (September 1978)

Hilado, C. J., Cumming, H. J., Machado, A. M., "Relative Toxicity of Pyrolysis Gases from Materials: Specific Toxicants and Special Studies", Fire and Materials, Vol. 2, No. 4, 141-153 (October 1978)

Hilado, C. J., Machado, A. M., Murphy, R. M., "Smoke Density Studies with the Arapahoe and NBS Chambers", Journal of Fire and Flammability, Vol. 9, No. 4, 421-425 (October 1978)

Hilado, C. J., Cumming, H. J., Schneider, J. E., "Relative Toxicity of Pyrolysis Gases from Materials: Effects of Temperature, Air Flow, and Criteria", Fire and Materials, Vol. 3, No. 4, 183-187 (December 1979)

Hilado, C. J., Huttlinger, P. A., "Toxic Hazards from Common Materials", Fire Technology, Vol. 17, No. 3, 177-182 (August 1981)

Hilado, C. J., "Flammability Handbook for Electrical Insulation", Technomic Publishing Company, Westport, Connecticut (1982)

Hilado, C. J., Huttlinger, P. A., "Review and Update of the Dome Chamber Toxicity Test Method", Proceedings of the International Conference on Fire Safety, Vol. 8, 169-189 (January 1983)

Hilado, C. J., Huttlinger, P. A., "Screening Materials by the NASA Dome Chamber Toxicity Test", Proceedings of the California Conference on Product Toxicity, Vol. 4, 20-60 (1983)

Hilado, C. J., "Toxicity of Off-Gases from Food and Plastic Products", Proceedings of the California Conference on Product Toxicity, Vol. 5, 49-55 (1984)

Hilado, C. J., "Flammability Handbook for Plastics", 4th Ed., Technomic Publishing Company, Lancaster, Pennsylvania (1990)

Hill, R. G., Eklund, T. I., Sarkos, C. P., "Aircraft Interior Panel Test Criteria Derived from Full-Scale Fire Tests", DOT/FAA/CT-85/12, Federal Aviation Administration, Atlantic City, New Jersey (September 1985)

Hinderer, R. K., "A Comparative Review of the Combustion Toxicology of Polyvinyl Chloride", Journal of Fire Sciences, Vol. 2, No. 1, 82-97 (January/February 1984)

Hirsch, D. B., Bunker, R. L., Janoff, D., "Effects of Oxygen Concentration, Diluents, and Pressure on Ignition and Flame-Spread Rates on Nonmetals: A Review Paper", ASTM STP 1111, 179-190, American Society for Testing and Materials, Philadelphia (1991)

Hirsch, D. B., Bunker, R. L., "Effects of Diluents on Flammability of Nonmetals in High-Pressure Oxygen Mixtures", ASTM STP 1197, 74-80, American Society for Testing and Materials, Philadelphia (1993)

Hirsch, D. B., Hshieh, F.-Y., Beeson, H., Bryan, C., "Ignitability in Air, Gaseoys Oxygen, and Oxygen-Enriched Environments of Polymers Used in Breathing-Air Devices", ASTM STP 1319, American Society for Testing and Materials, Philadelphia (1997)

Hirschler, M. M., "Fire Hazard and Toxic Potency of the Smoke from Burning Materials", Journal of Fire Sciences, Vol. 5, No. 5, 289-307 (September/October 1987)

Hirschler, M. M., "Analysis of Test Results from a Variety of Smoke Corrosivity Test Methods", Proceedings of the International Conference on Fire Safety, Vol. 18, 360-392 (January 1993)

Hirschler, M. M., "Analysis of Heat Release and Other Data from a Series of Plastic Materials Tested in the Cone Calorimeter", Proceedings of the International Conference on Fire Safety, Vol. 20, 214-228 (January 1995)

Hirschler, M. M., "Tests on Plastic Materials for the Wire and Cable Industry, Using the Cone Corrosimeter and the Cone Calorimeter", Proceedings of the October 1995 FRCA Conference, 103-124 (1995)

Hirschler, M. M., Trevino, J. O., "Repeatability Study on Heat Release Testing of Stacking Chairs", Proceedings of the International Conference on Fire Safety, Vol. 21, 56-68 (January 1996)

Hshieh, F.-Y., Motto, S. E., Hirsch, D. B., "Flammability Testing Using a Controlled-Atmosphere Cone Calorimeter", Proceedings of the International Conference on Fire Safety, Vol. 18, 299-325 (January 1993)

Hshieh, F.-Y., Beeson, H. D., "Flammability Testing of Pure and Flame Retardant-Treated Cotton Fabrics", Fire and Materials, Vol. 19, 233-239 (1995)

Hshieh, F.-Y., Stoltzfus, J. M., Beeson, H. D., "Note: Autoignition Temperature of Selected Polymers at Elevated Oxygen Pressure and Their Heat of Combustion", Fire and Materials, Vol. 20, 301-303 (1996)

Hu, X., Clark, F. R. S., "The Use of the ISO/TC92 Test for Ignitability Assessment", Fire and Materials, Vol. 12, No. 1, 1-5 (March 1988)

Innes, J. D., "Advances in Measurement of Combustion Gases Generated by the Cone

Calorimeter", Proceedings of the International Conference on Fire Safety, Vol. 21, 173-178 (January 1996)

Innes, J. D., Cox, A. W., "Technology Advances in Cone Calorimeter Testing", 8th Annual BCC Conference on Flame Retardancy (June 1997)

Kaplan, H. L., Grand, A. F., Hartzell, G. F., "Combustion Toxicology: Principles and Test Methods", Technomic Publishing Company, Lancaster, Pennsylvania (1983)

Kaplan, H. L., Hirschler, M. M., Switzer, W. G., Coaker, A. W., "A Comparative Study of Test Methods Used to Determine the Toxic Potency of Smoke", Proceedings of the International Conference on Fire Safety, Vol. 13, 279-287 (January 1988)

Kessel, S. L., Bennett, J. G., Rogers, C. E., "Corrosivity Test Methods for Polymeric Materials. Part 1. Radiant Furnace Test Method", Journal of Fire Sciences, Vol. 12, No. 2, 109-133 (March/April 1994)

Kessel, S. L., Rogers, C. E., Bennett, J. G., "Corrosivity Test Methods for Polymeric Materials. Part 5. A Comparison of Four Test Methods", Proceedings of the International Conference on Fire Safety, Vol. 18, 326-359 (January 1993)

Kessel, S. L., Rogers, C. E., Bennett, J. G., "Corrosivity Test Methods for Polymeric Materials. Part 5. A Comparison of Four Test Methods", Journal of Fire Sciences, Vol. 12, No. 2, 196-233 (March/April 1994)

Klimisch, H. J., Doe, J. E., Hartzell, G. E., Packham, S. C., Pauluhn, J., Purser, D. A., "Bioassay Procedures for Fire Effluents: Basic Principles, Criteria, and Methodology", Journal of Fire Sciences, Vol. 5, No. 2, 73-104 (March/April 1987)

Lawson, J. R., "An Examination of Variability and Precision for ASTM E648 Standard Test Method for Critical Radiant Flux of Floor Covering Systems", Proceedings of the International Conference on Fire Safety, Vol. 18, 198-224 (January 1993)

Levin, B. C., Fowell, A. J., Birky, M. M., Paabo, M., Stolte, A., Malek, D., "Further Development of a Test Method for the Assessment of the Acute Inhalation Toxicity of Combustion Products", NBSIR 82-2352, National Bureau of Standards, Gaithersburg, Maryland (June 1982)

Levy, M. M., "A Simplified Method for Determining Flame Spread", Fire Technology, Vol. 3, No. 1, 38-46 (February 1967)

Malin, D. S., "Cone Corrosimeter Testing of Fire Retardant and Other Polymeric Materials for Wire and Cable Applications", Proceedings of the International Conference on Fire Safety, Vol. 19, 211-240 (January 1994)

Manka, M. J., Pierce, H., Huggett, C., "Studies of the Flash Fire Potential of Aircraft

Cabin Interior Materials", FAA-RD-77-47, Federal Aviation Administration, Washington, D.C. (December 1977)

Nelson, G. L., Bridgman, A. L., "Heat Release Calorimetry and Radiant Panel Testing: A Comparative Study", Proceedings of the International Conference on Fire Safety, Vol. 11, 128-139 (January 1986)

Nelson, G. L., "Effect of EMI Coatings on the Fire Performance of Plastics", Proceedings of the International Conference on Fire Safety, Vol. 13, 367-378 (January 1988)

Norris, J. C., "National Institute of Building Sciences Combustion Toxicity Hazard Test", Proceedings of the March 1988 FRCA Conference, 146-155 (1988)

Norris, J. C., "Investigation of the Dual LC_{50} Values in Woods Using the University of Pittsburgh Combustion Toxicity Apparatus", ASTM STP 1082, 57-71, American Society for Testing and Materials, Philadelphia (1990)

Nurbakhsh, S., Damant, G. H., "Development of a Test Method for the Flammability of Stacking Chairs", Proceedings of the International Conference on Fire Safety, Vol. 19, 32-58 (January 1994)

Ohlemiller, T. J., Villa, K. M., "Material Flammability Test Assessment for Space Station Freedom", NISTIR 4591, National Institute of Standards and Technology, Gaithersburg, Maryland (June 1991)

O'Neill, T. J., "Flame and Heat Response of Halogenated and Non-Halogenated Insulating and Jacketing Materials for Electrical Wires and Cables", Proceedings of the International Conference on Fire Safety, Vol. 14, 229-236 (January 1989)

O'Neill, T. J., "Assessing the Corrosion Risk of Plastics in Fires", Fire Safety Journal, Vol. 15, 45-56 (1989)

Ostman, B. A., Svensson, I. G., Blomqvist, J., "Comparison of Three Test Methods for Measuring Rate of Heat Release", Fire and Materials, Vol. 9, No. 4, 176-184 (December 1985)

Paul, K. T., Christian, S. D., "Standard Flaming Ignition Sources for Upholstered Composites, Furniture, and Bed Assembly Tests", Journal of Fire Sciences, Vol. 5, No. 3, 178-211 (May/June 1987)

Pauluhn, J., Kimmerle, G., Martins, T., Prager, F., Pump, W., "Toxicity of the Combustion Gases from Plastics: Relevance and Limitations of Results Obtained in Animal Experiments", Journal of Fire Sciences, Vol. 12, No. 1, 63-104 (January/February 1994)

Peacock, R. D., Braun, E., "Fire Tests of Amtrak Passenger Rail Vehicle Interiors", NBS Technical Note 1193, National Bureau of Standards, Gaithersburg, Maryland (May 1984)

Prager, F. H., Einbrodt, H. K., Hupfeld, J., Muller, B., Sand, H., "Risk Oriented Evaluation of Fire Gas Toxicity Based on Laboratory Scale Experiments - The DIN 53436 Method", Journal of Fire Sciences, Vol. 5, No. 5, 308-325 (September/October 1987)

Quintiere, J. G., Harkleroad, M., "New Concepts for Measuring Flame Spread Properties", NBSIR 84-2943, National Bureau of Standards, Gaithersburg, Maryland (November 1984)

Richardson, L. R., Cornelissen, A. A., "Measurement of Smoke Generation by Building Materials", Proceedings of the International Conference on Fire Safety, Vol. 16, 278-286 (January 1991)

Richardson, L. R., Brooks, M. E., "Combustibility of Building Materials", Proceedings of the International Conference on Fire Safety, Vol. 16, 287-299 (January 1991)

Richardson, L. R., "Assessing Fire Performance of Claddings Using the ICAL (Intermediate-Scale Calorimeter)", Proceedings of the International Conference on Fire Safety, Vol. 23, 240-252 (January 1997)

Riley, R.E., Fishback, T. L., Yu-Hallada, L. C., "Flammability Study of Hydrocarbon-Blown Isocyanurate Foams", Proceedings of the 35th Annual Polyurethane Conference, 561-567 (October 1994)

Riley, R.E., Fishback, T. L., Yu-Hallada, L. C., "Flammability Study of Hydrocarbon-Blown Isocyanurate Foams", Journal of Cellular Plastics (September/October 1996)

Robertson, A. F., Gross, D., and Loftus, J. J., "A Method for Measuring Surface Flammability of Materials Using a Radiant Energy Source", ASTM Proceedings, Vol. 56, 1437-1453 (1956)

Robertson, A. F., "A Flammability Test Based on an ISO Spread of Flame Test", IMO Report FT-215 (1979)

Rodak, E. M., Taylor, R. J., Hirsch, D. B., Linley, L. J., "Effects of Sample and Test Variables on Electrical Wire Insulation Flammability", Journal of Testing and Evaluation, Vol. 22, No. 5, 449-452 (September 1994)

Rogers, C. E., Bennett, J. G., Kessel, S. L., "Corrosivity Test Methods for Polymeric Materials: CNET Test Method", Proceedings of the International Conference on Fire Safety, Vol. 17, 392-403 (January 1992)

Rogers, C. E., Bennett, J. G., Kessel, S. L., "Corrosivity Test Methods for Polymeric

Materials. Part 2. CNET Test Method", Journal of Fire Sciences, Vol. 12, No. 2, 134-154 (March/April 1994)

Sanders, D. C., Crane, C. R., Endecott, B. R., "Inhalation Toxicology: V. Evaluation of the Relative Toxicity to Rats of Thermal Decomposition Products from Two Aircraft Seat Fire-Blocking Materials", DOT/FAA/AM-86-1, Civil Aeromedical Institute, Federal Aviation Administration, Oklahoma City, Oklahoma (November 1985)

Sarkos, C. P., Filipczak, R. A., Abramowitz, A., "Preliminary Evaluation of an Improved Flammability Test Method for Aircraft Materials", DOT/FAA/CT-84/22, Federal Aviation Administration, Atlantic city, New Jersey (December 1984)

Scudamore, M. J., Briggs, P. J., Prager, F. H., "The Cone Calorimeter as a Test for Plastics. A European Plastics Industry Evaluation", Proceedings of the International Conference on Fire Safety, Vol. 16, 259-270 (January 1991)

Smith, E. E., "Heat Release Rate of Building Materials", ASTM Spec. Tech. Publ. No. 502, 119-134, American Society for Testing and Materials, Philadelphia (1972)

Smith, E. E., The Ohio State University, private communication (August 3, 1989)

Steiner, A. J., "Fire Hazard Classification of Building Materials", Underwriters Laboratories Research Bulletin No. 32 (September 1944)

Stevens, M. G., Voruganti, V., Rose, R., "Evaluation of Small Scale Screening Tests for ASTM E-84 Tunnel Test", Proceedings of the International Conference on Fire Safety, Vol. 21, 245-255 (January 1996)

Tewarson, A., Macaione, D. P., "Polymers and Composites: An Examination of Fire Spread and Generation of Heat and Fire Products", Journal of Fire Sciences, Vol. 11, No. 5, 421-441 (September/October 1993)

Tewarson, A., "Flammability Parameters of Materials: Ignition, Combustion, and Fire Propagation", Journal of Fire Sciences, Vol. 12, No. 4, 329-356 (July/August 1994)

Tewarson, A., "Fire Properties of Materials for Model-Based Assessments for Hazards and Protection Needs", in "Fire and Polymers II", Ed. G. L. Nelson, ACS Symposium Series 599, 450-497 (1995)

Tewarson, A., "Effectiveness of Fire Retardants in Reducing/Eliminating Non-Thermal Damage", Proceedings of the October 1995 FRCA Conference, 79-102 (1995)

Tomann, J., "Comparison of Nordtest FIRE 007, CEN Draft Proposal (Radiant Panel), and Cone Calorimeter Methods in the Fire Testing of Floor Coverings", Proceedings of the International Conference on Fire Safety, Vol. 17, 26-37 (January 1992)

Trabold, E. L., "Study to Develop Improved Fire Resistance Aircraft Passenger Seat Materials. Phase I", NASA CR-152056, National Aeronautics and Space Administration, Ames Research Center, Moffett Field, California (1977)

Tran, H. C., White, R. H., "Burning Rate of Solid Wood Measured in a Heat Release Rate Calorimeter", Fire and Materials, Vol. 16, 197-206 (1992)

Trevino, J. O., Grand, A. F., "Fire Test Methods for Fire Retardant Polymers", Proceedings of the March 1996 FRCA Conference, 31-46 (1996)

Truax, T. R., Harrison, C. A., "A New Test for Measuring the Fire Resistance of Wood", ASTM Proceedings, Vol. 29 (II), 973-989 (1929)

Tu, K.-M., Aaronson, A., Hombeck, R., "A Study on the Combustion Characteristics of Fire Resistant Industrial Fluids Using the Cone Calorimeter", Proceedings of the International Conference on Fire Safety, Vol. 21, 42-55 (January 1996)

Urbas, J., Shaw, J. R., "Testing of Wall Assemblies on Intermediate Scale Calorimeter (ICAL)", Proceedings of the International Conference on Fire Safety, Vol. 18, 225-240 (January 1993)

Urbas, J., "Nondimensional Heat of Gasification Measurements in the Intermediate Scale Rate of Heat Release Apparatus", Fire and Materials, Vol. 17, 119-123 (1993)

Urbas, J., Luebbers, G. E., "The Intermediate Scale Calorimeter Development", Fire and Materials, Vol. 19, 65-70 (1995)

Urbas, J., Parker, W. J., "Impact of Air Velocity on Ignition in the Intermediate Scale Calorimeter (ICAL), Fire and Materials, Vol. 21 (1997)

Vanspeybroeck, R. S. L., Sewell, R. A., Thoen, J. A., "Horizontal Burn Test at Different Oxygen Levels: A Small Scale Screening Test for Flexible Polyurethane Foam with Excellent Correlation to the BS 5852 Part 2 Source 5", Journal of Fire Sciences, Vol. 8, No. 6, 421-454 (November/December 1990)

Villa, K. M., "Textile Test Methods for Protective Clothing Standards", 4th Annual Conference on Protective Clothing, Clemson University (April 1990)

Villa, K. M., Krasny, J. F., "Small-Scale Vertical Flammability Testing for Fabrics", Fire Safety Journal, Vol. 16, 229-241 (1990)

Way, D. H., Hilado, C. J., "The Performance of Rigid Cellular Plastics in Fire Tests for Industrial Insulation", Journal of Cellular Plastics, Vol. 4, No. 6, 221-228 (June 1968)

White, R. H., "Oxygen Index Evaluation of Fire-Retardant-Treated Wood", Wood Science, Vol. 12, No. 2, 113-121 (October 1979)

White, R. H., Nordheim, E. V., "Charring Rate of Wood for ASTM E 119 Exposure", Fire Technology, Vol. 28, No. 1, 5-30 (February 1992)

Wilson, J. A., "Surface Flammability of Materials: A Survey of Test Methods and Comparison of Results", ASTM Spec. Tech. Publ. No. 301, 60-82, American Society for Testing and Materials, Philadelphia (February 1961)

Wortman, P. S., Williams, S. S., Damant, G. H., "Development of a Fire Test for Furniture for High Risk and Public Occupancies", Proceedings of the International Conference on Fire Safety, Vol. 9, 55-67 (January 1984)

Woyneroski, S. P., "Testing for Inhalation Toxicity of Combustion Products", Proceedings of the International Conference on Fire Safety, Vol. 14, 203-209 (January 1989)

Yarbrough, D. W., "Design of Improved Flammability Tests for Cellulosic Insulations", Proceedings of the International Conference on Fire Safety, Vol. 22, 30-49 (July 1996)

Table 4.1. Some International Standards Relevant to Fire Safety and Plastics Issued by International Standardization Organization (ISO), ISO Standards.

ISO 181	Flammability, rigid plastics, small specimens, incandescent rod
ISO 871	Temperature of evolution of flammable gases
ISO 1210	Flammability, small specimens, small flame
ISO 3582	Cellular plastics and rubber, horizontal, small specimens, small flame
ISO 4589	Plastics, oxygen index
ISO 5659-2	Plastics, smoke density, single chamber
ISO 9772	Cellular plastics, horizontal, small specimens, small flame
ISO 9773	Plastics, vertical, small flame
ISO 10351	Combustibility, 125 mm flame
ISO 11907-2	Plastics, smoke corrosivity, part 2. Static method
ISO/WD 1182	Building materials, noncombustibility
ISO/CD 1716	Building materials, calorific potential
ISO/DTR 5924	Smoke generated, building products
ISO/DTR 14697	Fire tests, use of substrates
ISO/NP 14934	Calibration of heat flux meters
ISO/DIS 5657	Ignitability, radiant heat source
ISO/DTR 11925-1	Ignitability, direct flame, theory
ISO/DIS 11925-2	Ignitability, direct flame, single source
ISO/DIS 11925-3	Ignitability, direct flame, multi-source
ISO/DTR 5658-1	Fire spread, guidance
ISO 5658-2	Fire spread, lateral spread, vertical
ISO/DTR 5658-3	Fire spread, lateral ignition and spread (LIFT)
ISO/CD 5658-4	Fire spread, intermediate scale vertical
ISO/DIS 9239-1	Floor coverings, horizontal spread. Part 1: radiant heat ignition
ISO/CD 9239-2	Floor coverings, horizontal spread. Part 2: higher heat flux levels
ISO/WD 5660-1	Heat, smoke, and mass loss rate, building products. Part 1: heat release rate
ISO/WD 5660-2	Heat, smoke, and mass loss rate, building products. Part 2: smoke release rate, dynamic measurement
ISO/NP 5660-3	Heat, smoke, and mass loss rate, building products. Part 3: mass loss rate
ISO/NP 5660-4	Heat, smoke, and mass loss rate, building products. Part 4: guidance on heat and smoke release rate
ISO/CD 8337	Small corner test
ISO/WD 9705-2	Full-scale room test: Part 2: theory
ISO/WD 13784	Large scale test for industrial sandwich panels
ISO/CD 13785-1	Facades. Part 1: intermediate scale tests
ISO/WD 13785-2	Facades. Part 2: large scale tests
ISO/DTR 14696	Intermediate scale heat release calorimeter
ISO/DTR 11696-1	Fire tests. Part 1: mathematical modelling
ISO/DTR 11696-2	Fire tests. Part 2: fire hazard analysis
ISO 4736	Fire tests, small chimneys
ISO/WD 14803	Fire doors and shutter assemblies
ISO/WD 14805	Contribution of structural elements
ISO/CD 14832-1	Loadbearing elements. Part 1: internal walls
ISO/CD 14832-5	Loadbearing elements. Part 5: beams
ISO/CD 14832-6	Loadbearing elements. Part 6: columns
ISO/FDIS 834-1	Fire resistance. Part 1: general requirements
ISO/WD 834-2	Fire resistance. Part 2: specific requirements
ISO/WD 14804	Calibration of furnaces
ISO/CD 14831-1	Non-loadbearing elements. Part 1: vertical separating elements
ISO/CD 14831-2	Non-loadbearing elements. Part 2: horizontal separating elements
ISO/CD 14831-3	Non-loadbearing elements. Part 3: vertical elements
ISO/CD 14832-2	Loadbearing elements. Part 2: horizontal elements

continued

Table 4.1 (continued). Some International Standards Relevant to Fire Safety and Plastics Issued by International Standardization Organization (ISO), ISO Standards.

ISO/CD 14832-3	Loadbearing elements. Part 3: beams
ISO/CD 14832-4	Loadbearing elements. Part 4: columns
ISO/WD 12469	Analytical determination
ISO/CD 12470	Interpolation and extrapolation
ISO/WD 12471	Input for analytical fire design
ISO/NP 15655	Thermal and mechanical properties
ISO/NP 15656	Calculation models
ISO/NP 15657	Computational structural fire design
ISO/NP 15658	Full scale structural fire tests
ISO/CD 3008	Fire doors and shutters
ISO/NP 3009	Glazed elements
ISO/CD 5925-1	Smoke control door and shutter assemblies. Part 1: leakage test
ISO/DTR 5925-2	Smoke control door and shutter assemblies. Part 2: commentary
ISO/CD 12472	Intumescent seals for fire doors
ISO/CD 6944	Fire resisting ducts
ISO 10294-1	Fire dampers for air distribution systems. Part 1: test method
ISO/DIS 10294-2	Fire dampers for air distribution systems. Part 2: classification
ISO/DIS 10294-3	Fire dampers for air distribution systems. Part 3: explanatory
ISO/CD 12468-1	External fire exposure to roofs. Part 1: small burning brands
ISO/CD 12468-2	External fire exposure to roofs. Part 2: burning brands
ISO/CD 12468-3	External fire exposure to roofs. Part 3: large burning brands
ISO/CD 10295-1	Part 1: penetration seals
ISO/CD 10295-2	Part 2: linear gap seals
ISO/WD 11067	Toxic gases in blood
ISO/NP 12473	Fire hazard analysis
ISO 13344	Lethal toxic potency of fire effluents
ISO/CD 13571	Life threat components of fire
ISO/WD 13387	Design objectives
ISO/WD 13388	Characterization of buildings
ISO/CD 13389	Mathematical fire models
ISO/WD 13390	Initiation and development of fire and fire effluents
ISO/WD 13391	Movement of fire effluents
ISO/WD 13392	Fire spread beyond enclosure of origin
ISO/WD 13393	Detection, activation, and suppression
ISO/WD 13394	Life safety

Table 4.2. *Some American Standards Relevant to Fire Safety and Plastics Issued by the American Society for Testing and Materials (ASTM).*

ASTM C 209 (insulating board)
ASTM C 541 (hot surface performance, high temperature insulation)
ASTM C 542 (lock-strip gaskets)
ASTM C 864 (compression seal gaskets, setting blocks, spacers)
ASTM C 1166 (dense and cellular elastomeric gaskets, accessories)

ASTM D 56 (liquids, Tag closed tester)
ASTM D 92 (liquids, Cleveland open cup)
ASTM D 93 (liquids, Pensky-Martens closed tester)
ASTM D 229 (electrical insulation, rigid sheet and plate)
ASTM D 240 (heat of combustion, bomb calorimeter)
ASTM D 350 (electrical insulation, flexible treated sleeving)
ASTM D 378 (flat rubber belting)
ASTM D 461 (felt)
ASTM D 470 (electrical insulation and jacket, thermosetting)
ASTM D 568 (plastics, vertical)
ASTM D 635 (plastics, horizontal)
ASTM D 757 (plastics, horizontal, incandescence, Globar)
ASTM D 876 (electrical insulation, nonrigid PVC tubing)
ASTM D 1000 (electrical insulation, adhesive tape)
ASTM D 1230 (apparel textiles, 45-degree angle)
ASTM D 1310 (flash point and fire point of liquids, Tag open cup)
ASTM D 1360 (paints, cabinet method)
ASTM D 1361 (paints)
ASTM D 1433 (plastics, 45-degree angle)
ASTM D 1692 (cellular plastics, horizontal)
ASTM D 1929 (plastics, ignition, Setchkin furnace)
ASTM D 2389 (minimum pressure for vapor phase ignition)
ASTM D 2512 (compatibility with liquid oxygen)
ASTM D 2584 (ignition loss, cured reinforced resins)
ASTM D 2633 (electrical insulation and jacket, thermoplastic)
ASTM D 2671 (electrical insulation, heat-shrinkable tubing)
ASTM D 2843 (smoke, Rohm and Haas chamber)
ASTM D 2859 (floor covering, methenamine tablet)
ASTM D 2863 (plastics, minimum oxygen concentration, oxygen index)
ASTM D 3014 (cellular plastics, vertical, Butler chimney)
ASTM D 3065 (aerosol products)
ASTM D 3119 (hydraulic fluids, mist spray flammability)
ASTM D 3278 (flash point of liquids, small scale closed cup)
ASTM D 3659 (apparel fabrics, semi-restraint method)
ASTM D 3675 (flexible cellular materials, radiant panel)
ASTM D 3713 (plastics, ignition, small flame)
ASTM D 3801 (plastics, vertical)
ASTM D 3806 (fire retardant paints, 2-foot tunnel)
ASTM D 3814 (combustion test methods for plastics)
ASTM D 3828 (flash point, Setaflash closed tester)
ASTM D 3874 (ignition by hot wire sources)
ASTM D 3894 (rigid cellular plastics, small corner)
ASTM D 3934 (flash/no flash test, equilibrium method, closed cup)
ASTM D 3941 (flash point, equilibrium method, closed cup)
ASTM D 4100 (plastics, smoke, gravimetric, Arapahoe chamber)
ASTM D 4108 (protective clothing, open flame method)
ASTM D 4151 (blankets, flammability)
ASTM D 4205 (rubber, flammability and combustion)

continued

ASTM D 4206 (sustained burning of liquids, small open cup)
ASTM D 4207 (sustained burning of low viscosity liquids, wick test)
ASTM D 4372 (camping tentage, flame resistant materials)
ASTM D 4433 (treated paper and paperboard)
ASTM D 4723 (textile heat and flammability test methods)
ASTM D 4804 (non-rigid solid plastics, UL 94HB)
ASTM D 4986 (cellular polymeric materials, horizontal, UL 94HF)
ASTM D 5025 (laboratory burner for small-scale burning tests)
ASTM D 5048 (burn-through of solid plastics, 125 mm flame, UL 94V5)
ASTM D 5132 (horizontal burning rate, motor vehicles, MVSS 302)
ASTM D 5207 (calibration of 20 and 125 mm test flames)
ASTM D 5238 (smoldering combustion potential, cotton-based batting)
ASTM D 5424 (electrical cables, smoke, vertical tray)
ASTM D 5425 (electrotechnical products, fire hazard assessment)
ASTM D 5485 (corrosion by combustion products, cone corrosimeter)
ASTM D 5537 (electrical cables, heat release, vertical tray)

ASTM D9 Z5215Z (electrical cables, heat release, cone calorimeter)
ASTM D9 Z5494Z (electrical cables, fire/smoke, horizontal, UL 910)
ASTM D9 Z6520Z (electrical insulation, toxicity of gas)

ASTM D13 Z6614Z (fabrics, vertical flame resistance)
ASTM D13 Z6615Z (textiles after exposure to vertical flame)
ASTM D13 Z6621Z (heat transfer through fabrics exposed to heat)

ASTM E 69 (treated wood, FPL fire tube)
ASTM E 84 (building materials, surface burning, UL 25-foot tunnel)
ASTM E 108 (roof coverings)
ASTM E 119 (standard fire tests)
ASTM E 136 (combustibility, vertical tube furnace, 750°C)
ASTM E 152 (door assemblies)
ASTM E 160 (treated wood, crib)
ASTM E 162 (surface flammability, radiant panel)
ASTM E 163 (window assemblies)
ASTM E 176 (terminology of fire standards)
ASTM E 286 (surface flammability, FPL 8-foot tunnel)
ASTM E 535 (preparation of fire test response standards)
ASTM E 603 (room fire experiments)
ASTM E 648 (floor covering systems, radiant panel)
ASTM E 662 (smoke, NBS chamber)
ASTM E 681 (chemicals)
ASTM E 789 (dust explosions)
ASTM E 800 (fire gases)
ASTM E 814 (through-penetration fire stops)
ASTM E 906 (heat release, OSU apparatus)
ASTM E 918 (combustible chemicals)
ASTM E 970 (attic floor insulation, radiant panel)
ASTM E 1226 (combustible dusts)
ASTM E 1317 (marine surface finishes, flammability)
ASTM E 1321 (ignition and flame spread, LIFT)
ASTM E 1352 (cigarette ignition resistance, furniture mock-up)
ASTM E 1353 (cigarette ignition resistance, furniture components)
ASTM E 1354 (heat release, cone calorimeter)
ASTM E 1355 (fire models)
ASTM E 1472 (computer software for fire models)
ASTM E 1474 (heat release, furniture components, cone calorimeter)

Table 4.2 (continued). Some American Standards Relevant to Fire Safety and Plastics Issued by the American Society for Testing and Materials (ASTM).

ASTM E 1529 (large hydrocarbon pool fires)
ASTM E 1537 (real scale upholstered furniture)
ASTM E 1546 (fire hazard assessment standards)
ASTM E 1590 (real scale mattresses)
ASTM E 1591 (data for fire models)
ASTM E 1623 (intermediate scale calorimeter)
ASTM E 1678 (smoke toxicity, NIBS)
ASTM E 1725 (barrier systems for electrical system components)
ASTM E 1740 (wallcovering composites, cone calorimeter)
ASTM E 1776 (fire risk assessment standards)
ASTM E 1822 (stacked chairs)

ASTM E5 Z5859Z (fire-resistive joint systems)
ASTM E5 Z6712Z (resistance of elements of building construction)
ASTM E5 Z6856Z (fire-resistive joint systems)
ASTM E5 Z2536Z (room fire test of wall and ceiling materials)
ASTM E5 Z6379Z (oxygen consumption calorimetry fire tests)
ASTM E5 Z5467Z (furniture in health care facilities)
ASTM E5 Z5871Z (floor coverings in health care properties)
ASTM E5 Z5469Z (school bus seating)
ASTM E5 Z5493Z (rail transportation vehicles)
ASTM E5 Z6857Z (rail transportation vehicles)
ASTM E5 Z3314Z (corrosivity of smoke)
ASTM E5 Z4175Z (acute lethality)
ASTM E5 Z5616Z (smoke toxicity in post-flashover fires)
ASTM E5 Z6390Z (heat release, conical radiant heater)
ASTM E5 Z6779Z (life threat components of fire)
ASTM E5 Z6880Z (fire propagation apparatus, FM)
ASTM E5 Z6829Z (wires and cables in air handling spaces)
ASTM E5 Z6173Z (deterministic fire models)

ASTM F 371 (compatibility with liquid oxygen)
ASTM F 495 (gasket materials)
ASTM F 501 (aerospace materials, vertical)
ASTM F 764 (compatibility with high-energy propellants)
ASTM F 776 (aerospace materials, horizontal)
ASTM F 777 (electrical wire insulation, 60-degree angle)
ASTM F 814 (aerospace materials, smoke, NBS chamber)
ASTM F 828 (aircraft evacuation slide and raft materials)
ASTM F 1103 (aerospace materials, 45-degree angle)
ASTM F 1550 (correctional facility furnishings, cone calorimeter)
ASTM G 72 (autoignition temperature, oxygen enriched environments)

Table 4.3. Some American Standards Relevant to Fire Safety and Plastics Issued by Underwriters Laboratories (UL).

UL 9 (window assemblies, similar to ASTM E 163)
UL 10B (door assemblies, similar to ASTM E 152)
UL 44 (rubber-insulated wires and cables)
UL 83 (thermoplastic-insulated wires and cables)
UL 94 (plastic materials for parts in devices and appliances)
   Sec. 7. (horizontal test, HB)
   Sec. 8. (vertical test, V-0, V-1, or V-2)
   Sec. 9. Method A (vertical bar, 5VA)
   Sec. 9. Method B (horizontal plaque, 5VB)
   Sec. 10. (flame spread index, similar to ASTM E 162)
   Sec. 11. (vertical test, VTM-0, VTM-1, or VTM-2)
   Sec. 12. (horizontal test for foamed materials, HBF, HF-1, or HF-2, similar to
     ASTM D 4986)
UL 181 (air ducts and connectors)
UL 214 (fabrics and films)
UL 224 (VW1 tubing)
UL 263 (standard fire tests, similar to ASTM E 119)
UL 510 (insulating tape)
UL 580 (roof assemblies)
UL 651 (heavy wall PVC conduit)
UL 651A (Type EB and A rigid PVC conduit and HDPE conduit)
UL 719 (non-metallic sheathed cables)
UL 723 (building materials, surface burning, 25-foot tunnel, similar to ASTM E 84)
UL 746A (polymeric materials, short term properties)
   Sec. 30. (hot wire ignition, similar to ASTM D 3874)
   Sec. 31. (high-current arc ignition)
   Sec. 32. (high-voltage arc ignition)
   Sec. 33. (glow wire ignition, similar to IEC 695-2-1/3)
UL 746B (polymeric materials, long term properties)
UL 746C (polymeric materials, electrical equipment)
   Sec. 51. (flammability, 12 mm flame test, similar to IEC 695-1-2-2)
   Sec. 52. (flammability, 3/4 in. flame test)
   Sec. 53. (flammability, 5 in. flame test)
   Sec. 54. (enclosure flammability, 746-5VS test)
UL 746D (polymeric materials, fabricated parts)
UL 746E (polymeric materials, industrial laminates, filament wound tubing,
   vulcanized fibre, and materials used in printed wiring boards)
UL 790 (roof coverings, similar to ASTM E 108)
UL 854 (service entrance cables)
UL 910 (electrical cables, horizontal, plenum, 25-foot tunnel)
UL 992 (flooring and floor covering materials)
UL 1040 (insulated wall construction)
UL 1056 (upholstered furniture)
UL 1063 (machine tool wires and cables)
UL 1114 (marine use flexible fuel line)
UL 1256 (roof deck constructions)
UL 1441 (VW-1 sleeving)
UL 1479 (through-penetration firestops)
UL 1581 (electrical wires, cables, and flexible cords, vertical tray, similar to IEEE 383)
   Sec. 1060. vertical test
   Sec. 1061. vertical cable flame test

Table 4.3 (continued). Some American Standards Relevant to Fire Safety and Plastics Issued by Underwriters Laboratories (UL).

UL 1666 (electrical cables, vertical shaft or riser)
UL 1685 (electrical cables, vertical, fire and smoke)
UL 1692 (polymeric materials, coil forms)
UL 1694 (small polymeric component materials)
UL 1709 (protection materials for structural steel)
UL 1715 (interior finish material)
UL 1820 (pneumatic tubing, flame and smoke characteristics)
UL 1821 (thermoplastic sprinkler pipe and fittings)
UL 1853 (plastic containers)
UL 1887 (plastic sprinkler pipe, flame and smoke characteristics)
UL 1895 (mattresses)
UL 1897 (roof covering systems)
UL 1975 (foamed plastics used for decorative purposes)
UL 2043 (heat and smoke release, air-handling spaces)
UL 2060 (mattresses and bedding)
UL 2079 (building joint systems)
UL 2154 (surgical fabrics)

Table 4.4. Some American Standards Relevant to Fire Safety and Plastics Issued by National Fire Protection Association (NFPA).

NFPA 53 (fire hazards in oxygen enriched atmospheres)
NFPA 70 (National Electrical Code)
NFPA 123 (underground bituminous coal mines)
NFPA 130 (fixed guideway transit systems)
NFPA 203 (roof coverings and roof deck constructions)
NFPA 221 (fire walls and fire barrier walls)
NFPA 251 (standard fire tests, similar to ASTM E 119, UL 263)
NFPA 252 (door assemblies, similar to ASTM E 152, UL 10B)
NFPA 253 (flooring, radiant panel, similar to ASTM E 648)
NFPA 255 (surface burning, 25-foot tunnel, similar to ASTM E 84, UL 723)
NFPA 256 (roof coverings, similar to ASTM E 108, UL 790)
NFPA 257 (window assemblies, similar to ASTM E 163, UL 9)
NFPA 258 (smoke, NBS chamber, similar to ASTM E 662)
NFPA 259 (potential heat, oxygen bomb calorimeter)
NFPA 260 (cigarette ignition, furniture components)
NFPA 261 (cigarette ignition, furniture mock-up)
NFPA 262 (wire and cable, horizontal, plenum, similar to UL 910)
NFPA 263 (heat release, OSU apparatus, similar to ASTM E 906)
NFPA 264 (heat release, cone calorimeter)
NFPA 264A (heat release, upholstered furniture and mattresses, cone calorimeter)
NFPA 265 (textile wall coverings)
NFPA 266 (flaming ignition, upholstered furniture)
NFPA 267 (flaming ignition, mattresses and bedding)
NFPA 268 (exterior wall assemblies)
NFPA 269 (toxic potency data)
NFPA 555 (potential for room flashover)
NFPA 701 (flame-resistant textiles and films, similar to UL 214)
NFPA 703 (fire retardant impregnated wood and fire retardant coatings)
NFPA 705 (field flame test for textiles and films)

Table 4.5. Some American Standards Relevant to Fire Safety and Plastics Issued by International Conference of Building Officials (ICBO).

UBC 2-1 (noncombustibility, similar to ASTM E 136)
UCB 4-1 (proscenium fire safety curtains)
UBC 7-1 (standard fire tests, similar to ASTM E 119)
UBC 7-2 (door assemblies, similar to UL 10B)
UBC 7-3 (tinclad fire doors)
UBC 7-4 (window assemblies, similar to ASTM E 163)
UBC 7-5 (through-penetration fire stops, similar to ASTM E 814)
UBC 7-6 (spray applied fire resistive material)
UBC 7-7 (methods for calculating fire resistance)
UBC 7-8 (horizontal sliding fire doors)
UBC 8-1 (building materials, surface burning, 25-foot tunnel, similar to ASTM E 84)
UBC 8-2 (textile wall covering)
UBC 15-2 (roof assemblies, similar to UL 790)
UBC 23-5 (fire retardant treated wood)
UBC 26-1 (building materials, potential heat)
UBC 26-2 (thermal barriers)
UBC 26-3 (foam plastic systems)
UBC 26-4 (wall panels using foam plastic insulation)
UBC 26-5 (smoke from burning or decomposition of plastics, Rohm and Haas chamber, similar to ASTM D 2843)
UBC 26-6 (ignition properties of plastics, similar to ASTM D 1929)
UBC 26-7 (light-transmitting plastics, similar to ASTM D 635)
UBC 26-8 (garage doors using foam plastic insulation)
UBC 26-9 (intermediate scale multistory test)
UBC 31-1 (flame retardant membranes)

Table 4.6. Some Canadian Standards Relevant to Fire Safety and Plastics Published by Canadian Standards Association (CSA).

CAN/CSA-C22.2 No. 0.17-92	Evaluation of properties of polymeric materials
C22.2 No. 38-95	Thermoset insulated wires and cables
C22.2 No. 75-M1983	Thermoplastic-insulated wires and cables
C22.2 No. 174-M1984	Cables and cable glands for use in hazardous locations
CAN/CSA-M422-M87	Fire performance and antistatic requirements for conveyor belting
CAN/CSA-M423-M87	Fire resistant hydraulic fluids
CAN/CSA-M424.1-88	Flameproof non-rail-bound diesel-powered machines for use in gassy underground coal mines
CAN/CSA-M417-M91	Fire performance and antistatic requirements for ventilation materials

Table 4.7. Some Canadian Standards Relevant to Fire Safety and Plastics Published by Underwriters Laboratories of Canada (ULC).

CAN/ULC-S101-M89	Fire endurance tests of building construction and materials
CAN/ULC-S102-M88	Surface burning characteristics of building materials and assemblies
CAN/ULC-S-102.2-M88	Surface burning characteristics of flooring, floor covering, and miscellaneous materials and assemblies
ULC-S102.3-M1982	Fire test of light diffusers and lenses
ULC-S102.4-M1987	Test for fire and smoke characteristics of electrical wiring and cable
CAN4-S104-M80	Fire tests of door assemblies
CAN4-S106-M80	Fire tests of window and glass block assemblies
CAN/ULC-S107-M87	Fire tests of roof coverings
CAN/ULC-S109-M87	Flame tests of flame-resistant fabrics and films
CAN/ULC-S110-M86	Fire tests for air ducts
ULC-S111-95	Fire tests for air filter units
CAN4-S114-M80	Tests for determination of non-combustibility in building materials
CAN4-S115-M85	Fire tests of firestop systems
CAN4-S117.1-M85	Methenamine tablet test for textile floor coverings
CAN4-S124-M85	Test for evaluation of protective coverings for foamed plastic
CAN/ULC-S126-M86	Test for fire spread under roof deck assemblies
ULC-S127-M1988	Flammability characteristics of non-melting building materials
ULC-S129-95	Smoulder resistance of loose fill insulation (basket method)
CAN/ULC-S130-M87	Ignition resistance of loose fill insulation (cigarette method)
ULC-S132-93	Standard for emergency exit and emergency fire exit hardware
CAN/ULC-S133-M90	Standard for door closers intended for use with swinging doors
CAN/ULC-S134-92	Standard method of fire test of exterior wall assemblies
CAN/ULC-S135-92	Test for determination of degrees of combustibility of building materials using an oxygen consumption calorimeter (cone calorimeter)
CAN/ULC-S701-97	Standard for thermal insulation, polystyrene, boards and pipe covering

Table 4.8. Some Australian Standards Relevant to Fire Safety and Plastics Issued by the Standards Association of Australia (SAA).

AS 1176. Textiles. Methods of test for combustion properties
   Part 1. Determination of ease of ignition of certain textile materials in a horizontal plane
   Part 2. Determination of burning time of textile materials
   Part 3. Determination of surface burning properties of certain textile materials
AS 1248. Fabrics for domestic apparel of the low fire hazard type
AS 1266. Fire control plans for ships
AS 1441. Methods of test for coated products
   Part 13. Method of determination of flammability
AS 1530. Methods of fire tests on building materials and structures
   Part 1. Combustibility test for materials
   Part 2. Test for flammability of materials
   Part 3. Test for early fire hazard properties of materials
   Part 4. Fire-resistance tests of elements of construction
AS 1647. Children's toys (safety requirements)
   Part 4. Flammability requirements
AS 2111. Method of test for textile floor coverings
   Part 18. Method for the determination of fire propagation properties. Fire propagation of the use-surface using a small ignition source
AS 2122. Combustion propagation properties of plastics
   Part 1. Determination of flame propagation following surface ignition of vertically oriented specimens of cellular plastics (similar to ASTM D 3014)
   Part 2. Determination of the minimum oxygen concentration for flame propagation following top surface ignition of vertically oriented specimens (similar to ASTM D 2863)
AS 2288. Guide to the selection and use of fire tests for plastics materials and products
AS 2404. Textile floor coverings. Fire propagation of the use-surface using a small ignition source
AS 2420. Fire test methods for solid insulating materials and non-metallic enclosures used in electrical equipment
AS 2755. Textile fabrics. Determination of burning behaviour
   Part 1. Determination of the ease of ignition of vertically oriented specimens
   Part 2. Measurement of flame spread properties of vertically oriented specimens
AS 3000. Electrical installations. Buildings, structures and premises (known as the SAA Wiring Rules)

Table 4.9. Some British Standards Relevant to Fire Safety and Plastics Issued by the British Standards Institution (BSI).

BS 229. Flameproof enclosure of electrical apparatus
BS 476. Fire tests on building materials and structures
  Part 3. Roof test
  Part 4. Noncombustibility test
  Part 5. Ignitability test
  Part 6. Fire propagation test
  Part 7. Surface flame spread test
  Part 8. Fire resistance tests
  Part 10. Guide to fire testing
  Part 13. Ignitability by thermal irradiance (ISO 5657)
  Part 15. Heat release, cone calorimeter (ISO 5660-1)
BS 738. Insulating materials for electrical products
BS 889. Flameproof electric lighting fittings
BS 1547. Flameproof industrial clothing
BS 1763. Thin PVC sheeting
BS 2000. Petroleum and petroleum products
  Part 10. Burning test
BS 2011. Basic environmental testing procedures
  Part 2. Flammability test
BS 2050. Conducting and antistatic products made from flexible polymeric material
BS 2655. Lifts, elevators, passenger conveyors
  Part 6. Fire resistance requirements
BS 2782. Methods of testing plastics
  Part 1. Method 140D. Flammability of thin PVC sheeting
  Part 1. Method 140E. Flammability with alcohol flame
  Part 1. Method 141. Flammability by oxygen index (ISO 4589, similar to ASTM D 2863)
BS 2848. Electrical products
BS 2963. Flammability of fabrics
BS 3119. Materials for flameproof clothing
BS 3120. Performance of flameproof clothing
BS 3121. Fabrics of low flammability
BS 3289. Conveyor belting for use underground
BS 3379. Flexible urethane foam for loadbearing applications
BS 3456. Household electrical appliances
BS 3497. Electrical products
BS 3791. Clothing for protection against intense heat
BS 3900. Part A-11. Combustibility of paints
BS 4145. Electrical products
BS 4422. Glossary of terms associated with fire
BS 4584. Printed circuits
BS 4790. Ignition of textile floor coverings (hot metal nut method)
BS 4808. PVC insulation and PVC sheath for telecommunication cables and wires
BS 4840. Rigid polyurethane foam in slab form
BS 4841. Rigid polyurethane foam in laminated boards
BS 5173. Hoses and hose assemblies
BS 5287. Textile floor coverings tested to BS 4790
BS 5438. Flammability of vertically oriented textile fabrics and fabric assemblies
  subjected to a small igniting flame
BS 5502. Section 1.3. Fire protection of agricultural buildings and structures
BS 5576. Camping tents, awnings, trailer tents
BS 5588. Fire precautions in design and construction of buildings
BS 5651. Flammability of textile fabrics and fabric assemblies

Table 4.9 (continued). Some British Standards Relevant to Fire Safety and Plastics Issued by the British Standards Institution (BSI).

BS 5665. Toys, flammability requirements
BS 5722. Flammability performance of fabrics and fabric assemblies used in sleepwear
BS 5724. Medical electrical equipment
BS 5852. Fire tests for furniture
  Part 1. Ignitability of upholstered assemblies by either a smouldering cigarette or a lighted match
  Part 2. Ignitability of upholstered composites by flaming sources
  Section 1. General and guidance
  Section 2. Smouldering ignition source
  Section 3. Flaming ignition sources
  Section 4. Methods of test for ignitability of upholstery composites
  Section 5. Methods of test for ignitability of complete items of furniture
BS 5867. Flammability of fabrics for curtains and drapes
BS 6249. Clothing for protection against heat and flame
BS 6307. Ignition of textile floor coverings (methenamine tablet test) (ISO 6925)
BS 6334. Flammability of solid electrical insulating materials (IEC 707)
BS 6341. Fabrics for camping tents
BS 6373. Glossary of terms relating to burning behaviour of textiles and textile products (ISO 4880)
BS 6401. Test method for smoke
BS 6458. Fire hazard testing of electrotechnical products
  Section 2.2. Needle-flame test
BS 6807. Methods of test for ignitability of mattresses
BS 6853. Fire precautions for railway passenger rolling stock
BS 7176. Resistance to ignition of upholstered furniture
BS 9400. Integrated circuits

Table 4.10. Smolder Susceptibility of Materials as Measured by the USF Smolder Susceptibility Test Method.

Upholstery Fabric Material	Smolder Time Minutes	Foam Weight Loss Grams
100% Polyolefin (furniture)	25.1 ± 1.7	0
100% Wool (aircraft)	28.2 ± 1.7	0
100% Wool (aircraft)	29.4 ± 3.9	0
83% Wool, 17% nylon (aircraft)	29.6 ± 0.8	0
49% Wool, 51% PVC (aircraft)	31.0 ± 3.7	0
100% Cotton (furniture)	63.1 ± 8.0	2.8 ± 1.5
100% Rayon (aircraft)	67.6 ± 1.6	1.4 ± 0.7

Table 4.11. Ignitability Data on Some Materials Using the ASTM D 5485 Method at 50 kW/m².

	Material	Time to Ignition Seconds Average
1	XL olefin elastomer, metal hydrate filler	89
2	Blend of HDPE and chlorinated PE elastomer	51
3	Chlorinated PE, fillers	38
4	EVA polyolefin, ATH filler	87
5	Blend of polyphenylene oxide and polystyrene	55
6	Polyetherimide	123
7	Polyetherimide/siloxane copolymer	73
8	Polypropylene, intumescent	25
9	Nylon, mineral filler	
10	Polyolefin copolymer, mineral filler	128
11	XL polyolefin copolymer, mineral filler	120
12	XL polyolefin copolymer, ATH filler	146
13	XL polyolefin copolymer, ATH filler	70
14	EVA polyolefin, mineral filler	58
15	Polyolefin, mineral filler	88
16	XL polyethylene copolymer, chlorinated additive	73
17	Polyvinylidene fluoride material	820
18	Polytetrafluoroethylene material	208
19	PVC material	38
20	PVC building wire compound	
21	polyethylene homopolymer	64
22	Douglas fir	28
23	EVA polyolefin copolymer	51
24	Nylon 6/6	93
25	XL polyethylene copolymer, brominated additive	99

Ref. Bennett, J. G., Kessel, S. L., Rogers, C. E., "Corrosivity Test Methods for Polymeric Materials. Part 4. Cone Corrosimeter Test Method," Journal of Fire Sciences, Vol. 12, No. 2, 175–195 (March/April 1994).

Table 4.12. Flash-Fire Propensity of Materials as Measured by the NBS Flash-Fire Test.

Material	Minimum Mass, Grams/Liter
Polyphenylene oxide, modified	0.23
Polyester urethane flexible foam, 1	0.41
Wool carpet, polyester backing, FR latex backcoating	0.55
Wool carpet, polyester backing, FR latex backcoating, polyurethane foam pad, nylon scrim	0.55
Wool fabric, FR	0.62
Polyvinyl chloride flexible foam	0.67
Polyester urethane flexible foam, 2	0.72
Polyester-fiberglass sheet	0.84
Polycarbonate, molded	1.20
Polycarbonate, transparency	1.25
Epoxy-fiberglass faces, aramid honeycomb	2.91
PVF/PVC finish, phenolic-fiberglass faces, epoxy adhesive, aramid honeycomb	3.30
PVF perforated finish, aramid-epoxy face, aramid honeycomb core, fiberglass-epoxy face	4.30
FR coating, cotton/rayon fabric	greater than 1.37
PVF finish, aramid-phenolic finish	1.41
Wool carpet, phenolic/epoxy-fiberglass face, epoxy adhesive, aramid honeycomb, epoxy-fiberglass face	1.70
Fiberglass batting, melamine binder	1.83
Epoxy-fiberglass sheet	2.04
Aramid fabric	2.09
PVF finish, fiberglass screen, aramid honeycomb core with fiberglass batting, phenolic/epoxy-fiberglass face	2.35
Fiberglass batt, silicone-treated, phenolic-impregnated	2.35
PVF finish, epoxy-fiberglass faces, aramid honeycomb	2.77
PVC finish, phenolic/epoxy-fiberglass face, epoxy adhesive, aramid honeycomb, epoxy-fiberglass face	3.67
Polyether urethane foams	
ungrafted, no FR	0.1917
polyacrylonitrile grafted, no FR	0.2385
polyacrylonitrile grafted, FR (Br,P)	0.3674
ungrafted, FR (Br,P)	0.3888

Table 4.13. Flash-Fire Propensity of Materials as Measured by the Douglas Flash-Fire Test.

Material	Time to Flash Fire, min	Sample Temp. at Flash, °C
Nylon fabric	1.2	570
Polyurethane flexible foam	1.36	600
Polyurethane-coated nylon fabric	1.36	450
Wool/nylon fabric	1.56	275
Polychloroprene flexible foam, 1	1.6	740
Phenolic fabric	1.6	850
Phenolic/aramid fabric, 1	1.72	750
Polyamide-imide fabric	2.86	910
Silicone flexible foam, 1	2.96	930
Silicone flexible foam, 2	3.0	825
Phenolic/aramid fabric, 2	3.1	940
Phenolic/aramid fabric, 3	3.28	810
Polybenzimidazole fabric	4.18	940
Polyphosphazene flexible foam	none	
Silicone flexible foam, 3	none	
Polychloroprene flexible foam (4 samples)	none	
Polyamide-imide/wool fabric	none	
Aramid fabric (2 samples)	none	

Table 4.14. Explosibility of Dusts of Various Polymers.

Polymer	Minimum Ignition Temp., °C Cloud	Layer	Minimum Explosive Conc., oz/cu. ft.	Minimum Ignition Energy, Joule	Ignition Sensitivity	Explosion Severity	Explosibility Index
Polyethylene							
Type D	450		0.025	0.080	2.2	1.1	2.4
High pressure, sample 1	450	380	0.020	0.030	7.5	1.4	>10
High pressure, sample 2	410					0.8	
Low pressure, sample 1	420		0.020	0.060	4.0	1.0	4.0
Low pressure, sample 2	450		0.020	0.010	22.4	2.3	>10
Low pressure, melt index 0.4	430					1.4	
Low pressure, melt index 6.0	420					2.2	
Wax, low molecular weight	400		0.020	0.035	7.2	0.8	5.8
Polypropylene							
Molecular weight 1.8 million	460		0.035	0.400	0.3	0.2	0.1
Molecular weight 1.1 million	460		0.030	0.025	5.8	0.4	2.3
Molecular weight 0.6 million	460		0.055	0.400	0.2	0.1	<0.1
Linear	420		0.020	0.030	8.0	1.0	8.0
No antioxidant	420		0.020	0.030	8.0	2.0	>10
0.3–0.4% antioxidant						1.7	
Polyvinyl acetate	550		0.040	0.160	0.6	0.4	0.2
Polyvinyl acetate alcohol	520	440	0.035	0.120	0.9	1.2	1.1
Polyvinyl butyral	390		0.020	0.010	25.8	0.9	>10
Polyvinyl chloride							
Sample 1	730	290	>2.00	>8.32	<0.1	<0.1	<0.1
Sample 2, fine	660	400	>2.00	>8.32	<0.1	<0.1	0.1
Sample 3, coarse	690	480	>2.00	>8.32	<0.1	<0.1	<0.1
Sample 4, powdered	680	400	>2.00	>8.32	<0.1	<0.1	<0.1
Copolymer	720	500	>2.00	>8.32	<0.1	<0.1	0.1
Binder for fiber batting	670		>2.00	>8.32	<0.1	<0.1	<0.1

continued

165

Table 4.14 (continued). Explosibility of Dusts of Various Polymers.

Polymer	Minimum Ignition Temp., °C		Minimum Explosive Conc., oz/cu. ft.	Minimum Ignition Energy, Joule	Ignition Sensitivity	Explosion Severity	Explosibility Index
	Cloud	Layer					
Vinyl chloride-vinyl acetate copolymer							
Sample 1	690		>2.00	>8.32	<0.1		<0.1
Sample 2	750		>2.00	>8.32	<0.1		<0.1
Sample 3	710		>2.00	>8.32	<0.1		<0.1
Molding compound, mineral filler	690		>2.00	>8.32	<0.1		<0.1
Vinyl chloride-acrylonitrile							
60/40 copolymer, water emuls. prod.	570	470	0.045	0.025	3.1	0.6	1.9
33/67 copolymer, water emuls. prod.	530	470	0.035	0.015	7.2	2.0	>10
Vinyl chloride-polyoctyl acrylate							
79/21 copolymer	500	430	0.100	0.960	<0.1	<0.1	<0.1
Vinyl chloride-diisopropyl fumarate							
70/30 copolymer	580		0.060	0.060	1.0	0.9	0.9
Polyvinyl chloride-dioctyl phthalate							
67/33 mixture	320		0.035	0.050	3.6	0.8	2.9
Polyvinyl chloride-Hycar rubber							
Copolymer, sample 1	490		0.025	0.030	5.5	1.7	9.4
Copolymer, sample 2, more resin	550	460	0.070	0.060	0.9	0.3	0.3
Vinyl-vinylidene chloride copolymer							
Mainly vinyl	780	450	>2.00	>8.32	<0.1		<0.1
Mainly vinylidene	>1000	420	>2.00	>8.32	<0.1		<0.1

Table 4.14 (continued). Explosibility of Dusts of Various Polymers.

Polymer	Minimum Ignition Temp., °C		Minimum Explosive Conc., oz/cu. ft.	Minimum Ignition Energy, Joule	Ignition Sensitivity	Explosion Severity	Explosibility Index
	Cloud	Layer					
Vinylidene copolymer, 10% plasticizer	830	390	>2.00	>8.32	<0.1		<0.1
Vinylidene chloride polymer mold. cpd.	900		>2.00	>8.32	<0.1		<0.1
Vinyl multipolymer Monomeric vinylidene cyanide	500	510	0.030	0.015	8.9	3.0	>10
Vinyl toluene-acrylonitrile-butadiene							
58/19/23 copolymer, sample 1	530		0.020	0.020	9.5	1.6	>10
58/19/23 copolymer, sample 2	530		0.020	0.020	9.5	2.2	>10
Polyvinyl toluene, sulfonated	540	330	1.000	2.880	<0.1		<0.1
Polyvinyl benzyl trimethyl ammonium chloride,							
flake yellow	420	240	0.045	0.140	0.8	0.3	0.2
divinyl benzene	410	220	0.035	0.100	1.4	0.7	1.0
Polystyrene							
clear	490		0.020	0.120	1.7	0.5	0.9
special grind	500		0.020	0.040	5.0		
molding compound	560		0.015	0.040	6.0	2.0	>10
beads	500	470	0.025	0.060	2.7	1.5	4.1
latex, spray dried	500	500	0.020	0.015	13.4	3.3	>10
Styrene-hydrocarbon monomer polymer							
85/15 copolymer, sample 1	460	450	0.020	0.035	6.3	2.1	>10
85/15 copolymer, sample 2	460	470					
Styrene-acrylonitrile 70/30 copolymer	500		0.035	0.030	3.8	0.5	1.9

continued

Table 4.14 (continued). Explosibility of Dusts of Various Polymers.

Polymer	Minimum Ignition Temp., °C		Minimum Explosive Conc., oz/cu. ft.	Minimum Ignition Energy, Joule	Ignition Sensitivity	Explosion Severity	Explosibility Index
	Cloud	Layer					
Polystyrene–Buna N rubber coprecipitate	510	500	0.020	0.080	2.5	2.3	5.8
Styrene–butadiene latex copolymer							
Less than 4% zinc stearate blend	470		0.030	0.060	2.4	0.6	1.4
over 75% styrene, alum coagulated	440		0.025	0.025	7.3	1.7	>10
Methyl methacrylate polymer							
Sample 1	480		0.030	0.020	7.0	0.9	6.3
Sample 2, molding compound, fines	440		0.020	0.015	15.3	1.0	>10
Sample 3, molding compound, fines	440		0.030	0.020	7.6	0.8	6.1
Methyl methacrylate–ethyl acrylate							
Copolymer, sample 1	480		0.030	0.010	14.0	2.7	>10
Copolymer, sample 2, spray dried	500		0.035	0.025	4.6	2.0	9.2
Methyl methacrylate–ethyl acrylate–styrene copolymer	440		0.025	0.020	9.2	1.7	>10
Methyl methacrylate–styrene–butadiene–acrylonitrile copolymer	480		0.025	0.020	8.4	1.4	>10
Methyl methacrylate–styrene–butadiene–ethyl acrylate copolymer	480		0.025	0.025	6.7	1.5	>10
Methacrylic acid polymer, modified	450	290	0.045	0.100	1.0	0.6	0.6
Isobutyl methacrylate polymer	500	280	0.020	0.040	5.0	1.0	5.0
Acrylamide polymer	410	240	0.040	0.030	4.1	0.6	2.5

Table 4.14 (continued). Explosibility of Dusts of Various Polymers.

Polymer	Minimum Ignition Temp., °C Cloud	Minimum Ignition Temp., °C Layer	Minimum Explosive Conc., oz/cu. ft.	Minimum Ignition Energy, Joule	Ignition Sensitivity	Explosion Severity	Explosibility Index
Acrylamide-vinyl benzyl trimethyl ammonium chloride copolymer	810	500	1.000	8.000	<0.1		<0.1
Acrylonitrile polymer	500	460	0.025	0.020	8.1	2.3	>10
Acrylonitrile-vinyl pyridine copolymer	510	240	0.020	0.025	7.9	2.4	>10
Acrylonitrile-vinyl chloride-vinylidene chloride copolymer 70/20/10	650	210	0.035	0.015	5.9	3.0	>10
Cellulose acetate							
Sample 1	420		0.040	0.015	8.0	1.6	>10
Sample 2	420		0.035	0.030	4.6	0.9	4.1
Sample 3	420	420	0.040	0.045	2.7		
Sample 4	420	420	0.050	0.045	2.1		
Sample 5	470	400	0.045	0.025	3.8	3.0	>10
Sample 6	460	430	0.045	0.040	2.4	3.2	7.7
Sample 7	400	400	0.040	0.050	2.5	2.2	5.5
Sample 8	460	430	0.040	0.035	3.1	2.1	6.5
Sample 9	480	380	0.040	0.045	2.3	2.9	6.7
Sample 10	420	400	0.040	0.050	2.4	2.4	5.8
Sample 11	430		0.045	0.030	3.5	3.2	>10
Sample 12	440	340	0.055	0.020	4.2	3.1	>10
Sample 13	430		0.035			1.5	
Sample 14, 5–10 micron	450	390	0.035	0.020	6.4	3.7	>10
Sample 15, molding compound	410		0.035	0.040	3.5	0.9	3.2
Sample 16, 54.5% acetyl, spinning	450		0.040	0.080	1.4	2.5	3.5
Sample 17, 53.0% acetyl, molding	470		0.035	0.060	2.0	2.3	4.6 continued

Table 4.14 (continued). Explosibility of Dusts of Various Polymers.

Polymer	Minimum Ignition Temp., °C		Minimum Explosive Conc., oz/cu. ft.	Minimum Ignition Energy, Joule	Ignition Sensitivity	Explosion Severity	Explosibility Index
	Cloud	Layer					
Cellulose triacetate							
Sample 1	430		0.035	0.030	4.5	1.2	5.4
Sample 2	430		0.040	0.030	3.9	1.9	7.4
Cellulose acetate butyrate							
Sample 1	410		0.035	0.030	4.7	1.2	5.6
Sample 2	440		0.035			1.5	
Sample 3, molding compound	370		0.025	0.030	7.3	1.1	8.0
Cellulose acetate butyrate–cellulose acetate mixture	430		0.030			2.8	
Cellulose propionate, 0.3% free OH	460		0.025	0.060	2.9	2.6	7.5
Cellulose tripropionate, 0% free OH	460		0.025	0.045	3.9	1.8	7.0
Ethyl cellulose							
Sample 1	370		0.020				
Sample 2, 5–10 micron	370	350	0.025	0.010	21.8	3.4	>10
Sample 3, no fill. or plast.	340	330	0.025	0.015	15.8	3.6	>10
Sample 4, molding compound	320		0.025	0.010	25.2	3.2	>10
Methyl cellulose, no fill. or plast.	360	340	0.030	0.020	9.3	3.1	>10
Carboxy methyl cellulose, sample 1	350	290				0.2	
low visc., 0.3–0.4% subs.	450	290	0.165	0.180	0.2	0.2	<0.1
low visc., 0.3–0.4% subs., acid prod.	460	310	0.060	0.140	0.5	2.7	1.4
med visc., 0.84% subs.	370	260	0.150	0.560	0.1	0.5	<0.1
0.65–0.95% subs.	360		0.400	1.920	<0.1	0.2	<0.1
0.65–0.95% subs.	330		0.350	0.800	<0.1	0.2	<0.1
0.2–0.3% subs.	400		0.300	1.920	<0.1	<0.1	<0.1
0.98% subs., 56.4% active agent	400	380	0.340	>8.32	<0.1	0.2	<0.1

Table 4.14 (continued). Explosibility of Dusts of Various Polymers.

Polymer	Minimum Ignition Temp., °C		Minimum Explosive Conc., oz/cu. ft.	Minimum Ignition Energy, Joule	Ignition Sensitivity	Explosion Severity	Explosibility Index
	Cloud	Layer					
Carboxy methyl hydroxyethyl cellulose							
Sample 1	400	330	0.250	1.280	<0.1	0.1	<0.1
Sample 2, 0.65–0.85% subs.	380		0.200	0.960	<0.1	0.3	<0.1
Hydroxyethyl cellulose-monosodium phosphate sizing compound	390	340	0.070	0.035	2.1	0.8	1.7
Acetal, linear (polyformaldehyde)	440		0.035	0.020	6.5	1.9	>10
Nylon (polyhexamethylene adipamide)							
Sample 1	500	430	0.030	0.020	6.7	1.8	>10
Sample 2	510		0.050	0.030	2.6	3.3	8.6
Sample 3, chemically precipitated	540		0.035	0.030	3.6	1.1	4.0
Chlorinated polyether alcohol	460		0.045	0.160	0.6	0.3	0.2
Ethylene oxide polymer	350		0.030	0.030	6.4	0.9	5.8
Polycarbonate	710		0.025	0.025	4.5	1.9	8.6
Phenol formaldehyde							
Sample 1	580		0.025	0.015	9.3	1.4	>10
Sample 2	670		0.035	0.025	3.4		
Sample 3	730		0.035	0.080	1.0	1.9	1.9
Sample 4	700		0.025	0.035	3.3	1.7	5.6
Sample 5	580		0.175	0.020	6.9	3.9	>10
Sample 6, powdered	630		0.175	>8.32	<0.1	0.1	<0.1
Sample 7, spray dried	580		0.175	3.840	<0.1	0.1	<0.1
Sample 8, spray dried	660	320	0.200	6.000	<0.1	0.1	<0.1
Sample 9, alkaline	620		0.040	0.030	2.7		
Sample 10, 1-step	640		0.040	0.010	7.9	5.3	10
Sample 11, 2-step	580		0.025	0.010	13.9	4.0	10
Sample 12, 2-step	580		0.030	0.015	7.7	4.1	10
Sample 13, 2-step	590		0.035	0.030	3.2		
Sample 14, novalac, angular part.	620		0.030	0.020	5.4	2.5	>10

continued

171

Table 4.14 (continued). Explosibility of Dusts of Various Polymers.

Polymer	Minimum Ignition Temp., °C Cloud	Layer	Minimum Explosive Conc., oz/cu. ft.	Minimum Ignition Energy, Joule	Ignition Sensitivity	Explosion Severity	Explosibility Index
Phenol formaldehyde (cont.)							
Sample 15, novalac, spherical part.	650		0.035	0.030	2.9	0.8	2.3
Sample 16, hollow spherical part.	500	190	0.250	>8.32	<0.1	0.2	<0.1
Sample 17, hollow spherical part.	490	180	0.250	>8.32	<0.1	0.3	<0.1
Sample 18, infusible, insoluble	500	210	0.200	1.300	<0.1	<0.1	<0.1
Sample 19, 1.5% zinc stearate and 1% oxalic acid	550	300	0.025	0.010	14.6	2.7	>10
Sample 20, fiber batting binder	600		0.025	0.025	5.4	3.5	>10
Sample 21, semi resinous	460		0.235	>8.32	<0.1	<0.1	<0.1
Sample 22, glass and fibers	590						
Sample 23, more fibers than above	540						
Sample 24, 20% cellulosic extender	500		0.120	3.840	<0.1	0.1	<0.1
Phenol formaldehyde molding compound							
Sample 1, cotton flock filler	490		0.030	0.010	13.7	5.3	>10
Sample 2, wood flour filler	500		0.030	0.015	8.9	4.7	>10

Table 4.14 (continued). Explosibility of Dusts of Various Polymers.

Polymer	Minimum Ignition Temp., °C		Minimum Explosive Conc., oz/cu. ft.	Minimum Ignition Energy, Joule	Ignition Sensitivity	Explosion Severity	Explosibility Index
	Cloud	Layer					
Phenol formaldehyde, C stage, modified by hydrophilic groups							
Sample 1	440						
Sample 2	590						
Sample 3	500						
Phenol formaldehyde, amine modified	510		0.070			0.1	
Phenol formaldehyde, polyalkylene polyamine modified	420	290	0.020	0.015	16.0	2.8	10
Phenol anhydro formaldehyde anilin, 2-step	570		0.035	0.010	10.1	5.1	10
Phenol formaldehyde derivative, calcium							
Sample 1, spray dried	460	480	0.030	0.025	5.8	1.0	5.8
Sample 2, spray and drum dried	460	480	0.025	0.020	8.8	3.6	>10
Phenol formaldehyde, sulfonated	640	430	>2.00	>8.32	<0.1		<0.1
Melamine formaldehyde, no plasticizer	810		0.085	0.320	0.1	0.2	<0.1
Melamine formaldehyde, plasticizer	790		0.065	0.050	0.8	0.9	0.7
Urea formaldehyde							
Sample 1	530		0.135	1.280	0.1	0.1	0.1
Sample 2, glue	510		0.070	0.640	0.1	0.4	0.1
Sample 3, glue	510		0.075	0.960	0.1	0.8	0.1
Sample 4, glue	470		0.070	0.080	0.8	0.6	0.5
Urea formaldehyde molding compound							
Sample 1, granular	480		0.165	0.080	0.3	0.2	0.1
Sample 2	530		0.090	0.080	0.5	1.1	0.6
Sample 3, grade I, medium fine	450		0.075	0.160	0.4	0.9	0.4
Sample 4, grade II, fine	460		0.085	0.080	0.6	1.7	1.0
Sample 5, wood flour filler	490	530	0.075	0.160	0.3	0.9	0.3

continued

Table 4.14 (continued). Explosibility of Dusts of Various Polymers.

Polymer	Minimum Ignition Temp., °C		Minimum Explosive Conc., oz/cu. ft.	Minimum Ignition Energy, Joule	Ignition Sensitivity	Explosion Severity	Explosibility Index
	Cloud	Layer					
Urea formaldehyde-phenol formaldehyde molding compound, wood flour filler	530	240	0.085	0.120	0.4	0.6	0.2
Epoxy, one part anhydride type, 1% cat.	530		0.030	0.035	3.6	2.0	7.2
Epoxy, no cat., modif., or addit.	540		0.020	0.015	12.4	2.7	>10
Epoxy-bisphenol A mixture	510		0.030	0.035	3.8	0.5	1.9
Polyethylene terephthalate	500		0.040	0.035	2.9	2.6	7.5
Polyurethane foam, TDI, not fire ret.	510	440	0.030	0.020	6.6	1.5	10
Polyurethane foam, TDI, fire ret.	550	390	0.025	0.015	9.8	1.7	10
Allyl alcohol derivative							
Sample 1, CR-39	510		0.035	0.020	5.6	3.6	>10
Sample 2, CR-39	500		0.035	0.060	1.9	6.7	>10
Sample 3, CR-149, glass (65/35)	540		0.345	1.600	0.1	0.2	<0.1
Phenol furfural	530		0.025	0.010	15.2	3.9	>10
Phenol furfural, 1.5% glycerol monooleate and 1% K_2CO_3	520	310	0.025	0.010	15.5	4.0	>10
Alkyd molding compound, mineral filler							
Sample 1, not self extinguishing	500	270	0.155	0.120	0.2	<0.1	<0.1
Sample 2, self extinguishing	510	270	>2.00	>8.32	0.1	<0.1	<0.1

Table 4.15. Flame Spread Characteristics of Materials as Measured by the NBS Flooring Radiant Panel Test.

Material	Weight, oz/yd²	Critical Radiant Flux, W/cm²
Residential carpets (no preheat)		
Nylon, level loop		0.76 ± 0.13
Wool, plush		0.64 ± 0.10
Nylon 6, cut pile		0.59 ± 0.09
Acrylic, level loop		0.47 ± 0.06
Polyester		0.40 ± 0.10
Acrylic		0.23 ± 0.05
Contract carpets (2-min preheat)		
Nylon 6/6, level loop, tufted, jute backing	28	did not ignite
Wool, level loop, velvet, latex backing	46	0.83 ± 0.09
Nylon 6/6, cut loop, woven, latex backing	34.7	0.67 ± 0.13
Polyester, level loop, tufted, jute backing	42	0.66 ± 0.08
Acrylic/nylon, level loop, tufted, foam backing	37	0.64 ± 0.12
Acrylic, level loop, tufted, jute backing	42	0.60 ± 0.08
Nylon 6, level loop, tufted, foam backing	16	0.19 ± 0.04
Polypropylene, level loop, tufted, jute backing*	28	0.10

*With carpet cushion pad.

Table 4.16. Flame Spread and Smoke Characteristics of Materials as Measured by the ASTM E 162 Test.

Material	Thickness, mils	Flame Spread Index
Red oak	750	99
Hardboard	218	136
Fiberboard, unfinished	500	236
Polystyrene, extruded	66	355
Polystyrene, tile	68	224
Polymethyl methacrylate, FR	125	376
Polyvinyl chloride, rigid	147	9.6
Polyvinyl chloride, rigid, FR	147	3.2
Polyvinyl chloride, flexible, on cotton	21	89
Polyvinyl chloride, flexible, FR, on cotton	18	4.5
Phenolic, laminate	63	107
Polyester, 21% glass reinforced	62	239
Polyester, 27% glass reinforced	85	154
Polyester, 27% glass reinforced, FR	95	66

Table 4.17. Flame Spread Characteristics of Materials as Measured by the ASTM E 162 Test.

Material	Thickness, in	Flame Spread Factor, F	Heat Evol. Factor, Q	Flame Spread Index, I
Hardboard, NBS 1002b	0.222	4.3	43	186
Hardboard	0.25	4.9	40.0	185
Plywood	0.25	10.6	15.9	169
Particle board	0.625	5.1	14.8	75
Oak flooring	0.75	5.5	12.3	65
Polystyrene, FR	0.125	5.3	11.1	59
Polystyrene, FR, structural foam	0.250	4.7	10.5	85
ABS, structural foam	0.250	6.3	21.0	131
Acrylic, 1	0.050	10.2	53.2	544
	0.060	12.0	19.0	228
	0.135	5.5	40.5	223
	0.232	5.2	60.5	316
Acrylic, 2	0.25	5.9	71.0	416
Acrylic, FR	0.125	5.2	60.5	316
Polycarbonate	0.080	5.3	11.0	65
	0.125	5.8	13.3	88
	0.250	3.3	17.2	55
Polycarbonate structural foam				
FR, 1	0.25	1.87	14.8	27.5
FR, 2	0.25	1.88	16.8	31.5
FR, 3	0.250	2.9	5.2	18
Lexan FL 1800	0.250	3.9	4.6	18
Lexan FL 900	0.250	5.6	8.3	47
Polycarbonate, FR, I	0.060	7.3	11.3	81
	0.125	5.7	19.9	115
Polycarbonate, FR, II	0.125	5.2	12.7	65
	0.25	4.0	18.4	73
Polycarbonate, FR, III	0.25	1.1	2.6	2.8
Polycarbonate, Lexan GF 3025	0.050	5.5	8.0	44
	0.060	3.5	4.0	14.3
	0.083	3.0	5.3	16.1
	0.090	3.3	5.8	21.8
	0.097	3.0	4.2	12.5
	0.123	2.7	6.1	16.6
Polycarbonate, Lexan F2004	0.080	6.8	8.8	60.1
	0.115	4.0	9.2	36.6
	0.125	5.2	12.7	65.1
	0.231	3.9	17.7	68.1
	0.250	4.0	18.4	72.6

Table 4.17 (continued). Flame Spread Characteristics of Materials as Measured by the ASTM E 162 Test.

Material	Thickness, in	Flame Spread Factor, F	Heat Evol. Factor, Q	Flame Spread Index, I
Polycarbonate, Lexan F6000	0.023	7.3	10.3	74.8
	0.060	7.3	11.3	81.1
Polycarbonate, Lexan 9030	0.121	5.7	12.7	72.9
Polycarbonate, Lexan 9600	0.063	5.4	2.9	16.8
	0.122	2.1	4.6	10.3
PPO, modified, structural foam	0.25	2.77	30.6	84.4
Polyetherimide, Ultem	0.0035	1.1	1.8	1.9
	0.005	1.1	1.1	1.2
	0.058	1.3	3.0	4.1
	0.101	1.1	2.3	2.5
	0.110	1.2	2.8	3.4
	0.126	1.6	3.4	5.2
Polyethersulfone, Victrex	0.064	2.2	2.4	5.4
	0.072	1.3	1.6	2.0
Polyethersulfone	0.125	2.2	2.4	5.4
Polybutylene terephthalate				
Valox 341	0.125	6.9	6.5	57.7
Valox 357	0.125	4.5	9.0	39.7
Valox 420	0.125	5.2	27.0	136
Polyurethane, RIM, structural foam	0.50	6.07	28.5	173.3
Polyester, glass reinforced				
1	0.022	21.4	10.9	361
3	0.060	10.9	17.4	190

Ref. Bridgman, A. L., Nelson, G. L., "Radiant Panel tests on Engineering Plastics," Proceedings of the International Conference on Fire Safety, Vol. 8, 191–226 (January 1983).

Nelson, G. L., Bridgman, A. L., "Heat Release Calorimetry and Radiant Panel Testing: A Comparative Study," Proceedings of the International Conference on Fire Safety, Vol. 11, 128–139 (January 1986).

Nelson, G. L., "Effect of EMI Coatings on the Fire Performance of Plastics," Proceedings of the International Conference on Fire Safety, Vol. 13, 367–378 (January 1988).

Table 4.18. Heat and Smoke Release Characteristics of Materials as Measured by the Ohio State University (OSU) Test (ASTM E 906).

Material	Thickness, in.	Heat Flux, W/sq.cm	Maximum Heat Release, kW/sq.m	Maximum smoke Release, smoke/sq.m.
Hardboard	0.22	2.8	444	279
		4.2	504	306
		7.0	545	376
Plywood	0.25	2.8	81	8
		4.2	140	26
Yellow pine, vertical	0.85	2.8	76	9
		4.2	117	43
Polystyrene, FR	0.125	4.2	325	1310
ABS	0.12	1.4	746	1300
		2.8	770	1360
Acrylic, FR	0.125	1.4	593	216
Polycarbonate, Lexan 9030	0.080	2.8	260	735
		4.2	445	1100
	0.125	2.8	144	333
		4.2	257	640
	0.250	1.4	42	72
		2.8	284	740
		4.2	381	999
		7.0	290	994
Polycarbonate, Lexan MR4000	0.250	4.2	333	780
		7.0	263	870
Polycarbonate, Lexan F2004	0.080	2.8	115	230
		4.2	199	938
Polycarbonate, Lexan MR4604	0.060	2.8	13	50
		4.2	153	910
Polycarbonate, Lexan F6000	0.060	2.8	15	5
		4.2	165	470
Polycarbonate, Lexan TF6000	0.060	2.8	165	303
		4.2	170	480
Polycarbonate, Lexan F6006	0.080	4.2	177	480
Polycarbonate, FR, I	0.060	4.2	177	429
Polycarbonate, FR, II	0.125	2.8	105	344
		4.2	179	701
	0.250	4.2	235	652
Polycarbonate, FR, III	0.235	4.2	90	268
		7.0	137	248
PPO, modified, Noryl EN185	0.13	4.2	325	1310
Polyetherimide, Ultem	0.055	4.2	18	9
		5.6	122	232
		7.0	127	298
	0.13	7.0	122	229
Polyether sulfone	0.125	4.2	17	13
		7.0	121	375
Polyester, glass reinforced, 1	0.022	1.4	178	588
		2.8	348	901
Polyester, glass reinforced, 2	0.036	1.4	187	608
		2.8	268	855
Polyester, glass reinforced, 3	0.060	2.8	297	765

Ref.: Bridgman, A. L., Nelson, G. L., Journal of Fire and Flammability, Vol. 13, No. 2, 114–134 (April 1982).

Nelson, G. L., Bridgman, A. L., Proceedings of the International Conference on Fire Safety, Vol. 11, 128–139 (1986).

Table 4.19. Heat and Smoke Release Characteristics of Materials as Measured by the Ohio State University (OSU) Test (FAA Method).

Material	Maximum Heat Release	2 Min. Heat Release	Maximum Smoke Release	2 Min. Smoke Release
Pre-1987 Materials at 3.5 W/sq.cm				
Phenolic Kevlar, no dec.	71	75	43	29
Phenolic glass, no dec.	53	42	20	10
Epoxy glass, no dec.	83	87	447	156
Epoxy Kevlar, no dec.	60	66	423	125
Phenolic Kevlar, 2 mil wht. Tedlar	64	83	35	27
Phenolic glass, 2 mil wht. Tedlar	64	72	43	21
Epoxy glass, 2 mil wht. Tedlar	89	92	402	158
Epoxy Kevlar, 2 mil wht. Tedlar	66	63	364	124
Phenolic Kevlar, TXT dec.	100	116	80	59
Phenolic glass, TXT dec.	112	99	81	48
Epoxy glass, TXT dec.	106	114	353	199
Epoxy Kevlar, TXT dec.	109	104	368	177
ABS, FR	288	280	327	322
Polycarbonate	225	199	360	35
Polyetherimide	75	54	78	1
1989 Materials at 3.5 W/sq.cm				
Thermoplastic A, 60 mil	36	0	65	5
Thermoplastic, B, 60 mil	45	0	57	3
Modified phenolic glass, low HR dec.	44	40	80	25
Modified phenolic glass, paint	48	35	43	15
Pre-1987 Materials at 5.0 W/sq.cm				
Phenolic Kevlar, no dec.	124	119	151	99
Phenolic glass, no dec.	87	78	67	33
Epoxy glass, no dec.	104	82	771	215
Epoxy Kevlar, no dec.	103	120	824	251
Phenolic Kevlar, 2 mil wht. Tedlar	106	122	143	106
Phenolic glass, 2 mil wht. Tedlar	91	93	171	61
Epoxy glass, 2 mil wht. Tedlar	111	96	767	241
Epoxy Kevlar, 2 mil. wht. Tedlar	101	119	881	266
Phenolic Kevlar, TXT dec.	175	161	237	149
Phenolic glass, TXT dec.	150	116	184	90
Epoxy glass, TXT dec.	138	114	698	285
Epoxy Kevlar, TXT dec.	164	138	650	314

Units: Maximum heat release in kW/sq.m.; 2-min. heat release in kW-min/sq.m.; maximum smoke release in smoke/min-sq.m.; and 2-min. smoke release in smoke/sq.m.

Ref.: Smith, E. E., private communication (August 3, 1989).

Table 4.20. Heat Release Characteristics of Some Materials Using the Cone Calorimeter Test at 50 kW/m².

	Material	Peak HRR, kW/m²	Avg HRR, kW/m²
1	XL olefin elastomer, metal hydrate filler	107.8	88.7
2	Blend of HDPE and chlorinated PE elastomer	172.1	138.5
3	Chlorinated PE, fillers	108.2	100.3
4	EVA polyolefin, ATH filler	152.0	123.6
5	Blend of polyphenylene oxide and polystyrene	486.1	342.6
6	Polyetherimide	120.4	79.5
7	Polyetherimide/siloxane copolymer	160.3	105.3
8	Polypropylene, intumescent	86.0	45.3
9	Nylon, mineral filler		
10	Polyolefin colpolymer, mineral filler	115.1	98.3
11	XL polyolefin copolymer, mineral filler	127.9	112.0
12	XL polyolefin copolymer, ATH filler	285.3	164.6
13	XL polyolefin copolymer, ATH filler	195.3	155.0
14	EVA polyolefin, mineral filler	305.9	239.6
15	Polyolefin, mineral filler	117.0	104.7
16	XL polyethylene copolymer, chlorinated additive	485.3	365.0
17	Polyvinylidene fluoride material	30.3	17.9
18	Polytetrafluoroethylene material	104.6	75.7
19	PVC material	54.3	25.0
20	PVC building wire compound		
21	Polyethylene homopolymer	806.5	310.0
22	Douglas fir	156.3	84.3
23	EVA polyolefin copolymer	1163.0	356.3
24	Nylon 6/6	864.5	354.3
25	XL polyethylene copolymer, brominated additive	422.0	336.4

Ref. Hirschler, M. M. "Analysis of Heat Release and Other Data from a Series of Plastic Materials Tested in the Cone Calorimeter," Proceedings of the International Conference on Fire Safety, Vol. 20, 214–228 (January 1995).

Table 4.21. Smoke-Producing Characteristics of Materials as Measured by the ASTM D 2843 Test.

Material	Thickness, mils	Maximum Light Absorption, pct
Polyethylene, UCC DXM-100	250	9
Polyethylene, UCC DFDA-6311	125	82
Polyethylene, UCC DHDA-1811	125	79
Polyethylene, UCC DMDA-7075	125	49
Polypropylene, UCC JMD-8500	250	20
Polyvinyl chloride, UCC QCA-2460	15	39
	20	40
	40	87
Polyvinyl chloride, UCC QYTQ	250	100
Polyvinyl butyral, UCC XYHL	250	11
Polystyrene, UCC SMD-3500	250	98
Styrene-acrylonitrile, UCC C-11	250	100
ABS, Cycolac	250	100
Polysulfone, UCC P-1700	250	59
Polycarbonate, Lexan 113-T	250	90
Polyester, Paraplex P-43	125	100
Polyester, Hetron 92	125	100
Polyurethane rigid foam	250	50
Polyurethane rigid foam, FR	250	24
Polystyrene rigid foam, FR	250	56
Polyvinyl chloride rigid foam	250	51

Table 4.22. Smoke-Producing Characteristics of Materials as Measured by the OSU Release Rate Test.

Material	Orientation V—Vertical H—Horizontal	Applied Heat Flux, W/cm²	Maximum Smoke Release Rate, units/min-m³	Total Smoke Release, units/m³	
				3 min	10 min
Oak, 1 in	V	1.0	0.9	0.9	2.3
		2.0	5.5	1.8	4.6
		2.5	6.9	9.1	11.4
Pine, 1 in	V	1.5	2.3	6.9	9.1
		2.5	27.4	32.0	36.6
Red oak, flooring	H	1.0	0.9	0.9	2.3
		2.0	2.3	1.4	4.6
		2.75	6.9	9.1	11.4
Hardboard, 0.25 in	V	1.0	13.7	0	13.7
		2.0	36.6	0.9	73.1
Particle board, 0.5 in	V	1.0	0.9		
		2.5	22.8	9.1	27.4
Particle board, FR, 0.5 in	V	1.5	0.9		
		2.5	4.6	0.9	4.6
Fiberboard, low density, 0.625 in	V	1.0	27.4	45.7	45.7
		2.0	45.7	80.0	80.0
Exterior plywood, 0.5 in	V	1.0	3.4	1.4	4.6
		2.0	4.6	3.7	6.9
Polypropylene, sheet, 0.125 in	H	1.0	228	4.6	686
Polyvinyl chloride, rigid, pipe	V	1.0	4.6	0.9	22.8
		2.6	114	91.4	457
Polyvinyl chloride, flexible, sheet 90 mil	V	1.0	228	228	640
		2.0	366	640	731
Polystyrene, light diffuser	H	0	77.7	16.0	366
		1.0	137	45.7	457
ABS, sheet, 0.125 in	H	0	274	2.3	137
		1.0	366	155	457
		2.5	366	457	548
Polymethyl methacrylate, light diffuser	H	0	6.9	1.8	9.1
		1.0	4.6	2.3	13.7
Polysulfone, sheet, 95 mil	H	1.0	0	0	0
		2.5	183	32.0	320
Polyurethane rigid foam, 1 in	V	0	18.3	9.1	9.1
		1.0	27.4	9.1	9.1
Polyurethane rigid foam, FR, 1 in	V	0.5	0.9	0	0.9
		1.0	137	114	114
		2.5	183	137	137
Polyisocyanurate rigid foam, 1 in	V	1.0	9.1	2.3	2.3
		2.75	228	27.4	32.0

Table 4.23. Smoke-Producing Characteristics of Materials as Measured by the Arapahoe Smoke Test.

Material	Percent Smoke Based On:	
	Initial Weight	Weight Loss
Hardwoods		
Beech	0.05	0.08
Yellow birch	0.09	0.14
Red oak	0.11	0.18
Aspen poplar	0.13	0.20
Softwoods		
Southern yellow pine	0.08	0.12
Douglas fir	0.14	0.24
Eastern white pine	0.17	0.27
Western hemlock	0.20	0.33
Western red cedar	0.23	0.35
Cellulosic boards		
Hardboard	0.06	0.08
Medium density hardboard	0.11	0.16
Cellulose fiberboard, core board	0.57	0.75
Plastics		
Polymethyl methacrylate	0.08	0.19
Polyvinyl chloride flooring	0.21	6.23
Acrylic, unidentified	0.33	0.47
Linoleum	0.52	1.53
Polychloroprene rubber, filled, FR	0.80	8.72
Polycarbonate, 2	0.89	13.34
Polyvinyl chloride, rigid	1.33	10.52
Polycarbonate, 1	1.34	11.96
Polyvinyl chloride/acrylic	1.38	7.60
Polypropylene, FR	1.64	13.42
Polyester, brominated, fiberglass reinforced	1.70	14.52
Polyvinyl chloride, flexible, FR	2.36	12.74
Acrylonitrile-butadiene-styrene, FR	4.02	20.54
Polystyrene	4.86	12.71
Stytene-butadiene rubber, filled, FR	6.63	10.51

Table 4.24. Toxicity Data on Some Materials Using the Polish Fire Safety Centre Method.

	Material	Mode: p—pyrolysis, c—combustion	LC_{50}, mg/liter
1	Polyacrylonitrile	p	13.5
2	Polyurethane foam, stiff	p	18.3
3	Polyester	p	33.0
4	Wallboard	p	38.0
5	Polyacrylonitrile	c	40.3
6	Polystyrene foam	c	41.0
7	Polyurethane foam, stiff	c	41.5
8	Forsan 752	p	43.0
9	Polyamide	p	43.4
10	Wool, natural	p	43.3
11	Wallboard	c	45.3
12	Polyurethane foam, elastic	p	46.0
13	Polystyrene foam	p	47.0
14	Unilam	p	47.0
15	Forsan 752	c	54.5
16	Polyester	c	56.0
17	Wool, natural	c	58.0
18	Polyurethane foam, elastic	c	60.0
19	Hardboard	p	64.0
20	Polyamide	c	65.5
21	Unilam	c	68.9
22	Rubber insulation	c	71.4
23	Hardboard	c	80.5
24	Rubber insulation	p	82.7

Ref. Czerczak, S., Stetkiewicz, J., "Toxicity Classification of Thermal Degradation Products of Chemical Materials Used in Construction," Journal of Fire Sciences, Vol. 14, 367–378 (September/October 1996).

Table 4.25. Toxic Gas Evolution from Materials as Measured by the University of Pittsburgh Method (New York State Protocol).

Material	LC$_{50}$, grams
White pine	41.8–65.4
Cellulose loose fill insulation	8.3–35.0
Gypsum wallboard	97.0–119.9
Vinyl coated gypsum wallboard	88.5–107.6
Vinyl wall covering	15.7–37.5
Polyester flexible duct	8.7–16.5
Fiberglass reinforced polyester duct	11.2–16.4
Fiberglass reinforced epoxy duct	15.7–21.7
Fiberglass reinforced PVC flexible duct	12.3–65.0
Rigid polyvinyl chloride pipe	6.5–13.1
Polystyrene molding	5.5–15.7
Polyurethane foam	2.9–7.9
Fluoroplastic wire insulation	2.0–12.6
Polyvinyl chloride wire insulation	8.3–16.3
Nylon wire insulation	5.8–7.0

Ref. Woynerowski, S. P., Proceedings of the International Conference on Fire Safety, Vol. 14, 203–209 (1989).

Table 4.26. Corrosivity Data on Some Materials Using the Modified NIBS Apparatus.

	Material	Metal Loss, angstroms 2,500 Angstrom Probe, Average
1	XL olefin elastomer, metal hydrate filler	205
2	Blend of HDPE and chlorinated PE elastomer	cc
3	Chlorinated PE, fillers	cc
4	EVA polyolefin, ATH filler	627
5	Blend of polyphenylene oxide and polystyrene	182
6	Polyetherimide	66
7	Polyetherimide/siloxane copolymer	184
8	Polypropylene, intumescent	663
9	Nylon, mineral filler	
10	Polyolefin copolymer, mineral filler	1023+
11	XL polyolefin copolymer, mineral filler	568
12	XL polyolefin copolymer, ATH filler	76
13	XL polyolefin copolymer, ATH filler	219
14	EVA polyolefin, mineral filler	415
15	Polyolefin, mineral filler	663
16	XL polyethylene copolymer, chlorinated additive	cc
17	Polyvinylidene fluoride material	cc
18	Polytetrafluoroethylene material	cc
19	PVC material	cc
20	PVC building wire compound	cc
21	Polyethylene homopolymer	291
22	Douglas fir	177
23	EVA polyolefin copolymer	93
24	Nylon 6/6	198
25	XL polyethylene copolymer, brominated additive	2362+

cc—probe completely corroded, off-scale at 2500+.

Ref. Kessel, S. L., Bennett, J. G., Rogers, C. E., "Corrosivity Test Methods for Polymeric Materials. Part 1. Radiant Furnace Test Method," Journal of Fire Sciences, Vol. 12, No. 2, 109–133 (March/April 1994).

Table 4.27. Corrosivity Data on Some Materials Using the Modified DIN Method.

	Material	pH, min. ave.	Conduct, max. ave.
1	XL olefin elastomer, metal hydrate filler	3.81	30.6
2	Blend of HDPE and chlorinated PE elastomer	2.21	3560.8
3	Chlorinated PE, fillers	2.20	3378.7
4	EVA polyolefin, ATH filler	3.91	13.1
5	Blend of polyphenylene oxide and polystyrene	3.50	24.3
6	Polyetherimide	3.69	212.1
7	Polyetherimide/siloxane copolymer	3.56	55.6
8	Polypropylene, intumescent	5.58	944.3
9	Nylon, mineral filler		
10	Polyolefin copolymer, mineral filler	3.60	16.0
11	XL polyolefin copolymer, mineral filler	3.59	18.1
12	XL polyolefin copolymer, ATH filler	4.10	15.0
13	XL polyolefin copolymer, ATH filler	3.70	16.8
14	EVA polyolefin, mineral filler	3.56	24.9
15	Polyolefin, mineral filler	3.73	13.5
16	XL polyethylene copolymer, chlorinated additive	2.52	1435.8
17	Polyvinylidene fluoride material	1.99	4540.3
18	Polytetrafluoroethylene material	2.42	1952.5
19	PVC material	1.82	7208.0
20	PVC building wire compound	1.75	10920.7
21	Polyethylene homopolymer	3.58	35.2
22	Douglas fir	3.49	82.7
23	EVA polyolefin copolymer	3.27	114.4
24	Nylon 6/6	4.07	956.6
25	XL polyethylene copolymer, brominated additive	3.69	154.7

Ref. Bennett, J. G., Kessel, S. L., Rogers, C. E., "Corrosivity Test Methods for Polymeric Materials. Part 3. Modified DIN Test Method," Journal of Fire Sciences, Vol. 12, No. 2, 155–174 (March/April 1994).

Table 4.28. Corrosivity Data on Some Materials Using the CNET Method.

	Material	% Corrosivity Factor, ohms	Data Spread %
1	XL olefin elastomer, metal hydrate filler	8.3	13
2	Blend of HDPE and chlorinated PE elastomer	13.7	108
3	Chlorinated PE, fillers	7.8	141
4	EVA polyolefin, ATH filler	2.6	105
5	Blend of polyphenylene oxide and polystyrene	5.9	98
6	Polyetherimide	5.4	111
7	Polyetherimide/siloxane copolymer	5.0	108
8	Polypropylene, intumescent	6.7	69
9	Nylon, mineral filler		
10	Polyolefin copolymer, mineral filler	5.1	120
11	XL polyolefin copolymer, mineral filler	6.6	100
12	XL polyolefin copolymer, ATH filler	5.5	72
13	XL polyolefin copolymer, ATH filler	3.2	135
14	EVA polyolefin, mineral filler	3.0	205
15	Polyolefin, mineral filler	4.5	77
16	XL polyethylene copolymer, chlorinated additive	4.6	97
17	Polyvinylidene fluoride material	9.6	57
18	Polytetrafluoroethylene material	5.4	76
19	PVC material	30.3	34
20	PVC building wire compound	47.8	27
21	Polyethylene homopolymer	7.4	104
22	Douglas fir	4.7	99
23	EVA polyolefin copolymer	6.8	72
24	Nylon 6/6	2.3	
25	XL polyethylene copolymer, brominated additive	6.2	99

Ref.: Rogers, C. E., Bennett, J. G., Kessel, S. L., "Corrosivity Test Methods for Polymeric Materials: CNET Test Method," Proceedings of the International Conference on Fire Safety, Vol. 17, 392–403 (January 1992).

Rogers, C. E., Bennett, J. G., Kessel, S. L., "Corrosivity Test Methods for Polymeric Materials. Part 2. CNET Test Method," Journal of Fire Sciences, Vol. 12, No. 2, 134–154 (March/April 1994).

CHAPTER 5

Prevention, Inhibition, and Extinguishment

The essential ingredient of a plastic material is an organic substance which will burn under sufficiently severe exposure to heat and oxygen. This ability to burn must be reduced under the specific conditions of the many applications for which plastics are designed, so that plastics will not represent a fire hazard.

The work of preventing, inhibiting, and extinguishing fire in plastics can be divided into three areas:

1. Design of the material
2. Design of the product
3. Design of the application

SECTION 5.1. DESIGN OF THE MATERIAL

Fire retardance involves disruption of the burning process at one or more stages so that the process is terminated within an acceptable period of time, preferably before ignition actually occurs. When dealing with plastic materials, the following general approaches are available.

1. Design of the basic polymer so that exposure to heat and oxygen will not produce combustion. This is achieved by producing thermally stable polymers with high decomposition temperatures and high fractions of solid residue after decomposition. Such polymers may require higher temperatures for processing and may lack versatility in processing and performance characteristics.

2. Modification of existing polymers so that they exhibit satisfactory performance upon exposure to fire. Examples of this approach are the chlorination of polyethylene, the substitution of chlorendic anhydride or tetrabromophthalic anhydride for other compounds in polyesters, and the use of phosphorus-containing polyols in polyurethanes. These substituted materials are known as reactive type fire retardants.

3. Incorporation of compounds so that the resulting plastic materials exhibit

189

satisfactory performance upon exposure to fire. Examples of this approach are the addition of tris(2,3-dibromopropyl) phosphate, and the addition of zinc borate. These added materials are known as additive type fire retardants.

The first approach provides little help in imparting fire retardance to the materials presently being used at the rate of several billion pounds annually. New high-temperature polymers lack sufficient versatility in processing and performance and sufficiently favorable economics to displace presently used materials to a major extent.

The third approach provides the most expeditious means of providing fire retardance, particularly at the end-use or compounding stages. The incorporation of additives, however, has limitations and disadvantages, and the second approach is favored when fire resistance can simply be added to the list of properties considered desirable when modifying existing polymers.

The general approaches involved in the use of additive type fire retardants are the following.

1. Incorporation of a compound that will redirect the decomposition and combustion reactions toward the evolution of gases which are non-combustible, or heavy enough to interfere with normal interchange of combustion gases and combustion air. Ammonium carbonate decomposes into ammonia, water, and carbon dioxide, and tetrabromoethane decomposes to evolve hydrogen bromide. Phosphorus compounds are believed to alter the decomposition reactions so as to increase the amount of carbonaceous residue at the expense of combustible gases. Halogen compounds are believed to inhibit the free radical chain reactions involved in decomposing the polymer into combustible gases. This approach essentially constitutes interference with Stages II and III of the burning process on the macro scale (Decomposition and Ignition), with effects on Stages IV and V (Combustion and Propagation).

2. Incorporation of a compound that will redirect the decomposition and combustion reactions toward reduced heat of combustion. Phosphorus compounds appear effective in this approach, in favoring the formation of carbonaceous char rather than carbon-containing combustible gases, and in inhibiting glowing oxidation of the char. One gram-mol of carbon produces 26,400 calories of heat in burning to carbon monoxide and 94,000 calories of heat in burning to carbon dioxide; this heat is not produced if carbon is kept in the form of char. This approach essentially constitutes interference with Stage IV of the burning process on the macro scale (Combustion), with effects on Stage V (Propagation).

3. Incorporation of a compound that will conserve the physical integrity of the material, in order to impede access of oxygen and heat, and reduce disintegration of the structure. Fillers and reinforcing glass fibers contribute to physical integrity by assuring certain minimum levels of solid residue. Phosphorus compounds appear effective in this approach, in favoring retention of the carbon as char, because carbon monoxide and carbon dioxide are gases and contribute nothing to physical integrity, and may even

disrupt the structure of the solid residue during thermal expansion. This approach essentially constitutes interference with Stage II of the burning process on the macro scale (Decomposition), with effects on Stages III, IV, and V (Ignition, Combustion, and Propagation).

4. Incorporation of a compound that will increase the specific heat or thermal conductivity of a material, or otherwise increase absorption or dissipation of heat, to reduce the heat available for every stage of the burning process. High-density fillers tend to increase heat absorption and dissipation, and inorganic hydrates such as hydrated aluminum oxide require energy for dehydration. This approach essentially constitutes interference with Stage I of the burning process on the macro scale (Heating), with effects on succeeding stages. However, it is not always desirable for plastic materials whose primary function is thermal insulation, such as some rigid cellular plastics.

The general approaches involved in the use of reactive type fire retardants are the following.

1. Modification of the polymer to favor the decomposition and combustion reactions producing gases which are non-combustible, or heavy enough to interfere with normal interchange of combustion gases and combustion air. Phosphorus is introduced into polyurethane polymers in the form of phosphonate, phosphate, and phosphite polyols. Halogens are introduced into polyester polymers in the form of halogenated anhydrides such as chlorendic anhydride and tetrabromophthalic anhydride, into epoxy polymers in the form of halogenated compounds such as tetrabromobisphenol-A, and into polyolefins in the form of chlorinated paraffins and olefins. This approach essentially constitutes interference with Stages II and III of the burning process on the macro scale (Decomposition and Ignition), with effects on Stages IV and V (Combustion and Propagation).

2. Modification of the polymer to favor decomposition and combustion reactions producing reduced heat of combustion. Phosphorus compounds appear effective in this approach. This approach essentially constitutes interference with Stage IV of the burning process on the macro scale (Combustion), with effects on Stage V (Propagation).

3. Modification of the polymer to increase the amount of solid residue, to maintain structural integrity and to impede access of oxygen and heat. Phosphorus compounds appear effective in this approach. Char formation in polyurethanes is improved by substituting polymeric polyaryl isocyanates for tolylene diisocyanate, and by employing sucrose-based polyethers. This approach essentially constitutes interference with Stage II of the burning process on the macro scale (Decomposition), with effects on Stages III, IV, and V (Ignition, Combustion, and Propagation).

4. Modification of the polymer to reduce ease of ignition, by increasing minimum decomposition temperature or minimum ignition temperature, or by increasing the energy required for decomposition or ignition. Phosphorus and halogen compounds appear effective in this approach. This approach essentially constitutes interference with

Stage III of the burning process on the macro scale (Ignition), with effects on Stages IV and V (Combustion and Propagation).

The promotion of char formation is perhaps the most important overall concept in fire-retarding plastic materials. Retention of carbon in the char has the following important benefits:

1. Reduction of heat of combustion, from 94,000 calories per gram mol of carbon dioxide, to 26,400 calories per gram mol of carbon monoxide, to zero for carbon retained in the solid phase
2. Conservation of mechanical integrity, because retained carbon is a solid, and carbon monoxide and carbon dioxide are gases which can damage the char structure during thermal expansion
3. Reduction of smoke evolution
4. Reduction of oxygen depletion
5. Reduction of toxic gas evolution, particularly carbon monoxide

Reactive type fire retardants appear to be most versatile in thermosetting plastic materials, in which substitution of compounds can be effected much closer to end-use than might be the case with thermoplastics. Additive type fire retardants appear to be most versatile in thermoplastics.

SECTION 5.2. FLAME RETARDANCE MECHANISMS

Much of the development of flame retardants has been based on empirical work, because the mechanisms of flame extinguishment are not completely understood. On the basis of empirical results, various theories have been postulated to describe the mechanisms involved in flame inhibition and extinguishment.

The compounds which have been found to be most effective in producing flame retardance are compounds containing phosphorus, bromine, or chlorine, or two or more of these elements. Other elements which have exhibited some flame retardant effect are antimony, boron, nitrogen, silicon, and zinc. Because phosphorus, bromine, and chlorine have proven so much more effective than the other elements cited, considerable work has gone into the study of the possible mechanisms involved in their inhibiting and extinguishing action.

In the discussion of the mechanisms involved in the performance of various elements considered to be flame retardants, four concepts should be borne in mind:

1. The phase in which retardance occurs (gaseous or condensed) and the stage of the burning process at which retardance occurs are important.
2. The flame retardant must be available and active at the right time and in the right place.
3. To impart a specific level of flame retardance to a specific material using a specific

flame retardant, a specific amount of flame retardant is required, and amounts exceeding this optimum are unnecessary and sometimes undesirable.
4. A flame retardant is undesirable for a given material if the amount required has an intolerable effect on other properties, such as mechanical strength or thermal stability.

Flame retardant action is attributed to a variety of compounds which contain elements found in certain groups of elements in the Period Table of the Elements:

Group VII B, which includes fluorine, chlorine, bromine, and iodine
Group VI B, which includes sulfur
Group V B, which includes nitrogen, phosphorus, arsenic, antimony, and bismuth
Group IV B, which includes carbon, silicon, tin, and lead
Group III B, which includes boron and aluminum
Group II B, which includes zinc, cadmium, and mercury
Group I B, which includes copper, silver, and gold
Group VIII A, which includes iron, cobalt, nickel, and platinum
Group VII A, which includes manganese
Group VI A, which includes chromium, molybdenum, and tungsten
Group V A, which includes vanadium
Group II A, which includes magnesium, calcium, strontium, and barium

Section 5.2.1. Group VII B Elements

The Group VII B elements include the halogens: fluorine, chlorine, bromine, and iodine.

The lightest element in this group shows great strength in bonds to carbon and low density in its gaseous compounds. It appears to be more effective in the condensed phase.

Fluorine can be considered a flame retardant element in that its presence in the basic polymer contributes to fire resistance in many cases. Its bond to carbon is so strong that it is difficult to render the attached carbon available for oxidation. Where the hydrogen atoms in the polymer have been replaced with fluorine atoms, as in the case of polytetrafluoroethylene, decomposition temperatures are high and the decomposition products are noncombustible.

Fluorine is more effective as an inherent flame retardant in the basic polymer than as a reactive flame retardant or additive flame retardant employed in later stages of the product. As part of the basic polymer, it functions essentially in the condensed phase. It is effective in Stage II (Decomposition) and Stage III (Ignition), and increases the possibility of arresting the burning process at an earlier stage. Fluorine-containing gases are lighter than their chlorine and bromine containing analogs, and their smothering or blanketing action would be less effective in the gaseous phase.

The most important fluorine-containing compounds which function as additive flame retardants have been the chlorofluorocarbons, which contributed to fire resistance when used as blowing agents in polyurethane and polystyrene foams. Chlorofluorocarbons have also been useful as fire extinguishing agents.

Chlorine is believed to perform its flame retardant function in both the gaseous and condensed phases. In the gaseous phase, it employs the chemical mechanism of redirection or termination of the chemical reactions involved in combustion, and the physical mechanism of evolution of heavy chlorine-containing gases to protect the condensed phase by inhibiting access of oxygen and heat. In the condensed phase, it redirects the chemical reactions involved in decomposition.

The chemical reactions may involve halogenation followed by dehydrohalogenation, to yield a polymeric residue rich in double bonds which is then converted to carbon.

The action of chlorine-containing compounds appears to be primarily effective in Stage III (Ignition), by blocking access to oxygen and heat, and in Stage IV (Combustion), by influencing the combustion reactions. Their influence on decomposition indicates some effectiveness in Stage II (Decomposition). It is necessary for the burning process to proceed through Stage II (Decomposition) and release chlorine-containing compounds into the gaseous phase for flame retardant action to occur.

Aliphatic chlorine is more effective than aromatic chlorine in both polyesters and polyurethanes. The form in which chlorine is available influences its effectiveness. In polypropylene, perchlorofulvalene is more efficient than perchloropentacyclodecane, perhaps because perchlorofulvalene decomposes above 250°C. into hexachlorocyclopentadiene and hydrogen chloride.

For epoxy resins containing no other flame retardants, self-extinguishing behavior requires about 16 to 20 per cent chlorine.

For polyesters containing no other flame retardants, self-extinguishing behavior requires about 21 per cent alicyclic chlorine or 28 to 31 per cent aromatic chlorine (in forms such as hexahalobenzenes and chlorinated biphenyls).

Large quantities of halogen can produce poor weathering resistance in polyesters. Highly chlorinated paraffins are preferred over lower chlorinated waxes because of the latter's plasticizing effect. If tetrachlorophthalic anhydride, which contains only 49.5 per cent chlorine, is employed in sufficient quantity to bring chlorine content in the polyester to the 30 per cent level, the resulting polyester is inferior in heat distortion and mechanical properties.

Bromine is believed to perform its flame retardant function in both the gaseous and condensed phases. In the gaseous phase, it employs the chemical mechanism of redirection or termination of the chemical reactions involved in combustion, and the physical mechanism of evolution of heavy bromine-containing gases to protect the condensed phase by inhibiting access of oxygen and heat.

The chemical reactions may involve halogenation, which is followed by dehydrohalogenation, to yield a polymeric residue rich in double bonds which is then converted to carbon.

The action of bromine-containing compounds appears to be primarily effective in Stage III (Ignition), by blocking access to oxygen and heat, and in Stage IV (Combustion), by influencing the combustion reactions. It is necessary for the burning process to proceed through Stage II (Decomposition) and release bromine-containing compounds into the gaseous phase for flame retardant action to occur.

Aliphatic and alicyclic bromine compounds have been found to be more effective than aromatic bromine compounds in flame retarding polystyrene. This difference is attributed to the higher bond energy of aromatic bromine, which makes it less readily available for flame retardant action.

For polyesters containing no other flame retardants, self-extinguishing behavior requires about 17 per cent bromine.

The concentrations of bromine that can be employed are limited in various polymers by its effect on flexibility, mechanical properties, heat distortion, or durability.

Iodine has such a weak bond to carbon and to organic compounds in general that it is likely to initiate decomposition in normal service, and its relative inactivity does not make it likely to influence combustion reactions. Iodine has been used in iodinated polystyrene.

Section 5.2.2. Group VI B Elements

The Group VI B elements include oxygen, sulfur, and selenium.

Sulfur can be considered to be a flame retardant in that its presence in the basic polymer contributes to fire resistance in many cases. Many sulfur-containing polymers are resistant to high temperatures. In such instances, sulfur functions in the condensed phase, and appears to be primarily effective in Stage II (Decomposition) and Stage III (Ignition).

Section 5.2.3. Group V B Elements

The Group V B elements include nitrogen, phosphorus, arsenic, antimony, and bismuth.

The elements in Group V B cross the boundary between non-metals and low-melting heavy metals. Nitrogen, phosphorus, and arsenic are non-metals, and antimony and bismuth are low-melting heavy metals.

The lightest element in this group shows great strength in bonds to carbon and low density in its gaseous compounds, and is more effective in the condensed phase.

Nitrogen can be considered to be a flame retardant in that its presence in the basic polymer contributes to fire resistance in many cases. Many nitrogen-containing polymers are resistant to high temperatures. In such instances, nitrogen functions in the condensed phase, and appears to be primarily effective in Stage II (Decomposition) and Stage III (Ignition).

Some nitrogen-containing compounds which function as reactive or additive flame retardants appear to act in the condensed phase and to be effective in Stage II (Decomposition) and Stage III (Ignition). Melamine is effective in various plastics, and especially in polyurethane foams.

Phosphorus is believed to perform most of its flame retardant function in the condensed phase (including both the solid and liquid phases, because various degrees of melting are involved at fire temperatures). Phosphorus-containing compounds increase the amóunt of carbonaceous residue or char formed, by one or both of two mechanisms: redirection of the chemical reactions involved in decomposition in favor of reactions yielding carbon rather than carbon monoxide or carbon dioxide; and formation of a surface layer of protective char which inhibits access of oxygen and escape of oxides of carbon by physical blockade, thus preventing gasification of the carbon by simply denying the oxygen required.

The effect of phosphorus on the chemical reactions involved in decomposition may involve partial oxidation followed by dehydration to yield water, a noncombustible gas with relatively low heat of formation, and carbon.

The effect of phosphorus on the surface characteristics of the char may involve the formation of nonvolatile oxides of phosphorus which act as a flux for the carbonaceous residue.

The action of phosphorus-containing compounds is particularly desirable for the following reasons:

1. It is effective in Stage II (Decomposition) rather than in the later Stages III (Ignition) and IV (Combustion), increasing the probability of arresting the burning process at an earlier stage.
2. It is effective in the condensed phase, which is the fraction of the material which may have some residual value, rather than in the gaseous phase, which contains the fraction of the material which has no further practical value.
3. It interferes with the oxidation reactions which are responsible for the one major source of heat in the burning of plastic materials: carbon oxidation.

4. It interferes with gasification, which is largely responsible for increased heat transfer and flame propagation.

For greatest effectiveness, a flame retardant should be made available and exhibit its greatest activity at the precise conditions of time and space at which the burning process is most vulnerable to its particular mechanism of extinguishment. A flame retardant which is made available and active at optimum conditions will prove significantly more effective than a similar, supposedly flame retardant, compound which becomes available and active at a less propitious time or in a less desirable place.

Phosphorus content alone is not an adequate measure of the effectiveness of flame retardants. The occasionally reported cases where a phosphorus compound appears less effective than another despite its higher phosphorus level are manifestations of this phenomenon.

A particular combination of fire conditions and material properties requires a specific amount of flame retardant action to provide a specific level of flame retardance. Once the necessary amount of flame retardant action is provided, any increase becomes superfluous and in some cases may be undesirable.

For this reason, increasing phosphorus content provides an increasing degree of flame retardance up to a certain optimum level, beyond which further increases in phosphorus content provide no further benefit. For polyethylene and polypropylene containing no other flame retardants, self-extinguishing behavior requires about 15 per cent phosphorus. For vinyl polymers containing no other flame retardants, such as polyvinyl alcohol, self-extinguishing behavior requires about 15 per cent phosphorus. For cellulose containing no other flame retardants, in the form of paper or cotton, self-extinguishing behavior requires 3 to 4 per cent phosphorus. For polyesters containing no other flame retardants, self-extinguishing behavior requires 2 to 8 per cent phosphorus. For polyurethanes containing no other flame retardants, self-extinguishing behavior requires 1.0 to 1.5 per cent phosphorus.

The physical structure of the plastic material can affect the phosphorus requirements for self-extinguishing behavior. Among the polyurethanes, the higher phosphorus requirements of flexible foams for self-extinguishing behavior (6 to 8 per cent) can be attributed to the large surface area exposed by the interconnected cell structure.

It is noteworthy that the polymers which tend to burn without formation of char (polyolefins and vinyls) require considerably more phosphorus for self-extinguishing behavior (about 15 per cent) than the polymers which tend to form char on burning (cellulose, polyesters, polyurethanes). Apparently that much additional phosphorus is required to provide sufficient phosphorus pentoxide (the most probable oxide form) for the dehydrating action required to promote char and the fluxing action required to stabilize it.

Many phosphorus esters are susceptible to hydrolysis. For this reason, phosphorus is used in pendant groups rather than in the main polymer chain in polyesters, and its use in polyurethanes is limited by humid aging behavior.

Arsenic is known to have some flame retardant action, but its toxicity has essentially eliminated it from consideration for general use.

Antimony appears to be ineffective when used by itself. Antimony trioxide melts at 1550°C., a temperature too high to render it an effective fluxing agent in the manner of phosphorus, and it does not have the dehydrating action of phosphorus pentoxide. It is principally used as a synergist with the halogens.

Bismuth is known to have some flame retardant action, but it is less active and less well known than antimony.

Section 5.2.4. Group IV B Elements

The Group IV B elements include carbon, silicon, germanium, tin, and lead.

The elements in Group IV B cross the boundary between non-metals and low-melting heavy metals. Carbon and silicon are non-metals, and tin and lead are low-melting heavy metals.

Carbon would appear unlikely to be a flame retardant because it is the principal fuel in the burning of plastics. However, the presence of carbonaceous char contributes to fire resistance in many cases. The chemical effect is reduction in heat of combustion. The physical effect is formation of a physical barrier of carbonaceous residue. In such instances, carbon functions in the condensed phase, and appears to be primarily effective in Stage II (Decomposition) and Stage III (Ignition).

One form of carbon used as a flame retardant is expandable graphite (EG), which exhibits expansion when it is rapidly heated above 300°C.

Another form of carbon used to increase fire resistance is carbonaceous fiber. Polyacrylonitrile (PAN) fibers oxidized at temperatures greater than 600°C. have yielded carbonaceous fibers with a carbon content greater than 65 per cent, useful in flame retardant and fire barrier structures.

Silicon can be considered to be a flame retardant in that its presence in the basic polymer contributes to fire resistance in many cases. Many silicon-containing polymers are resistant to high temperatures. In such instances, silicon functions in the condensed phase, and appears to be primarily effective in Stage II (Decomposition) and Stage III (Ignition).

Silicon can be used as a flame retardant in the form of a sodium aluminosilicate, montmorillonite clay, which can be combined with polymers to form nanocomposites.

Nylon-6 clay nanocomposites containing a clay mass fraction of only 5 per cent show a substantial reduction in peak heat release rate.

Tin is believed to perform most of its flame retardant function in the condensed phase. It may be more effective as a smoke retardant.

Lead is believed to perform most of its flame retardant function in the condensed phase. It may be more effective as a smoke retardant.

Section 5.2.5. Group III B Elements

The Group III B elements include boron and aluminum.

Boron is believed to perform most of its flame retardant function in the condensed phase. Boron-containing compounds increase the amount of char formed, by one or both of two mechanisms: redirection of the chemical reactions involved in decomposition in favor of reactions yielding carbon rather than carbon monoxide or carbon dioxide; and formation of a surface layer of protective char which inhibits access of oxygen and escape of oxides of carbon by physical blockade, thus preventing gasification of the carbon by simply denying the oxygen required.

The chemical influence of boron may involve removal of vulnerable hydroxyl groups by dehydration, causing increased formation of char. The physical influence of boron may involve the formation of nonvolatile boric oxide which acts as a flux for the carbonaceous residue.

The action of boron-containing compounds is desirable because:

1. It is effective in Stage II (Decomposition) rather than in the later Stages III (Ignition) and IV (Combustion), increasing the probability of arresting the burning process at an earlier stage.
2. It is effective in the condensed phase, which is the fraction of the material which may have some residual value, rather than in the gaseous phase, which contains the fraction of the material which has no further practical value.
3. It interferes with the oxidation reactions which are responsible for the one major source of heat in the burning of plastic materials: carbon oxidation.
4. It interferes with gasification, which is largely responsible for increased heat transfer and flame propagation.

Hydrated boron compounds, such as boric acid, appear to be generally more effective than the anhydrous compounds at equal boron concentration. This is believed due to the removal of heat required to effect dehydration.

Aluminum is believed to perform most of its flame retardant function in the condensed phase.

The action of aluminum-containing compounds is desirable because:

1. It is effective in Stage II (Decomposition) rather than in the later Stages III (Ignition) and IV (Combustion), increasing the probability of arresting the burning process at an earlier stage.
2. It is effective in the condensed phase, which is the fraction of the material which may have some residual value, rather than in the gaseous phase, which contains the fraction of the material which has no further practical value.
3. It interferes with the oxidation reactions which are responsible for the one major source of heat in the burning of plastic materials: carbon oxidation.
4. It interferes with gasification, which is largely responsible for increased heat transfer and flame propagation.

Hydrated aluminum compounds, such as hydrated aluminum oxide, appear to be generally more effective than the anhydrous compounds at equal aluminum concentration. This is believed due to the removal of heat required to effect dehydration.

Section 5.2.6. Group II B Elements

The Group II B elements include zinc, cadmium, and mercury.

Zinc is believed to perform most of its flame retardant function in the condensed phase. It may be more effective as a smoke retardant. It is used with boron in the form of zinc borate, and with molybdenum in the form of zinc molybdate.

Cadmium has not been widely considered as a flame retardant.

Mercury has not been considered for use as a flame retardant because of its known toxicity.

Section 5.2.7. Group I B Elements

The Group I B elements include copper, silver, and gold.

Copper is believed to perform any flame retardant function in the condensed phase. It may be more effective as a smoke retardant.

Silver has not been considered for use as a flame retardant because of its high cost.

Gold has not been considered for use as a flame retardant because of its even higher cost.

Section 5.2.8. Group VIII A Elements

The Group VIII A elements include iron, cobalt, nickel, and platinum.

Iron is believed to perform most of its flame retardant function in the condensed phase. It may be more effective as a smoke retardant. Iron compounds may act both as cross-linking agents to promote the formation of stable char, and as oxidation catalysts to convert polymer carbon into carbon monoxide and carbon dioxide.

Cobalt is believed to perform any flame retardant function in the condensed phase. It may be more effective as a smoke retardant.

Nickel is believed to perform any flame retardant function in the condensed phase. It may be more effective as a smoke retardant.

Platinum has not been considered for use as a flame retardant because of its high cost.

Section 5.2.9. Group VII A Elements

The Group VII A elements include manganese.

Manganese is believed to perform any flame retardant function in the condensed phase. It may be more effective as a smoke retardant.

Section 5.2.10. Group VI A Elements

The Group VI A elements include chromium, molybdenum, and tungsten.

Chromium is believed to perform any flame retardant function in the condensed phase. It may be more effective as a smoke retardant.

Molybdenum is believed to perform most of its flame retardant function in the condensed phase. It may be more effective as a smoke retardant. Over 90% of the molybdenum remains in the char. Molybdenum compounds may act primarily as cross-linking agents to promote the formation of stable char.

Tungsten has not been widely considered as a flame retardant.

Section 5.2.11. Group V A Elements

The Group V A elements include vanadium.

Vanadium is believed to perform any flame retardant function in the condensed phase. It may be more effective as a smoke retardant.

Section 5.2.12. Group II A Elements

The Group II A elements include magnesium, calcium, strontium, and barium.

Magnesium is believed to perform any flame retardant function in the condensed phase. It may be more effective as a smoke retardant.

Calcium is believed to perform any flame retardant function in the condensed phase. It may be more effective as a smoke retardant. Calcium carbonate would perform as a flame retardant in the condensed phase.

Strontium has not been widely considered as a flame retardant.

Barium is believed to perform any flame retardant function in the condensed phase. It may be more effective as a smoke retardant. Barium carbonate would perform as a flame retardant in the condensed phase.

Section 5.2.13. Synergism

Synergism is the term applied to the observed ability of combinations of compounds containing certain elements to provide a greater degree of flame retardance than equal amounts of each used separately.

Phosphorus-bromine synergism is the term applied to the observed ability of combinations of phosphorus and bromine containing compounds to provide a greater degree of flame retardance than equal amounts of each used separately. This phenomenon is utilized to reduce the amounts of any of the two required. In polyolefins, for example, the use of phosphorus alone is not feasible because of the high levels required, and phosphorus/bromine combinations are effective.

Comparisons of relative effectiveness of phosphorus and bromine indicate that 1 per cent of phosphorus is equivalent to about 6 per cent of bromine in polyolefins and polystyrene, about 4 to 5 per cent of bromine in polyesters, and about 5 to 7 per cent of bromine in polyesters.

The synergism can be explained by the mechanisms employed by phosphorus and bromine compounds. Without phosphorus, hydrogen bromide is believed to be the most active bromine compound. Since this compound boils at -67°C., only its heavy molecular weight (80.92) is responsible for its relatively long residence time in the combustion zone, where it participates in successive halogenation and dehydrohalogenation. The presence of phosphorus promotes the formation of char, which further restricts movements in the gaseous phase, and results in the formation of phosphorus tribromide, phosphorus pentabromide, and phosphorus oxybromide, which are less readily gasified and are heavier gases (molecular weights of 270.70, 430.52, and 286.70, respectively). Since these bromides are continually regenerated on the halogenation-dehydrohalogenation process, their effectiveness increases with their residence time in the combustion region, and keeping them in effective contact twice as long would tend to make them twice as effective.

The fluxing action of phosphorus is probably limited to the interface of the combustion and pyrolysis regions, so that any excess over the amount needed to stabilize the char is ineffective. This excess is available for the formation of bromides without any loss of fluxing action. Since phosphorus oxides can effect only so much dehydration while the continually regenerated bromides can repeat the dehydrohalogenated process, phosphorus/bromine combinations would be more effective that phosphorus oxides in promoting char formation.

Phosphorus-chlorine synergism is the term applied to the observed ability of combinations of phosphorus and chlorine containing compounds to provide a greater degree of flame retardance than equal amounts of each used separately.

Phosphorus-chlorine synergism is similar to phosphorus-bromine synergism. It is more extensively utilized because of the greater variety of available chlorine-containing compounds, but it appears to be less effective, to the same degree and for the same reasons that chlorine is not as effective on an equal weight basis as bromine. Phosphorus chlorides are lower-boiling and lighter gases than the corresponding phosphorus bromides, and can be expected to have a shorter residence time in the combustion zone.

Phosphorus-nitrogen synergism is the term applied to the observed ability of combinations of phosphorus and nitrogen containing compounds to provide a greater degree of flame retardance than equal amounts of each used separately. One phenomenon attributed to phosphorus-nitrogen synergism is intumescence.

Phosphorus-nitrogen synergism has been found effective with cellulose, and may be a factor in polyesters and polyurethanes. Acrylonitrile and triethyl phosphate exhibited some synergism in polyesters, and amino-phosphate polyols are employed in polyurethanes. Phosphorus-nitrogen combinations, however, have been found ineffective in polyolefins.

Antimony-halogen synergism is the term applied to the observed ability of combinations of antimony and halogen containing compounds to provide a greater degree of flame retardance than equal amounts of each used separately.

Antimony is principally used as a synergist with the halogens, particularly bromine and chlorine. Where polyesters containing no other flame retardant require about 17 per cent bromine for self-extinguishing behavior, similar results can be obtained with 13 per cent bromine and 1 per cent antimony trioxide, or 9 per cent bromine and 2 per cent antimony trioxide. The increased effectiveness is probably due to the longer residence time of bromide as antimony tribromide, relative to bromide as hydrogen bromide. Greater effectiveness of bromine compared to chlorine is observed here. For polyesters containing 5 per cent antimony trioxide, 6 per cent bromine produces the same level of flame retardance as 24 to 25 per cent chlorine.

Carbon-phosphorus synergism is the term applied to the observed ability of combinations of carbon and phosphorus containing compounds to provide a greater degree of flame retardance than equal amounts of each used separately.

Carbon-phosphorus synergism has been found effective with thermoplastics, when expandable graphite (EG) is used with red phosphorus (RP) or ammonium polyphosphate (APP).

SECTION 5.3. FLAME RETARDING VARIOUS POLYMERS

The flame retarding of polymers is both science and art, because flame retardance must be achieved with the minimum adverse impact on the properties which make the polymer desirable in the first place. Many flame retardant materials and compositions are proprietary, and many may not be mentioned in this section.

Polyethylene (PE) is combustible and requires modification for flame retardance. The approaches which have been used include:

1. Halogenation of the polymer, as in the case of chlorinated polyethylene
2. Addition of halogen-containing compounds, such as chlorinated paraffins. Aliphatic chlorine compounds appear to be more effective than aromatic chlorine compounds
3. Use of antimony oxide to increase effectiveness of halogen
4. Use of phosphorus to increase effectiveness of halogen. Phosphorus used alone requires high concentrations. Phosphorus/bromine appears to be more effective than phosphorus/chlorine.
5. Use of zinc borate

Polypropylene (PP) is combustible and requires modification for flame retardance. The approaches which have been used include:

1. Addition of halogen-containing compounds, such as chlorinated paraffins. Aliphatic chlorine compounds appear to be more effective than aromatic chlorine compounds
2. Use of antimony oxide to increase effectiveness of halogen
3. Use of phosphorus to increase effectiveness of halogen. Phosphorus used alone requires high concentrations. Phosphorus/bromine appears to be more effective than phosphorus/chlorine.

Ethylene-vinyl acetate (EVA) is combustible and requires modification for flame retardance. The approaches which have been used include:

1. Use of alumina trihydrate
2. Use of zinc borate
3. Use of magnesium carbonate

Polyvinyl chloride (PVC) as a pure polymer is inherently flame retardant because of its chlorine content of 56 per cent. Many applications, however, require the use of plasticizers to impart the desired degree of flexibility. Many plasticizers, such as phthalates, increase flammability by diluting the high-chlorine polymer with combustible material. The approaches which have been used include:

1. Use of phosphate plasticizers as primary plasticizers
2. Use of secondary plasticizers such as halogenated paraffins, biphenyls, and fatty acid esters
3. Use of antimony oxide in plasticized formulations
4. Use of zinc borate
5. Use of tin oxide
6. Use of zinc molybdate and calcium zinc molybdate
7. Use of magnesium hydroxide

Polyvinyl acetate, polyvinyl alcohol, and polyvinyl butyral are combustible and require modification for flame retardance. The approaches which have been used include:

1. Use of phosphate plasticizers as primary plasticizers
2. Use of secondary plasticizers such as halogenated paraffins
3. Copolymerization with halogen-containing monomers such as vinyl chloride
4. Copolymerization with phosphorus-containing monomers

Polystyrene (PS) is combustible and requires modification for flame retardance. The approaches which have been used include:

1. Addition of halogenated compounds. Bromine appears to be more effective than chlorine. Aliphatic and alicyclic bromine compounds appear to be more effective than aromatic bromine compounds.
2. Chemical bonding of halogenated compounds into the polystyrene chain
3. Improvement of efficiency of halogens by using small amounts of free radical initiators such as peroxides
4. Use of magnesium hydroxide

Acrylonitrile-butadiene-styrene (ABS) is combustible and requires modification for flame retardance. The approaches which have been used include:

1. Addition of halogenated compounds
2. Terpolymerization with halogen-containing monomers, such as bis(2,3-dibromopropyl)fumarate

Polymethyl methacrylate (PMMA) is combustible and requires modification for flame retardance. The approaches which have been used include:

1. Addition of compounds containing phosphorus and halogen
2. Combination with polyvinyl chloride (PVC) to form an alloy

Cellulose acetate, cellulose butyrate, and cellulose propionate are combustible and require modification for flame retardance. The approaches which have been used include:

1. Use of phosphate and halogenated phosphate plasticizers
2. Addition of phosphorus and halogen containing compounds

Nylons which are combustible may require modification for flame retardance. The approaches which have been used include the use of zinc borate.

Polyesters are combustible and require modification for flame retardance. The approaches which have been used include:

1. Use of relatively large amounts of inorganic fillers
2. Use of flame retardant plasticizers such as halogenated paraffins and phosphates
3. Substitution of intermediates that contain halogen, such as tetrachlorophthalic anhydride and tetrabromophthalic anhydride
4. Substitution of intermediates that are both halogen-containing and more temperature-resistant, such as chlorendic anhydride
5. Use of zinc borate

Polyurethanes (PUR) are combustible and require modification for flame retardance. The approaches which have been used include:

1. Use of phosphate plasticizers, where flexibility is desired
2. Use of polymeric diphenylmethane type isocyanates instead of toluene diisocyanate
3. Introduction of phosphorus-containing intermediates, such as phosphate, phosphite, phosphonate, and amino-phosphonate polyols
4. Introduction of halogen-containing intermediates, such as brominated isocyanates and chlorine-containing polymer/polyols
5. Use of chlorine and fluorine-containing compounds as blowing agents in foams. However, chlorofluorocarbons (CFC) are no longer widely used because of environmental concerns.
6. Use of relatively large amounts of inorganic fillers
7. Use of melamine

Epoxy resins are combustible and require modification for flame retardance. The approaches which have been used include:

1. Introduction of halogen into the bisphenol component, in the form of tetrachlorobisphenols and tetrabromobisphenols
2. Use of halogenated anhydrides as hardeners, such as chlorendic anhydride
3. Use of relatively large amounts of inorganic fillers
4. Use of zinc borate

SECTION 5.4. SMOKE RETARDANCE MECHANISMS

Smoke retardance mechanisms have not been as extensively studied as flame retardance mechanisms have been.

Certain compounds are believed to perform their smoke retardant function in the solid or condensed phase, by promoting char formation, diluting the polymer content, dissipating heat, and altering the chemical reactions in the condensed phase. This type of mechanism may be identified by an increase in the char residue.

Other compounds are believed to perform their smoke retardant function in the gas phase, by promoting oxidation of carbon or soot, and altering the chemical reactions in the gas phase.

Theoretically, a flame retardant which is effective in the condensed phase would be expected to perform a smoke retardant function, because increased retention of material in the solid phase or char would result in a reduction in combustible compounds and a reduction in compounds in the gaseous phases which would appear as smoke. However, a flame retardant may succeed only in preventing oxidation which produces heat and flame, and unoxidized material may escape into the gas phase as soot and tar, forming smoke.

A flame retardant which acts primarily in the gaseous phase would not prevent decomposition of the material, and by inhibiting ignition or combustion may ensure that the decomposition products are more likely to be visible as smoke.

Smoke retardant action is attributed to a variety of compounds which contain elements found in certain groups of elements in the Period Table of the Elements:

Group II A, which includes magnesium, calcium, strontium, and barium
Group V A, which includes vanadium
Group VI A, which includes chromium, molybdenum, and tungsten
Group VII A, which includes manganese
Group VIII A, which includes iron, cobalt, nickel, and platinum
Group I B, which includes copper, silver, and gold
Group II B, which includes zinc, cadmium, and mercury
Group III B, which includes boron and aluminum
Group IV B, which includes carbon, silicon, tin, and lead
Group V B, which includes nitrogen, phosphorus, antimony, and bismuth

Section 5.4.1. Group II A Elements

The Group II A elements include magnesium, calcium, strontium, and barium.

Magnesium is believed to perform most of its smoke retardant function in the condensed phase.

Calcium is believed to perform most of its smoke retardant function in the condensed phase.

Strontium has not been widely considered as a smoke retardant.

Barium is believed to perform most of its smoke retardant function in the condensed phase.

Section 5.4.2. Group V A Elements

The Group V A elements include vanadium.

Vanadium is believed to perform most of its smoke retardant function in the condensed phase. The vanadium compounds which have been used include the following:

vanadium oxide
vanadium (III) acetylacetonate
phthalocyanine complexes of vanadium

Section 5.4.3. Group VI A Elements

The Group VI A elements include chromium, molybdenum, and tungsten.

Chromium is believed to perform most of its smoke retardant function in the condensed phase. The chromium compounds which have been used include the following: hydroxyquinoline complexes of chromium.

Molybdenum is believed to perform most of its smoke retardant function in the condensed phase. Over 90% of the molybdenum remains in the char, and volatile molybdenum oxychloride has not been found in degradation and combustion studies. Molybdenum compounds may act primarily as cross-linking agents to promote the formation of stable char.

The molybdenum compounds which have been used include the following:

molybdenum trioxide
ammonium octamolybdate
melaminium beta-octamolybdate
calcium molybdate
zinc molybdate

Tungsten has not been widely considered as a smoke retardant.

Section 5.4.4. Group VII A Elements

The Group VII A elements include manganese.

Manganese is believed to perform most of its smoke retardant function in the condensed phase. The manganese compounds which have been used include the following:

hydroxyquinoline complexes of manganese
phthalocyanine complexes of manganese

Section 5.4.5. Group VIII A Elements

The Group VIII A elements include iron, cobalt, nickel, and platinum.

Iron is believed to perform most of its smoke retardant function in the condensed phase. Iron compounds may act both as cross-linking agents to promote the formation of stable char, and as oxidation catalysts to convert polymer carbon into carbon monoxide and carbon dioxide.

The iron compounds which have been used include the following:

dicyclopentadienyl iron (ferrocene)
ferrocene-1,1-dicarboxylic acid
ferrocene-1,1-dicarboxylic acid zinc salt
ferric oxide
ferric hydroxide
ferric orthophosphate
ferric potassium oxalate
ferrous oxide
ferrous oxalate
iron metal
hydroxyquinoline complexes of iron
phthalocyanine complexes of iron

Cobalt is believed to perform most of its smoke retardant function in the condensed phase. The cobalt compounds which have been used include the following: phthalocyanine complexes of cobalt.

Nickel is believed to perform most of its smoke retardant function in the condensed phase. The nickel compounds which have been used include the following: dicyclopentadienyl nickel (nickelocene).

Platinum has not been considered for use as a smoke retardant because of its high cost.

Section 5.4.6. Group I B Elements

The Group I B elements include copper, silver, and gold.

Copper is believed to perform most of its smoke retardant function in the condensed phase. The copper compounds which have been used include the following:

cuprous cyanide
cuprous thiocyanate
phthalocyanine complexes of copper

Silver has not been considered for use as a smoke retardant because of its high cost.

Gold has not been considered for use as a smoke retardant because of its even higher cost.

Section 5.4.7. Group II B Elements

The Group II B elements include zinc, cadmium, and mercury.

Zinc is believed to perform most of its smoke retardant function in the condensed phase. The zinc compounds which have been used include the following:

zinc borate
zinc molybdate

Cadmium has not been widely considered as a smoke retardant.

Mercury has not been considered for use as a smoke retardant because of its known toxicity.

Section 5.4.8. Group III B Elements

The Group III B elements include boron and aluminum.

Boron is believed to perform most of its smoke retardant function in the condensed phase. Boron-containing compounds increase the amount of char formed, by one or both of two mechanisms: redirection of the chemical reactions involved in decomposition in favor of reactions yielding carbon rather than carbon monoxide or carbon dioxide; and formation of a surface layer of protective char which inhibits access of oxygen and escape of oxides of carbon by physical blockade, thus preventing gasification of the carbon by simply denying the oxygen required.

The chemical influence of boron may involve removal of vulnerable hydroxyl groups by dehydration, causing increased formation of char. The physical influence of boron may involve the formation of nonvolatile boric oxide which acts as a flux for the carbonaceous residue.

The boron compounds which have been used include the following: zinc borate.

Aluminum is believed to perform most of its smoke retardant function in the condensed phase. The aluminum compounds which have been used include the following:

aluminum acetylacetonate
alumina trihydrate

Section 5.4.9. Group IV B Elements

The Group IV B elements include carbon, silicon, germanium, tin, and lead.

The elements in Group IV B cross the boundary between non-metals and low-melting heavy metals. Carbon and silicon are non-metals, and tin and lead are low-melting heavy metals.

Carbon would appear unlikely to be a smoke retardant because it is the principal fuel in the burning of plastics. However, an increase in the percentage of carbon remaining as carbonaceous char results in a decrease in the percentage of carbon appearing as smoke.

Silicon can be considered to be a smoke retardant in that its presence in the basic polymer contributes to fire resistance in many cases. Many silicon-containing polymers are resistant to high temperatures. In such instances, silicon functions in the condensed phase.

Tin is believed to perform most of its smoke retardant function in the condensed phase. The tin compounds which have been used include the following:

tin oxide
sodium stannate
calcium stannate
zinc stannate

Lead is believed to perform most of its smoke retardant function in the condensed phase. The lead compounds which have been used include the following: tetraphenyl lead.

Section 5.4.10. Group V B Elements

The Group V B elements include nitrogen, phosphorus, arsenic, antimony, and bismuth.

Nitrogen can be considered to be a smoke retardant in that its presence in the basic polymer results in reduced smoke in many cases. Many nitrogen-containing polymers are resistant to high temperatures. In such instances, nitrogen functions in the condensed phase.

Phosphorus is believed to perform most of its smoke retardant function in the condensed phase. A variety of phosphorus compounds have been used.

Arsenic may have some smoke retardant action, but its toxicity has essentially eliminated it from consideration for general use.

Antimony is believed to perform most of its smoke retardant function in the condensed phase. The antimony compounds which have been used include the following: antimony trioxide.

Bismuth may have some flame retardant action, but it is less active and less well known than antimony.

SECTION 5.5. SMOKE RETARDING VARIOUS POLYMERS

The smoke retarding of polymers, perhaps more than the flame retarding of polymers, is an art, because both flame retardance and smoke retardance must be achieved with the minimum adverse impact on the properties which make the polymer desirable in the first place. Many smoke retardant materials and compositions are proprietary, and many may not be mentioned in this section.

Polyethylene (PE) and polypropylene (PP) tend to burn with relatively low levels of smoke, but some flame retardant formulations can produce higher levels of smoke. The approaches which have been used include:

1. Use of aluminum compounds such as alumina trihydrate and aluminum acetylacetonate
2. Use of iron compounds such as dicyclopentadienyl iron (ferrocene)
3. Use of nickel compounds such as dicyclopentadienyl nickel (nickelocene)
4. Use of calcium carbonate
5. Use of zinc borate

Polyvinyl chloride (PVC) tends to produce relatively high levels of smoke over certain ranges of formulations and conditions, and may require modification for smoke retardance. The approaches generally use additives to dilute the polymer content, increase heat dissipation, and promote retention of material in the char. The approaches which have been used include:

1. Use of aluminum compounds such as:
         alumina
         alumina trihydrate
   Hydrated oxides such as alumina trihydrate dissipate more heat than anhydrous oxides.
2. Use of iron compounds such as:
         dicyclopentadienyl iron (ferrocene)
         ferrocene-1,1-dicarboxylic acid

ferrocene-1,1-dicarboxylic acid zinc salt
ferric oxide
ferric hydroxide
ferric orthophosphate
ferric potassium oxalate
ferrous oxide
ferrous oxalate
iron metal
3. Use of copper compounds such as:
cuprous cyanide
cuprous thiocyanate
4. Use of vanadium compounds such as:
vanadium oxide
vanadium (III) acetylacetonate
5. Use of molybdenum compounds such as:
molybdenum trioxide
melaminium beta-octamolybdate
calcium molybdate
zinc molybdate
calcium zinc molybdate
6. Use of tin compounds such as tin oxide
7. Use of zinc compounds such as zinc borate
8. Use of magnesium compounds such as magnesium hydroxide

Polystyrene (PS) tends to produce relatively high levels of smoke and may require modification for smoke retardance. The approaches which have been used include:

1. Use of inorganic additives such as silica, alumina trihydrate, magnesium hydroxide, and hydrated clay
2. Use of hydroxyquinoline complexes of iron, manganese, and chromium
3. Use of phthalocyanine complexes of iron, copper, manganese, cobalt, and vanadium

Acrylonitrile-butadiene-styrene (ABS) tends to produce relatively high levels of smoke and may require modification for smoke retardance. The approaches which have been used include:

1. Use of inorganic additives such as silica, alumina trihydrate, magnesium hydroxide, and hydrated clay
2. Use of hydroxyquinoline complexes of iron, manganese, and chromium
3. Use of phthalocyanine complexes of iron, copper, manganese, cobalt, and vanadium
4. Use of lead compounds such as tetraphenyl lead
5. Use of molybdenum compounds such as molybdenum trioxide

Polychloroprene and chlorosulfonated polyethylene may produce relatively high levels of smoke and may require modification for smoke retardance. The approaches which have been used include the use of molybdenum compounds such as molybdenum trioxide.

Polyesters may produce relatively high levels of smoke over certain ranges of formulations and may require modification for smoke retardance. The approaches which have been used include:

1. Use of molybdenum compounds such as molybdenum trioxide and ammonium octamolybdate
2. Use of tin compounds such as tin oxide, sodium stannate, calcium stannate, and zinc stannate

Epoxy resins may produce relatively high levels of smoke and may require modification for smoke retardance. The approaches which have been used include the use of zinc borate.

SECTION 5.6. SMOLDER RETARDANCE MECHANISMS

For continued smoldering, the oxygen supply rate to the region of smoldering appears to be more important than the chemical composition of the fuel. The most effective smolder retardance mechanism appears to be preventing or limiting access of oxygen to the smoldering material. A smolder retardant would be expected to perform its function in the solid phase, by forming a coating or acting as a flux to seal the solid from access to oxygen.

Compounds which act as cross-linking agents to promote the formation of a stable char are desirable, but the char should not have a level of porosity which would make it vulnerable to smoldering combustion. Access to oxygen can be preventing by fluxing action or formation of a coating.

Boron compounds as a class appear to show promise as smolder retardants. Boron and boric oxide are known to be good inhibitors for the oxidation of carbonaceous materials. Boric acid and borax have been found to hinder smoldering in cellulosic insulation.

SECTION 5.7. REDUCING TOXIC GAS EVOLUTION

The most effective and most straightforward method of reducing toxic gas evolution would be to prevent the evolution of any gases, by terminating the burning process in the heating or decomposition stage. A flame retardant which performs its function in the condensed phase during the heating or decomposition stage could also reduce toxic gas evolution.

Carbon monoxide is produced when burning most organic materials including plastics, and has been found to be the primary toxicant in many fire fatalities due to inhalation. Reducing the amount of carbon monoxide evolved would be a significant factor in reducing toxic gas evolution.

To minimize the amount of carbon which would be evolved as carbon monoxide, the carbon in the plastic should be retained in the condensed phase, either in the polymer or in carbonaceous char residue, or oxidized to carbon dioxide. The first approach is preferable, because retaining carbon in the polymer or in carbonaceous char residue would reduce the amount of heat produced by oxidation. The second approach, oxidation to carbon dioxide, would result in greater heat release, because heat is produced when carbon or carbon monoxide is oxidized to carbon dioxide.

Flame retardants which are effective in the condensed phase would be expected to reduce toxic gas evolution.

Nitrogen-containing plastics may produce hydrogen cyanide or nitrogen oxides when burning. Sulfur-containing plastics may produce hydrogen sulfide or sulfur oxides when burning. Chlorine-containing plastics may produce hydrogen chloride when burning. These volatile products of combustion may identify specific plastics as contributors to toxic effects.

Because most plastics produce carbon monoxide when burning, carbon monoxide can not be used to identify a specific plastic as a contributor to toxic effects.

SECTION 5.8. REDUCING CORROSIVE GAS EVOLUTION

The most effective and most straightforward method of reducing corrosive gas evolution would be to prevent the evolution of any gases, by terminating the burning process in the heating or decomposition stage. A flame retardant which performs its function in the condensed phase during the heating or decomposition stage could also reduce corrosive gas evolution.

Halogen-containing gases are considered to be among the most corrosive gases evolved from burning plastics. On this basis, non-halogen or low-halogen materials would appear to be the preferred materials for applications where corrosive gas evolution is a major concern. Non-halogen compounds such as zinc borate can be used as flame retardants and smoke retardants in halogen-free systems.

However, halogen-containing plastics offer advantages in performance and price for many applications. Polyvinyl chloride (PVC) may be the most versatile of all plastics because it can be blended with plasticizers and other materials to give a wide range of properties. Dehydrochlorination of PVC releases hydrogen chloride which is a corrosive gas. Compounds of certain heavy metals such as molybdenum, iron, tin, and antimony may act as sequestering agents, sometimes called scavengers, for hydrogen chloride and retain the chloride in the condensed phase. Such compounds should be carefully selected for each particular formulation to avoid undesirable effects. For example, zinc oxide may cause blackening of rigid PVC and release hydrogen chloride in the phenomenon called "zinc failure". Some heavy metal compounds are already used as flame retardants and smoke retardants for PVC.

SECTION 5.9. DESIGN OF THE APPLICATION

The use of plastic materials, to be safe from a flammability standpoint, must take into consideration their behavior on exposure to fire and the probability and nature of possible fire exposure. The following general principles should be followed.

1. All highly flammable or easily ignited material should be removed from a potential source of heat or flame, and materials employed near this possible fire source should be tested to ensure that they have the necessary degree of flame retardance. The main example of this type of material, cellulose nitrate, should be replaced with other materials or adequately formulated with flame retardants.

2. Where possible, a material which is relatively easy to ignite should be protected with a coating or covering which is more difficult to ignite and which prevents access of oxygen to the easily ignited material.

3. Where possible, a material which exhibits relatively high surface flammability should be protected with a coating or covering of low surface flammability. Because the thermal insulating qualities of many plastic foams prevent dissipation of heat and thus increase surface flame spread, plastic foams in general should not be applied without a protective coating or covering.

4. A material which exhibits relatively high surface flammability should not be applied continuously for great distances, and should not be applied in areas of high surface flame spread probability, such as ceilings. Regular firestops, either physical barriers or intervals of nonflammable surface, should be included in applications covering great distances.

5. A material which exhibits relatively high smoke production should not be exposed in large quantities or over large surface areas.

6. A material which exhibits relatively high toxic gas production should not be exposed in large quantities or over large surface areas.

Ten rules presented for fire endurance rating by Harmathy are cited here because of their relevance to plastic materials used in structures.

1. The thermal fire endurance of a construction consisting of a number of parallel layers is greater than the sum of the thermal fire endurances characteristic of the individual layers when exposed separately to fire. A sandwich panel consisting or rigid polyurethane foam between polyvinyl chloride sheet facings, for example, provides more fire endurance than the individual components tested separately. The polyvinyl chloride

sheet prevents the foam surface from being exposed to direct flame and limits access of oxygen, and the polyurethane foam provides support for the char from the polyvinyl chloride.

2. The fire endurance of a construction does not decrease with the addition of further layers. This rule is subject to various restrictions and should be carefully applied.

3. The fire endurance of constructions containing continuous air gaps or cavities is greater than the fire endurance of similar constructions of the same weight, but containing no air gaps or cavities. A polymer in cellular form exhibits greater fire endurance than the same weight of polymer in sheet form.

4. The farther an air gap or cavity is located from the exposed surface, the more beneficial is its effect on the fire endurance. Cellular plastics are more beneficial to fire endurance if they are not exposed to high heat, so that their thermal insulating qualities can be utilized without exposure to destructive temperatures.

5. The fire endurance of a construction cannot be increased by increasing the thickness of a completely enclosed air layer. This rule applies to gaps of 1/2 inch or greater.

6. Layers of materials of low thermal conductivity are better utilized on that side of the construction on which fire is more likely to happen. This rule is not applicable to materials such as plastics which undergo physical and chemical changes and evolve heat on burning.

7. The fire endurance of assymmetrical constructions depends on the direction of heat flow. This rule is related to rules 4 and 6.

8. The presence of moisture, if it does not result in explosive spalling, increases the fire endurance. Water vapor has a beneficial effect in providing a noncombustible gas as well as requiring heat for vaporization. Oven-dried plastic materials tend to be more combustible than the same materials with equilibrium moisture. An extension of this rule is the incorporation of hydrated fillers.

9. Load-supporting elements, such as beams, girders, and joists, yielded higher fire endurances when subjected to fire endurance tests as parts of floor, roof, or ceiling assemblies than they would when tested separately.

10. The load-supporting elements of a floor, roof, or ceiling assembly can be replaced by such other load-supporting elements which, when tested separately, yielded fire endurances not less than that of the assembly.

SECTION 5.10. FIRE CONTROL AND EXTINGUISHMENT

In order to put out a fire it is necessary to stop the combustion reaction causing the fire. Fires are generally controlled and extinguished by one or more of the following methods:

1. Cooling
2. Separation or replacement of oxidizing agent
3. Dilution or removal of fuel supply
4. Chemical extinguishment

Extinguishment by cooling is the most widely used means of extinguishment, and the most effective in the case of ordinary combustible materials. In order to extinguish a fire by cooling, it is only necessary to absorb a portion of the total heat, because extinguishment occurs when the surface of the burning material is cooled to the point where it no longer releases enough vapors to maintain a combustible mixture. The efficiency of an extinguishing agent as a cooling medium depends on its specific and latent heats, and in this regard water is particularly effective because its specific and latent heats are higher than those of most extinguishing agents.

Extinguishment by separation of the oxidizing agent is accomplished by blanketing or smothering a fire. Carbon dioxide, foam, and vaporizing liquids form a blanket which prevents oxygen from reaching the fire. If the blanket is maintained long enough to cool the material below its self-ignition temperature and there are no ignition sources, the fire will stay out. It is, however, difficult to maintain the blanket long enough to extinguish all smoldering sources of ignition, and extinguishment by separation cannot be accomplished with materials containing their own oxygen supply, such as cellulose nitrate.

Extinguishment by removal of the fuel supply is accomplished by removing from exposure to the fire those combustible materials which have not reached their ignition temperatures.

Chemical extinguishment is based on the observed effectiveness of certain halogenated hydrocarbons and inorganic salts as extinguishing agents. This effectiveness is greater than can be accounted for by cooling, smothering, or blowing out mechanisms. The commercially used extinguishing agents are strikingly similar in chemical nature to the flame retardants used in plastics.

REFERENCES

Andrews, C. R., Tarquini, M. E., "The Effects of Fillers on the Dehydrochlorination and Smoke Properties of Rigid PVC", Proceedings of the International Conference on Fire Safety, Vol. 14, 237-242 (January 1989)

Antia, F. K., Cullis, C. F., Hirschler, M. M., "The Combined Action of Aluminum Oxides and Halogen Compounds as Flame Retardants", European Polymer Journal, Vol. 17, 451-455 (1981)

Antia, F. K., Cullis, C. F., Hirschler, M. M., "Binary Mixtures of Metal Compounds as

Flame Retardants for Organic Polymers", European Polymer Journal, Vol. 18, 95-107 (1982)

Antia, F. K., Baldry, P. J., Hirschler, M. M., "Comprehensive Study of the Effect of Composition on the Flame-Retardant Activity of Antimony Oxide and Halogenated Hydrocarbons in Thermoplastic Polymers", European Polymer Journal, Vol. 18, 167-174 (1982)

Avento, J. M., Touval, I., "Flame Retardants: Antimony and Other Inorganic Compounds", Encyclopedia of chemical Technology, Vol. 10, 355-372, John Wiley and Sons (1980)

Benrashid, R., Nelson, G. L., Ferm, D. J., "Effect of Zinc and Zinc Borate on Fire Properties of Modified Polyphenylene Oxide", Journal of Fire Sciences, Vol. 11, No. 3, 210-231 (May/June 1993)

Benrashid, R., Nelson, G. L., Ferm, D. J., Chew, L. W., "Effect of Zinc, Zinc Oxide, and Zinc Borate on the Flammability of Polycarbonate", Journal of Fire Sciences, Vol. 13, No. 3, 224-234 (May/June 1995)

Braksmayer, D., "Brominated Flame Retardant Improves Property Profiles", Modern Plastics (October 1988)

Broadbent, J. R. A., Hirschler, M. M., "Red Phosphorus as a Flame Retardant for a Thermoplastic Nitrogen-Containing Polymer", European Polymer Journa;, Vol. 11, 1087-1093 (1984)

Buszard, D. L., Bentley, R. L., "Advances in the Reduction of Smoke and Toxic Gases from Burning Polyurethane Foam", Proceedings of the International Conference on Fire Safety, Vol. 16, 54-72 (January 1991)

Camps, M., Jebri, A., Drouet, J.-C., "Comparative Study of Thermal Stabilities and Limiting Oxygen Indexes of Chlorinated, Brominated, and Iodinated Polystyrenes", Journal of Fire Sciences, Vol. 14, 251-262 (July/August 1996)

Carty, P., Docherty, A., "Iron-Containing Compounds as Flame Retarding/Smoke Suppressing Additives for PVC", Fire and Materials, Vol. 12, No. 3, 109-113 (September 1988)

Christopher, A. J., "Effects of Additives on Fire Properties of Polyethylene", Fire and Materials, Vol. 12, No. 1, 7-18 (March 1988)

Coaker, A. W., "Developing Flame Retardant PVC Compounds Based on Information Furnished by Materials' Suppliers", Proceedings of the International Conference on Fire Safety, Vol. 24, 60-85 (July 1997)

Cullis, C. F., Hirschler, M. M., "Char Formation from Polyolefins: Correlationms with Low-Temperature Oxygen Uptake and wiuth Flammability in the Presence of Metal-Halogen Systems", European Polymer Journal, Vol. 20, No. 1, 53-60 (1984)

Cullis, C. F., Hirschler, M. M., Khattab, M. A. A. M., "The Flame Retardance of a Natural Polymer by a Sulphur-Aluminum-Bromine System", European Polmer Journal, Vol. 20, No. 6, 559-562 (1984)

Cullis, C. F., Gad, A. M. M., Hirschler, M. M., "Metal Chelates as Flame Retardants and Smoke Suppressants for Thermoplastic Polymers", European Polymer Journal, Vol. 20, No. 7, 707-711 (1984)

Cullis, C. F., Hirschler, M. M., Thevaranjan, T. R., "Combinations of Titanium(IV) Oxide, Iron (III) Oxide and Molybdenum (VI) Oxide as Flame Retardants and Smoke Suppressants for Thermoplastic Polymers", European Polymer Journal, Vol. 20, No. 9, 841-847 (1984)

Cullis, C. F., "Bromine Compounds as Flame Retardants", Proceedings of the International Conference on Fire Safety, Vol. 12, 307-323 (January 1987)

Cusack, P. A., Davis, P. E., "Tin-Based Flame Retardants and Smoke Suppressants", Proceedings of the International Conference on Fire Safety, Vol. 13, 326-336 (January 1988)

Deanin, R. D., Ali, M., "Aromatic Organic Phosphate Oligomers as Flame Retardants in Plastics", in "Fire and Polymers II", Ed. G. L. Nelson, ACS Symposium Series 599, 56-64 (1995)

Donaldson, J. D., Donbavand, J., Hirschler, M. M., "Flame Retardance and Smoke Suppression by Tin(IV) Oxide Phases and Decabromobiphenyl", European Polymer Journal, Vol. 19, 33-41 (1983)

Donaldson, J. D., Donbavand, J., Hirschler, M. M., "Effect of Dispersing and Binding Agents on the Flammability of, and Smoke Production from, Thermoplastic Polymers", European Polymer Journal, Vol. 20, No. 4, 323-327 (1984)

Duffy, J. J., Ilardo, C. S., "New Reduced Smoke Flame Retardant Formulations", 33rd International Wire and Cable Symposium (November 13, 1984)

Eicher, W. J., Campbell, M. A., "Fomox: A Unique Intumescent Fire Barrier", Proceedings of the International Conference on Fire Safety, Vol. 53, 319-327 (January 1990)

Ferm, D. J., Shen, K. K., "Zinc Borate as a Smoke Suppressant in PVC", Proceedings of the International Conference on Fire Safety, Vol. 17, 242-253 (January 1992)

Ferm, D. J., Shen, K. K., "The Effect of Zinc Borate on Heat Release Properties of Several Polymeric Systems", Proceedings of the International Conference on Fire Safety, Vol. 19, 258-265 (January 1994)

Ferm, D. J., Shen, K. K., "Zinc Borate/ATH in Flexible PVC", Plastics Compounding (November/December 1994)

Gilman, J. W., Kashiwagi, T., Lichtenhan, J. D., "Nanocomposites: A Revolutionary New Flame Retardant Approach", SAMPE Journal, Vol. 33, No. 4, 40-46 (July/August 1997)

Goin, C. L., "Property Enhancement Using Metallocene Polymers in High Filled Fire Retarded Polyolefins", Proceedings of the International Conference on Fire Safety, Vol. 23, 199-207 (January 1997)

Grayson, S. J., Smith, D. A., "Effect of Calcium Carbonate Filler on the Fire and Smoke Properties of Moulded Polypropylene", Fire and Materials, Vol. 8, No. 3, 125-136 (September 1984)

Green, J., "Brominated Phosphate Ester Flame Retardants for Engineering Thermoplastics", Proceedings of the March 1989 FRCA Conference (1989)

Green, J., "Phosphorus-Bromine Flame Retardant Synergy", Proceedings of the International Conference on Fire Safety, Vol. 18, 258-273 (January 1993)

Green, J., "A Review of Phosphorus-Containing Flame Retardants", Proceedings of the March 1996 FRCA Conference, 71-87 (1996)

Harmon, W. S., Ramirez, J. E., "Layered Silicate/PBI Papers for High Performance Thermal Applications", Proceedings of the International Conference on Fire Safety, Vol. 14, 168-180 (January 1989)

Harscher, M., "Brominated Flame Retardants", Proceedings of the March 1996 FRCA Conference, 105-115 (1996)

Hilado, C. J., "Flammability Handbook for Plastics", 4th Ed., Technomic Publishing Company, Lancaster, Pennsylvania (1990)

Hirschler, M. M., "Thermal Analysis and Flammability of Polymers: Effect of Halogen-Metal Additive Systems", European Polymer Journal, Vol. 19, No. 2, 121-129 (1983)

Hirschler, M. M., Tsika, O., "The Effect of Combinations of Aluminum(III) Oxides and Decabromobiphenyl on the Flammability of and Smoke Production from Acrylonitrile-Butadiene-Styrene Terpolymer", European Polymer Journal, Vol. 19, No. 5, 375-380 (1983)

Hochberg, A., "Review of Flame Retardant Engineering Thermoplastics", Proceedings of the October 1996 FRCA Conference, 159-169 (1996)

Horn, W. E., Clever, T. R., Mineral Hydroxides: Their Manufacture and Use as Flame Retardants, Proceedings of the March 1996 FRCA Conference, 147-159 (1996)

Hshieh, F.-Y., Buch, R. R., "Controlled-Atmosphere Cone Calorimeter, Intermediate-Scale Calorimeter, and Cone Corrosimeter Studies of Silicones", Proceedings of the International Conference on Fire Safety, Vol. 23, 213-239 (January 1997)

Hshieh, F.-Y., "Shielding Effects of Silica-Ash Layer on the Combustion of Silicones and Their Possible Applications on the Fire Retardancy of Organic Polymers", Proceedings of the International Conference on Fire Safety, Vol. 24, 139-152 (July 1997)

Huggard, M., "Antimony Free Fire Retardants Based on Halogens: Phosphorus Substitution of Antimony", Proceedings of the International Conference on Fire Safety, Vol. 20, 279-288 (January 1995)

Huggard, M., "Flame Retardant Polyolefins: Impact and Flow Enhancement Using Metallocene Polymers", Journal of Fire Sciences, Vol. 14, 393-408 (September/October 1996)

Ilardo, C. S., Markezich, R. L., "Reduced Smoke Flame Retardant Formulations for High Performance Applications", Wire Association International Conference (May 1988)

Ilardo, C. S., Markezich, R. L., "Polyolefin Wire and Cable Materials with Reduced Smoke", Proceedings of the International Conference on Fire Safety, Vol. 15, 197-206 (January 1990)

Innes, J. D., "Advances in Synergist Systems for Lower Cost Flame Retarded Products", Proceedings of the International Conference on Fire Safety, Vol. 20, 290-296 (January 1995)

Innes, J. D., "Non-Halogen Flame Retardants in HDPE: A Case Study", Proceedings of the International Conference on Fire Safety, Vol. 21, 88-99 (January 1996)

Innes, J. D., Cox, A. W., "The Mechanism of Smoke Suppression and Synergism of Molybdate Compounds", Proceedings of the International Conference on Fire Safety, Vol. 23, 183-198 (January 1997)

Innes, J. D., Cox, A. W., "Smoke: Test Standards, Mechanisms, and Suppressants", Journal of Fire Sciences, Vol. 15, No. 3, 227-239 (May/June 1997)

Innes, J. D., Cox, A. W., "Magnesium Hydroxide Flame Retardant and Smoke Suppressant", Proceedings of the International Conference on Fire Safety, Vol. 24, 127-138 (July 1997)

Innes, J. D., Cox, A. W., "Molybdenum Flame and Smoke Retardant Mechanisms", Proceedings of the 2nd Symposium, Bureau of Flame Retarded Materials, Shibura Institute of Technology, 90-102 (July 1997)

Jeng, J. P., Terranova, S. A., Bonaplata, E., Goldsmith, K., Williams, D. M., Wojchiechowski, R. J., Starnes, W. H., "Reductive Coupling Promoted by Zerovalent Copper: A Potential New Method of Smoke Suppression for Vinyl Chloride Polymers", in "Fire and Polymers II", Ed. G. L. Nelson, ACS Symposium Series 599, 118-125 (1995)

Kitano, Y., "The Development and Application of New Non-Halogen Inorganic Flame Retardant", Proceedings of the March 1997 FRCA Conference, 37-47 (1997)

Knauss, D. M., McGrath, J. E., Kashiwagi, T., "Copolycarbonates and Poly(arylates) Derived from Hydrolytically Stable Phosphine Oxide Comonomers", in "Fire and Polymers II", Ed. G. L. Nelson, ACS Symposium Series 599, 41-55 (1995)

Landry, S. D., "Solving the UV Resistance Problem in FR HIPS", Proceedings of the March 1994 FRCA Conference (1994)

Larsen, E. R., "Halogenated Flame Retardants", Kirk Othmer Encyclopedia of Chemical Technology, 3rd Ed., Vol. 10, 373-395 (1980)

Levendusky, T. L., Musselman, L. L., "Alumina Trihydrate as a Flame Retardant and Smoke Suppressant Additive for Polypropylene: Development of a Commercial Flame Retardant Polypropylene", 38th Annual Technical Conference, Society of Plastics Engineers (May 1983)

Londa, M., Gingrich, R. P., Kormelink, H. G., Proctor, M. G., "Development of Nonhalogenated Flame Retardant Aliphatic Polyketone Compounds", Proceedings of the October 1995 FRCA Conference, 157-167 (1995)

Markezich, R. L., "Effectiveness of Synergists with Halogenated Flame Retardants", 44th Annual Technical Conference, Society of Plastics Engineers (April 1986)

Markezich, R. L., "Recent Advances Using Halogenated Flame Retardants", Flame Retardants '87 Conference, London (November 1987)

Markezich, R. L., Ilardo, C. S., "Evaluation of Synergists with Halogen Flame Retardants", 46th Annual Technical Conference, Society of Plastics Engineers (April 1988)

Markezich, R. L., "A New Chlorinated Flame Retardant for High Impact Polystyrene", Proceedings of the International Conference on Fire Safety, Vol. 17, 205-213 (January 1992)

Markezich, R. L., "Chlorine Containing Flame Retardants", Proceedings of the March 1996 FRCA Conference, 89-104 (1996)

Markezich, R. L., Mundhenke, R. F., "The Use of Bromine/Chlorine Synergism to Flame Retard Plastics", Proceedings of the October 1996 FRCA Conference, 29-41 (1996)

Miyachi, Y., "Development of the Flame Retardant with Smoke Suppressant Effect", Proceedings of the March 1995 FRCA Conference, 49-56 (1995)

Molesky, F., "Use of Magnesium Hydroxide for Flame Retardant Polypropylene", Proceedings of the International Conference on Fire Safety, Vol. 16, 212-226 (January 1991)

Moore, F. W., Ference, R. A., "Molybdenum Compounds as Smoke Suppressants", Proceedings of the International Conference on Fire Safety, Vol. 12, 324-339 (January 1987)

Moore, F. W., Kennelley, W. J., "Developments in Molybdate Smoke Suppressants", Proceedings of the International Conference on Fire Safety, Vol. 15, 174-196 (January 1990)

Moore, F. W., Kennelley, W. J., "New Developments in Smoke Suppression with Molybdenum and Boron Based Polymer Additives", Proceedings of the International Conference on Fire Safety, Vol. 16, 227-238 (January 1991)

Moore, F. W., Musselman, L. L., "Development of Flame Retardant, Low Smoke Epoxy Formulations with Alumina Trihydrate and Other Additives", Proceedings of the International Conference on Fire Safety, Vol. 17, 214-224 (January 1992)

Mount, R. A., "Non-Halogen Flame-Retarded Polypropylene", Proceedings of the International Conference on Fire Safety, Vol. 14, 285-286 (January 1989)

Mount, R. A., Pysz, J. F., "Phosphates as Flame Retardant Additives", Proceedings of the International Conference on Fire Safety, Vol. 16, 203-211 (January 1991)

NaB, B., Wanzke, W., "Intumescent Flame Retardants: New Developments", Proceedings of the October 1996 FRCA Conference, 63-73 (1996)

Nelson, E. D., "Effect of Tin Dioxide on Smoke and Heat Release of PVC", Proceedings of the International Conference on Fire Safety, Vol. 12, 242-259 (January 1987)

Nelson, G. L., Chan, E., "Elucidating the Mechanism of the Effects of Coatings on the Flammability of Engineering Plastics", Proceedings of the International Conference on Fire Safety, Vol. 15, 207-219 (January 1990)

Nelson, G. L., Benrashid, R., "Zinc, Zinc Compounds, and the Fire Performance of

Modified Polyphenylene Oxide", Proceedings of the International Conference on Fire Safety, Vol. 18, 283-298 (January 1993)

Nishihara, H., Tanji, S., Kanatani, R., Maeda, K., "Non-Halogenated Flame Retardant Polystyrene with Excellent Melt Flow Characteristics", Proceedings of the March 1995 FRCA Conference, 79-96 (1995)

Nishizawa, H., Ookoshi, M., "The Study of Nonhalogen Lowsmoke Flame Retardant Materials for Wire and Cables", Proceedings of the International Conference on Fire Safety, Vol. 24, 111-119 (July 1997)

Okisaki, F., "Flamecut GREP Series: New Non-Halogenated Flame-Retardant Systems", Proceedings of the March 1997 FRCA Conference, 11-24 (1997)

Olson, D. R., "Inherently Flame Retardant Silicone Polyetherimide Resins", Proceedings of the International Conference on Fire Safety, Vol. 15, 328-336 (January 1990)

Page, W. C., Buch, R. R., "Recent Studies on the Fire Behavior of Silicones", Proceedings of the International Conference on Fire Safety, Vol. 21, 185-188 (January 1996)

Park, B. Y., "Reference to Apply Flame Retardants of Brominated Epoxy Oligomer Origins to Styrenics Polymer", Proceedings of the March 1997 FRCA Conference, 147-155 (1997)

Rees, T. C., "Suppression of Smoke and Toxic Gases from Burning PVC by Molybdenum and Zinc Compounds", The Sherwin-Williams Company, Coffeyville, Kansas (Agust 17, 1984)

Scharf, D., Nalepa, R., Heflin, R., Wusu, T., "Studies on Flame Retardant Intumescent Char", Proceedings of the International Conference on Fire Safety, Vol. 15, 306-318 (January 1990)

Shanley, L. A., Slaten, B. L., Shanley, P., Broughton, R., Hall, D., Baginski, M., "Thermal Properties of Novel Carbonaceous Fiber Battings", Journal of Fire Sciences, Vol. 12, No. 3, 238-245 (May/June 1994)

Shen, K. K., "Zinc Borate", Plastics Compounding, Vol. 8, No. 5, 66 (September/October 1985)

Shen, K. K., "The Use of Zinc Borate as a Fire Retardant in Halogen-Free Polymer Systems", Proceedings of the International Conference on Fire Safety, Vol. 12, 340-365 (January 1987)

Shen, K. K., "Zinc Borate as a Flame Retardant in Halogen-Free Wire and Cable Systems", Plastics Compounding (November/December 1988)

Shen, K. K., Ferm, D. J., "Use of Zinc Borate in Electrical Applications", Proceedings of the October 1995 FRCA Conference, 129-143 (1995)

Shen, K. K., Ferm, D. J., "Use of Zinc Borate in Electrical Applications", Proceedings of the International Conference on Fire Safety, Vol. 21, 224-238 (January 1996)

Shen, K. K., Ferm, D. J., "Boron Compounds as Fire Retardants", Proceedings of the March 1996 FRCA Conference, 137-146 (1996)

Shen, K. K., Ferm, D. J., "Recent Advances in the Use of Borates as Fire Retardants in Nylons", 8th Annual BCC Conference on Flame Retardancy (June 1997)

Skinner, G. A., Haines, P. J., "Molybdenum Compounds as Flame-Retardants and Smoke-Suppressants in Halogenated Polymers", Fire and Materials, Vol. 10, No. 2, 63-69 (June 1986)

Tarquini, M. E., Andrews, C. R., "Flame Retardant Thermosets Having Lower Smoke Generation", 43rd Annual Conference, Composites Institute, Society of the Plastics Industry (February 1988)

Touval, I., "Antimony Oxide: Synergism in Flame Retardants", Plastics Compounding (September/October 1982)

Touval, I., "Antimony Flame Retarder Synergists", Proceedings of the March 1996 FRCA Conference, 117-136 (1996)

Wan, I.-Y., McGrath, J. E., Kashiwagi, T., "Triarylphosphine Oxide Containing Nylon 6,6 Copolymers", in "Fire and Polymers II", Ed. G. L. Nelson, ACS Symposium Series 599, 29-40 (1995)

Weil, E. D., "Recent Advances in Phosphorus-Containing Polymers for Flame Retardant Applications", Proceedings of the International Conference on Fire Safety, Vol. 12, 210-218 (January 1987)

Yacomeni, C. W., "New Technology for Effective Control of Flammability of Polypropylene Fibers/Textiles", Proceedings of the International Conference on Fire Safety, Vol. 18, 274-282 (January 1993)

Yuxiang, O., Fuye, G., Xin, L., Xiaomei, W., Ruibin, L., "Synthesis and Applications of Some New Flame Retardants", Proceedings of the March 1997 FRCA Conference, 125-137 (1997)

Table 5.1. Properties of Possible Flame Retardant Compounds.

Material	Formula	Mol. Wt.	Melting Point, °C	Boiling Point, °C	Decomp. Temp., °C
Phosphorus trioxide	P_2O_3	109.95	23.8	173.8	
Phosphorus tetroxide	P_2O_4	125.95	>100		
Phosphorus pentoxide	P_2O_5	141.94	580–585	300 (s)	
Phosphorus trifluoride	PF_3	87.97	−151.5	−101.5	
Phosphorus pentafluoride	PF_5	125.97	−83	−75	
Phosphorus trichloride	PCl_3	137.33	−112	75.5	
Phosphorus pentachloride	PCl_5	208.24		162 (s)	166.8
Phosphorus tribromide	PBr_3	270.70	−40	172.9	
Phosphorus pentabromide	PBr_5	430.52			106
Phosphorus oxychloride	$POCl_3$	153.33	2	105.3	
Phosphorus oxybromide	$POBr_3$	286.70	56	189.5	
Hydrogen fluoride	HF	20.01	−83.1	19.54	
Hydrogen chloride	HCl	36.46	−114.8	−84.9	
Hydrogen bromide	HBr	80.92	−88.5	−67.0	
Hydrogen iodide	HI	127.91	−50.8	−35.38	
Antimony trioxide	Sb_2O_3	291.50	656	1550	
Antimony trifluoride	SbF_3	178.75	292	319 (s)	
Antimony pentafluoride	SbF_5	216.74	7	149.5	
Antimony trichloride	$SbCl_3$	228.11	73.4	283	
Antimony pentachloride	$SbCl_5$	299.02	28	79	
Antimony tribromide	$SbBr_3$	361.48	96.6	280	
Antimony oxychloride	SbOCl	173.20			170
Arsenic trioxide	As_2O_3	197.84	315		
Arsenic pentoxide	As_2O_5	229.84			315
Arsenic trifluoride	AsF_3	131.92	−8.5	−63	
Arsenic trichloride	$AsCl_3$	181.28	−8.5	63	
Arsenic tribromide	$AsBr_3$	314.65	32.8	221	
Arsenic triiodide	AsI_3	455.64	146	403	
Boron trioxide	B_2O_3	69.62	460	1860	
Boron trifluoride	BF_3	67.81	−126.7	−99.9	
Boron trichloride	BCl_3	117.17	−107.3	12.5	
Boron tribromide	BBr_3	250.54	−46	91.3	
Boron triiodide	BI_3	391.52	49.9	210	
Boron carbide	B_4C	55.26	2350	>3500	

Table 5.2. Strengths of Chemical Bonds.

Chemical Bond	Average Bond Energy, kcal.
C—H	98.2
C—F	102
C—Cl	78
C—Br	65
C—I	57
H—H	103.2
H—F	135
H—Cl	102.1
H—Br	86.7
H—I	70.6
F—F	37
Cl—Cl	57.1
Br—Br	45.4
I—I	35.6
C—C	77.7 ethane
	79.0 propane
	79.6 normal butane
	80.1 isobutane
	84.9 solid carbon
C=C	140.0 ethylene
C≡C	193.3 acetylene
C—C	123.8 benzene

Table 5.3. Chemical Extinguishing Agents.

	Fire Extinguishing Effectiveness, weight pct. based on CH_3Br	Explosion Suppression Effectiveness, weight pct. based on CCl_4
Carbon dioxide (CO_2)	124	
Carbon tetrachloride (CCl_4), tetrachloromethane, Halon 104	68–75	100
Methyl bromide (CH_3Br), Halon 1001	100	192
Chlorobromomethane (CH_2BrCl), bromochloromethane, Halon 1011	80	180
Dibromodifluoromethane (CBr_2F_2), Halon 1202, Freon 12B2	100–148	201
Bromotrifluoromethane ($CBrF_3$), Halon 1301, Freon FE 1301, Freon 13B1	105–146	195
Bromochlorodifluoromethane ($CBrClF_2$), Halon 1211, Freon 12B1	75	115
Dichlorodifluoromethane (CCl_2F_2), Halon 122	68–104	98
Methylene bromide (CH_2Br_2), Halon 1002		195
Bromodifluoromethane ($CHBrF_2$)		161
Ethyl bromide (C_2H_5Br), Halon 2001	50–96	
Dibromotetrafluoroethane ($CBrF_2CBrF_2$), Halon 2402	74–107	139
Dibromochlorotrifluoroethane ($CBrF_2CBrClF$), Halon 2312		59
Dibromodifluoroethane ($CH_2BrCBrF_2$), Halon 2202		116

229

CHAPTER 6

Market Acceptance Criteria

The factors which determine acceptance of any material into any market are many: these include economics, politics, salesmanship and marketing, and personal appeal. The principal factor, however, is performance. The material must meet the requirements of the application. If a material fails to meet performance requirements, blaming lack of acceptance on other factors may be soothing but definitely is irrelevant. The most favorable purchasing agent, the most receptive building code, the lowest raw material costs, the most advantageous product and process economics, can not make an unsatisfactory material work.

The fact that plastic materials are so thoroughly established in so many applications indicates that plastic materials do meet the requirements of those applications. In many cases, the plastic material was designed for the application, and the essential requirements of the application were not altered for the benefit of the plastic material. This situation holds true for fire retardance. Each market, and each application in that market, has a requirement for a specific level of fire retardance, depending on experience and judgement.

SECTION 6.1. MARKETS FOR THE PLASTICS INDUSTRY

Plastic materials have gained general acceptance because they contribute to the fulfillment of human needs and human desires. The markets for these materials can be divided into seven broad and sometimes overlapping areas:

1. Building
2. Transportation
3. Packaging
4. Electrical and electronic
5. Consumer goods
6. Industrial goods
7. International

Because of population growth and demographic changes, conditions in all of these markets are changing rapidly and generating pressures for changes in many requirements affecting plastic materials. These pressures may tend to make acceptance requirements more reasonable, but the requirement of adequate safety for human life and property will never be abolished, and government involvement is making certain that safety will not be neglected.

SECTION 6.2. TYPES OF ACCEPTANCE CRITERIA

For a plastic material to win acceptance in any market, it must meet three types of requirements: customer requirements, government requirements, and insurance requirements. These are not necessarily independent, and in some cases one type of requirement dominates and influences acceptance of the material.

Customer requirements are based on the basic tenets of the free-enterprise system. The customer is supposedly free to set his own requirements and to purchase materials from any supplier who meets those requirements. Because the customer is not always the ultimate consumer, customer requirements sometimes include not only his own requirements but also the requirements of his own customers. Where use of the material in a particular application is subject to government and insurance regulations, these additional requirements are often incorporated in the customer's specifications. Where the customer is part of an industry association or a trade association, the quality standards agreed upon by the members of the association will often be incorporated into the customer's requirements.

Government requirements are based on the government's function of promoting the safety and welfare of its citizens. Government requirements have increased greatly in importance for the following reasons:

1. The rise of the government as a large customer itself, because of its massive expenditures in the military and aerospace areas
2. The tendency toward centralization of standard-setting for government purchases of items in quantity, such as motor vehicles and office supplies
3. The widespread use of influence on specific sectors of government to promote special interests, such as past efforts by plumbers' unions and cast iron trade associations to block the acceptance of plastic pipe and fittings in local building codes
4. The failure of various industries to exhibit adequate concern for the public welfare, by placing on the market materials such as highly flammable fabrics, and interior finishes with high surface flame spread characteristics
5. The increasing importance of international trade, especially with countries in which governments have a major and sometimes decisive role

The Code of Federal Regulations (CFR) is a codification of the general and permanent rules published in the Federal Register by the Executive departments and agencies of the Federal Government of the United States. The Code is divided into 50 titles which represent broad areas subject to Federal regulation. Each title is divided into chapters which usually bear the name of the issuing agency. Each chapter is further subdivided into parts covering specific regulatory areas. The outline of the Code of

Federal Regulations, and some documents relevant to fire safety and plastics, are presented in Table 6.1.

Insurance requirements arise from the general practice of purchasing insurance for protection against catastrophic losses, and the consequent desire of the insurance companies to protect themselves against excessive risks, or risks not justified by the premiums charged. Because insurance premiums are part of the cost of doing business, insurance rates have a direct influence on profitability.

SECTION 6.3. THE BUILDING MARKET

The building market can be divided into six areas: residential (homes, apartments, mobile homes), institutional (schools, hospitals, nursing homes, correctional facilities), commercial (offices, business, mercantile, warehouses), public (theaters, stadiums, coliseums), industrial (plants, factories, pipe lines), and transportation (highways, bridges, harbors, airports). The nature of this market makes it difficult to divide plastics consumption according to these areas, because building materials suppliers usually serve two or more areas. The consumption of plastics in the building market is large, and the potential is great. The quantities of some plastics used in building and construction are shown in Table 6.2.

There are thousands of building materials supply companies in the United States, and they in effect control the selection of materials for this highly fragmented industry. The architects and engineers involved in major projects exercise some choice in specifying materials, but the overriding influence in customer requirements is the building code involved, because the building code restricts the architect's freedom in design.

Government requirements are to a large degree embodied in building codes enforced by local building officials. Because each building code body is independent, a material of construction must obtain approval within every locality within which it is to be used. To provide guidance for cities and towns which cannot support extensive research and testing facilities, and to encourage uniformity, several model building codes have been prepared by various organizations.

The most important model building codes are:

1. Uniform Building Code, International Conference of Building Officials, Whittier, California, most widely accepted in the western United States
2. BOCA National Building Code, Building Officials and Code Administrators International, Country Club Hills, Illinois, most widely accepted in the northeastern and midwestern United States
3. Standard Building Code, Southern Building Code Congress International, Birmingham, Alabama, most widely accepted in the southeastern United States

These model building codes have been adopted to a substantial extent but not always in their entirety by individual building code groups in each locality. Each building code group is independent and is influenced by local experience and by the most persuasive special-interest groups in its locality.

There is an ongoing effort to combine the different building codes into one building code which would cover the entire United States by the year 2000.

There are two ways of classifying buildings: according to construction (Table 6.3), and according to occupancy (Tables 6.4, 6.5, and 6.6). Depending on the code, and the edition of the code, there may be as many as six construction classifications, and the most highly-rated fire-resistance classifications involve the use of steel, iron, concrete, and masonry. These materials are noncombustible and backed by years of experience; their continued use is promoted by their respective trade associations, and guarded by their labor unions: ironworkers, cement workers, and bricklayers. The least fire-resistant classifications involve the use of wood, a material which is combustible but entrenched by centuries of use; its continued use is promoted by its trade associations and guarded by carpenters' unions.

Despite the generally conservative appearance of the building codes, they are more receptive to new materials and techniques than is immediately apparent, because they are to a substantial extent performance-oriented. The requirements are largely presented on a performance basis, and while these requirements may seem extremely demanding to some parts of the plastics industry, two features of these requirements must be pointed out: they are more or less justified by concern for public safety, and they are now being met by commercially available materials.

Any material, plastic or otherwise, can be classified as noncombustible if it meets one of the following requirements:

1. No part of it will ignite and burn when subjected to fire. A material meets this requirement if it conforms to UBC 2-1 (based on ASTM E 136).
2. It has a structural base of noncombustible material as defined in (1) and a surfacing not over 1/8 inch thick with a flame spread rating of 50 or less according to UBC 8-1 (based on ASTM E 84).

Fire resistance ratings are generally defined according to ASTM E 119. Since few commercially available plastic materials can provide significant resistance as a structural component in buildings, plastics find their greatest use in applications which do not require fire resistance ratings: non-structural uses in noncombustible construction, and both structural and non-structural uses in combustible construction: in short, direct competition with wood and glass.

The occupancy classifications of buildings are a measure of relative hazard to life, because they indicate both the number and mobility of the persons in the buildings. The formal classifications vary with the building code (Tables 6.4, 6.5, and 6.6).

Interior finish materials, a substantial application area for plastic materials, are presently classified for fire hazard on the basis of flame spread ratings by ASTM E 84. Classifications tend to be the same among the building codes (Table 6.7), but requirements are a function of occupancy (Table 6.8). The most demanding requirements are applied for exitways, and for occupants who are incapacitated or under restraint in institutions.

Fire safety requirements for buildings, or parts of buildings, are based on the ability of occupants to escape from the building, or that part of the building, in the event of fire. Fire safety requirements for buildings, or parts of buildings, increase as the ability of occupants to escape from the building, or that part of the building, in the event of fire decreases. Ability to escape is defined by terms such as ambulatory, nonambulatory, in supervised environment, under restraint, and not capable of self-preservation.

The emphasis on flame spread rating for exitways in particular and all occupancies in general is the result of several tragic fires involving great loss of life. One factor contributing to the death of 492 persons in the Coconut Grove night club fire in Boston in 1942 was rapid flame spread; the ceiling finish material in the room had a flame spread rating of 2,500. In general, a plastic material with a flame spread rating over 200 has a limited future in the building market; a rating under 200 may make a portion of the market accessible, a rating under 75 makes available another portion of the market, and a rating under 25 makes available still more of the market.

For applications other than structural members and interior finish materials, the "approved plastics" classification is pertinent.

The Uniform Building Code (UBC) defines an approved plastic material, other than a foam plastic, as one which has a self-ignition temperature of 650°F. (343°C.) or greater when tested according to UBC 26-6 (based on ASTM D 1929) and a smoke-density rating not greater than 450 when tested according to UBC 8-1 (based on ASTM E 84) in the way intended for use, or a smoke-density rating no greater than 75 when tested in the thickness intended for use by UBC 26-5 (based on ASTM D 2843). Approved plastics are classified as either CC1 or CC2 according to UBC 26-7 (based on ASTM D 635). Foam plastic insulation is required to have a flame-spread rating of not more than 75 and a smoke-developed rating of not more than 450 when tested in the maximum thickness intended for use according to UBC 8-1 (based on ASTM E 84).

The BOCA National Building Code defines an approved light-transmitting plastic as any thermoplastic, thermosetting, or reinforced thermosetting plastic material which has a self-ignition temperature of 650°F. (343°C.) or greater when tested according to ASTM D 1929 and a smoke-density rating not greater than 450 when tested according to ASTM E 84 in the manner intended for use, or a smoke-density rating no greater than 75 when tested in the thickness intended for use by ASTM D 2843. Approved plastics are classified as either C1 or C2 according to ASTM D 635.

Government regulations on the state level are generally within the jurisdiction of state fire marshals, who usually maintain close contact with local building code officials and insurance groups. State fire marshals are represented in the National Association of State Fire Marshals (NASFM). In the larger states such as California and New York, school officials may set requirements for the approval of materials to be used in schools.

Government requirements on the Federal level come from a variety of departments: the Department of Housing and Urban Development (HUD), particularly the Federal Housing Administration (FHA), for government-supported residential buildings; the Department of Health and Human Services (HHS), for government-supported buildings providing health care and human services; the Department of Transportation (DOT), for government-supported buildings used for transportation; the Department of Defense (DOD), for buildings on military installations; the General Services Administration (GSA), for government buildings in general.

The Code of Federal Regulations (CFR) is a codification of the general and permanent rules published in the Federal Register by the Executive departments and agencies of the Federal Government of the United States. The Code is divided into 50 titles which represent broad areas subject to Federal regulation. Each title is divided into chapters which usually bear the name of the issuing agency. Each chapter is further subdivided into parts covering specific regulatory areas. The outline of the Code of Federal Regulations, and some documents relevant to fire safety and plastics, are presented in Table 6.1.

Insurance requirements are set by individual underwriters, and the insurance companies generally rely on their technical experts for guidance in the field of fire hazard. Underwriters Laboratories (UL) has been associated with the Factory Insurance Association (FIA), while the Factory Mutual (FM) Research Corporation has been associated with the Associated Factory Mutual Insurance Companies.

SECTION 6.4. THE TRANSPORTATION MARKET

The transportation market can be divided into five general areas: automotive, principally passenger cars; highway transportation, including trailer and tank trucks; railway transportation, including passenger cars, freight cars, and tank cars; air and space transportation, including aircraft and aerospace applications; and marine transportation, including boats, barges, and ocean-going vessels.

The quantities of some plastics used in transportation are shown in Table 6.9.

Government requirements in the transportation market are to a substantial extent determined by the Department of Transportation (DOT). The National Highway Traffic Safety Administration (NHTSA) has the responsibility for motor vehicles and highway transportation, the Federal Transit Administration (FTA), formerly the Urban Mass Transit Administration (UMTA), for intercity railway transportation and urban mass transit systems, the Federal Railroad Administration (FRA) for interstate railway transportation, and Federal Aviation Administration (FAA) for air transportation and aircraft, and the U.S. Coast Guard (USCG) for ships and boats.

The Code of Federal Regulations (CFR) is a codification of the general and permanent rules published in the Federal Register by the Executive departments and agencies of the Federal Government of the United States. The Code is divided into 50 titles which represent broad areas subject to Federal regulation. Each title is divided into chapters which usually bear the name of the issuing agency. Each chapter is further subdivided into parts covering specific regulatory areas. The outline of the Code of Federal Regulations, and some documents relevant to fire safety and plastics, are presented in Table 6.1.

In the Code of Federal Regulations (CFR), government requirements in the transportation market are organized as follows:

Title 14. Aeronautics and Space
    Chapter I. Federal Aviation Administration
        Subchapter C. Aircraft
        Parts 23, 25, 27, 29. Airworthiness Standards
        Part 23. Normal, utility, and commuter airplanes
        Part 25. Transport category airplanes
        Part 27. Normal category rotorcraft
        Part 29. Transport category rotorcraft
        Subchapter G. Air Carriers
        Part 121. Domestic, flag, and supplemental carriers
        Part 125. Airplanes with 20 or more passengers or
        6000 lb. or more payload
    Title 46. Shipping
        Subchapter H. Passenger Vessels
        Subchapter I. Cargo and Miscellaneous Vessels
    Title 49. Transportation
        Chapter V. National Highway Traffic Safety Administration
        Part 571. Federal Motor Vehicle Safety Standards
        Chapter VI. Federal Transit Administration
        Chapter VII. National Railroad Passenger Corporation
        (AMTRAK)

The automotive area is the largest consumer of plastics in the transportation market.

Motor vehicles must comply with the requirements of Motor Vehicle Safety Standard (MVSS) 302, contained in 49 CFR 571.302.

Fire safety requirements for transportation vehicles are based on the ability of occupants to escape from the vehicle in the event of fire. Fire safety requirements for transportation vehicles increase as the ability of occupants to escape from the vehicle in the event of fire decreases.

Passenger coaches, food service cars, commuter cars, and light rail vehicles (LRV) used in intercity railway transportation and urban mass transit systems must comply with the requirements of the Urban Mass Transit Administration (UMTA), now the Federal Transit Administration (FTA), first published as guidelines in the Federal Register on August 14, 1984. They must meet the requirements of the individual mass transit system, such as the Bay Area Rapid Transit (BART) District, the Washington Metropolitan Transit Authority (WMTA), the Southeastern Pennsylvania Transit Authority (SEPTA), and the New York Metropolitan Transit Authority (MTA) which extends into Connecticut.

These requirements cover exterior plastic components such as end caps and roof housings, diaphragms between vehicles, and interiors including seats, wall and ceiling panels, flooring and floor coverings, and windows. The tests specified are FAR 245.853 for ignitability, ASTM D 3675 and ASTM E 162 for flame spread, ASTM E 119 for fire endurance, and ASTM E 662 for smoke evolution.

Individual mass transit systems may have more demanding requirements, and may specifically ban certain materials such as polyurethane.

Passenger cars, freight cars, and tank cars used in interstate railway transportation must comply with the requirements of the Federal Railroad Administration (FRA), and must meet the specifications on the individual railroad.

Passenger coaches, sleeping cars, diners, and food service cars used in interstate railway transportation must comply with the requirements of the Federal Railroad Administration (FRA), first published as guidelines in the Federal Register on January 17, 1989.

These requirements cover exterior plastic components such as end caps and roof housings, diaphragms between vehicles, and interiors including seats, wall and ceiling panels, flooring and floor coverings, windows, curtains, and mattresses. The tests specified are FAR 245.853 for ignitability, ASTM D 3675 and ASTM E 162 for flame spread, ASTM E 119 for fire endurance, and ASTM E 662 for smoke evolution.

Some fire test requirements for passenger rail vehicles are shown in Table 6.10.

Aircraft use a smaller total amount of plastics than motor vehicles, but have much more demanding fire safety requirements because they involve situations of limited or no egress in case of fire.

To be certified by the Federal Aviation Administration (FAA), aircraft must comply with the requirements of the Federal Aviation Regulations (FAR) in the Code of Federal Regulations (CFR). The sections of the Code involved are:

Airworthiness Standards
14 CFR 23. Normal, utility, and commuter category airplanes
14 CFR 25. Transport category airplanes
14 CFR 27. Normal category rotorcraft
14 CFR 29. Transport category rotorcraft
Certification and Operation
14 CFR 121. Domestic, flag, and supplemental air carriers
14 CFR 125. Airplanes of certain sizes and larger

Some sections of the Federal Aviation Regulations (FAR) applicable to fire performance are:

Airworthiness Standards
Normal, Utility, and Commuter Category Airplanes
     FAR 23.853. Compartment interiors
Transport Category Airplanes
     FAR 25.853. Compartment interiors
     FAR 25.855. Cargo and baggage compartments
     FAR 25.865. Powerplant components
     FAR 25.867. Powerplant components
     FAR 25.869. Electric wire insulation
     FAR 25.1191. Powerplant components
     FAR 25.1193. Powerplant components
     FAR 25.1359. Electric wire and cable
Normal Category Rotorcraft
     FAR 27.853. Compartment interiors
     FAR 27.855. Cargo and baggage compartments
Transport Category Rotorcraft
     FAR 29.853. Compartment interiors
     FAR 29.855. Cargo and baggage compartments
Certification and Operation
Domestic, Flag, and Supplemental Air Carriers
     FAR 121.215. Cabin interiors
Airplanes with at least 20 passengers or 6000 lb. payload
     FAR 125.113. Cabin interiors

The Federal Aviation Administration (FAA) requires a variety of fire tests for aircraft certification. These include:

1. Vertical bunsen burner test for cabin and cargo compartment materials (FAR 25.853 and FAR 25.855) (ASTM F 501)
2. 45-degree bunsen burner test for cargo compartment liners and waste stowage compartment materials (FAR 25)
3. Horizontal bunsen burner test for cabin, cargo compartment, and miscellaneous materials (FAR 25.853) (ASTM F 776)
4. 60-degree bunsen burner test for electric wire (FAR 25.869) (ASTM F 777)

5. Heat release rate test for cabin materials (FAR 25.853)
6. Smoke test for cabin materials (FAR 25.853)
7. Oil burner test for seat cushions (FAR 25.853)
8. Oil burner test for cargo liners (FAR 25.855)
9. Radiant heat test for evacuation slides, ramps, and rafts (TSO C69A) (ASTM F 828)
10. Fire containment test for combustible waste containers, carts, and compartments (FAR 25.853)
11. Fire test for aircraft thermal acoustical insulation
12. Fire resistance of powerplant hose assemblies (TSO C42, C53A, and C75)
13. Powerplant fire penetration test (FAR 25.865, 25.867, 25.1191, and 25.1193)
14. Oil burner test for repaired cargo compartment liners
15. Horizontal flammability test for aircraft blankets
16. Smoke test for insulated aircraft wire
17. Dry arc tracking test for wire insulation
18. Dry arc-propagation resistance test for wire insulation

Some FAA fire test requirements for aircraft materials are shown in Table 6.11.

The commercial transport aircraft industry in the United States and in most of the rest of the world is dominated by the Boeing Airplane Company and Airbus Industrie. Aircraft manufacturer specifications and guidelines have an importance and influence on acceptance of a material which is similar to government regulation.

Boeing Airplane Company specifies Boeing fire tests (BSS) similar to FAA fire tests.

BSS 7230. bunsen burner tests
BSS 7238. NBS smoke chamber test
BSS 7239. toxicity of gases from NBS smoke chamber
BSS 7303. oil burner test for seat cushions (FAR 25.853)
BSS 7304. radiant panel test (ASTM E 162)
BSS 7322. OSU heat release rate test for cabin materials (FAR 25.853)
BSS 7323. oil burner test for cargo liners (FAR 25.855)
BSS 7324. 60-degree bunsen burner test for electric wire (FAR 25.869) (ASTM F 777)
BSS 7357. thermal acoustical insulation

Some Boeing fire test requirements contained in Boeing specification D6-51377 for commercial aircraft are shown in Table 6.12.

The Joint Airworthiness Authorities (JAA) of Europe issue Joint Airworthiness Requirements (JAR) similar to the Federal Aviation Regulations (FAR) issued by the Federal Aviation Administration (FAA).

Airbus Industrie specifies Airbus Industrie test methods (AITM) similar to FAA fire tests.

AITM 2.0002. flammability of non-metallic materials, small burner test, vertical (JAR/FAR 25.853)

AITM 2.0003. flammability of non-metallic materials, small burner test, horizontal (JAR/FAR 25.853)

AITM 2.0004. flammability of non-metallic materials, small burner test, 45-degree (FAR/FAR 25.855)

AITM 2.0005. flammability of non-metallic materials, small burner test, 60-degree (JAR/FAR 25.1359)

AITM 2.0006. heat release and heat release rate of aircraft materials (JAR/FAR 25.853)

AITM 2.0007. smoke density of aircraft interior materials (JAR/FAR 25.853)

AITM 2.0008. smoke density of electrical wire/cable insulation

AITM 2.0009. fire resistance of aircraft seat cushion using high intensity open flame (JAR/FAR Part 25, Appendix F, Part II)

AITM 2.0010. fire resistance of aircraft cargo compartment lining materials using high intensity open flame (JAR/FAR Part 25, Appendix F, Part III)

AITM 2.0038. flammability of heat shrinkable tubings, small burner test, 60-degree

AITM 3.0005. gas components of smoke from aircraft interior materials

Some Airbus fire test requirements contained in Airbus specification ABD-0031 for commercial aircraft are shown in Table 6.13.

Ships and boats must comply with the regulations of the U.S. Coast Guard (USCG), contained in 46 CFR 72 for passenger vessels and 46 CFR 92 for cargo and miscellaneous vessels. In general, the regulations require noncombustible materials in compartments occupied by passengers and crew.

Some sections in the Code of Federal Regulations (CFR) relevant to ships and boats are:

Subchapter H. Passenger Vessels
46 CFR 72.05. Structural fire protection
    72.05-5(g). Standard fire test
    72.05-10. Fire control bulkheads and decks
    72.05-15. Ceilings, linings, trims, and decorations in accommodation spaces and safety areas
    72.05-55. Furniture and furnishings
Subchapter I. Cargo and Miscellaneous Vessels
46 CFR 92.05. General fire protection
45 CFR 92.07. Structural fire protection
    92.07-5(a). Standard fire test
    92.07-10. Construction

Fire safety requirements for transportation vehicles are based on the ability of occupants to escape from the vehicle in the event of fire. Fire safety requirements for transportation vehicles increase as the ability of occupants to escape from the vehicle in the event of fire decreases.

The most demanding fire safety requirements in transportation may be found in Military Standard MIL-STD-2031(SH) for composite material systems used in hull, machinery, and structural applications inside naval submarines. Some fire and toxicity test requirements in MIL-STD-2031(SH) are shown in Table 6.14.

SECTION 6.5. THE PACKAGING MARKET

The packaging market is the largest market for plastics, surpassing the more publicized building market. The packaging market consumed more than 18 million lb. of plastics in 1993, exceeding the 15 million lb. of plastics consumed by the building market. The packaging market can be divided into two areas, flexible packaging and rigid packaging.

Flexible packaging consumed 7995 million lb. of plastics in 1993, with low density polyethylene (LDPE) accounting for the largest share, 3630 million lb. (Table 6.15)

Rigid packaging consumed 10,120 million lb. of plastics in 1993, with high density polyethylene (HDPE) accounting for the largest share, 4625 million lb. (Table 6.16)

The complexity of this market makes knowledge of all flammability requirements difficult. In many cases, there are no flammability requirements other than those based on government or insurance requirements. Materials which are difficult to ignite appear to be generally acceptable if they meet other performance requirements.

SECTION 6.6. THE ELECTRICAL AND ELECTRONICS MARKET

The electrical and electronics market consumes large quantities of plastics for wire and cable, electrical controls and switchgear, computers, and communications equipment. This market consumed 3379 million lb. of plastics in 1994.

The electrical and electronics market includes the following areas which may overlap in many cases:

1. General low voltage (under 600V) wire and cable, such as household, appliance, automotive
2. High voltage (over 600V) wire and cable, such as power
3. Telecommunications and computers
4. Harnesses, splices, and terminations

The quantities of some plastics used in wire and cable are shown in Table 6.17. The quantities of some plastics used in electrical/electronics applications are shown in Table 6.18.

Government requirements on the Federal level are largely based on the Federal government's position as a major purchaser, and generally come from the General Services Administration (GSA). Various government agencies indirectly influence requirements.

The Code of Federal Regulations (CFR) is a codification of the general and permanent rules published in the Federal Register by the Executive departments and agencies of the Federal Government of the United States. The Code is divided into 50 titles which represent broad areas subject to Federal regulation. Each title is divided into chapters which usually bear the name of the issuing agency. Each chapter is further subdivided into parts covering specific regulatory areas. The outline of the Code of Federal Regulations, and some documents relevant to fire safety and plastics, are presented in Table 6.1.

The Mine Safety and Health Administration requires electrical cables to pass flame resistance tests described in Section 18.64 of Title 30, Part 18.

The military specifications pertinent to the electrical and electronics market have been assigned to the following groups in the Federal Supply Classification:

Group 58. Communication, Detection, and Coherent Radiation Equipment
Group 59. Electrical and Electronic Equipment Components
Group 61. Electric Wire, and Power and Distribution Equipment
Group 74. Office Machines, Visible Record Equipment, and Data Processing Equipment

Government requirements on the local level, and customer requirements throughout the United States, are to varying degrees based on the National Electrical Code (NEC) issued by the National Fire Protection Association (NFPA).

Some parts of the National Electrical Code which are relevant to fire safety and plastics include:

Article 310. Conductors for General Wiring
    Par. 310-13. Conductor constructions and applications
Article 318. Cable Trays
    Par. 318-5(f). Nonmetallic Cable Tray
Article 331. Electrical Nonmetallic Tubing
    Par. 331-1. Definition
    Par. 331-3. Uses Permitted
Article 336. Nonmetallic Sheathed Cable (Types NM, NMC, NMS)
    Par. 336-25. Construction
Article 338. Service Entrance Cable (Types SE, USE)
    Par. 338-1. Construction
Article 340. Power and Control Tray Cable (Type TC)
    Par. 340-3. Construction

Article 347. Rigid Nonmetallic Conduit
   Par. 347-17. General
Article 400. Flexible Cords and Cables
   Par. 400-4. Types
Article 402. Fixture Wire
   Par. 402.3. Types
Article 760. Fire Protective Signalling Systems
   Par. 760-28(c). Fire resistance of cables
Article 770. Optical Fiber Cables
   Par. 770-6. Fire resistance of cables
Article 800. Conductor Circuits
   Par. 800-3(b). Fire resistance of wires and cables

The types of conductor insulation listed in the 1996 National Electrical Code are listed in Table 6.19.

Insurance requirements on particular products are based on standards developed by Underwriters Laboratories (UL) and Factory Mutual (FM).

The test standards most often used in the United States include UL 94 for small specimens, UL 910 and NFPA 262 for horizontal wire and cable, UL 1581 and IEEE 383 for vertical tray wire and cable, and UL 1666 vertical riser wire and cable.

The electrical and electronics market is an international market, and the important fire safety standards for the electrical and electronics market include international standards.

Some international standards relevant to fire safety and plastics issued by the International Electrotechnical Commission (IEC) are shown in Table 6.20.

Canadian standards relevant to fire safety and plastics include the following:

C22.2 No.38-95. Thermoset Insulated Wires and Cables
C22.2 No.75-M1983. Thermoplastic-Insulated Wires and Cables
C22.2 No.174-M1984. Cables and Cable Glands for Use in Hazardous Locations

Australian standards relevant to fire safety and plastics include the following:

AS 2420. Fire test methods for solid insulating materials and non-metallic enclosures used in electrical equipment
AS 3000. Electrical installations. Buildings, structures and premises (known as the SAA Wiring Rules)

British standards relevant to fire safety and plastics include the following:

BS 229. Flameproof enclosure of electrical apparatus

BS 738. Insulating materials for electrical products
BS 889. Flameproof electric lighting fittings
BS 2848. Electrical products
BS 3456. Household electrical appliances
BS 3497. Electrical products
BS 4145. Electrical products
BS 4584. Printed circuits
BS 4808. PVC insulation and PVC sheath for telecommunication cables and wires
BS 5724. Medical electrical equipment
BS 6334. Solid electrical insulating materials
BS 6458. Electrotechnical products
BS 9400. Integrated circuits

Japanese standards relevant to fire safety and plastics include the following:

JIS C 0060. Fire hazard testing
 Part 2. Test methods, glow-wire test and guidance
JIS C 0061. Fire hazard testing
 Part 2. Needle flame test
JIS C 3521. Flame test method for flame retardant sheath of telecommunication cables

French standards relevant to fire safety and plastics include the following:

NF C 17-300. dielectric liquids
NF C 20-454. Basic environmental testing procedures. Test methods. Fire behavior. Analysis and titrations of gases evolved during pyrolysis or combustion of materials used in electrotechnics. Exposure to abnormal heat or fire. Tube furnace method.
NF C 20-455. Test methods. Fire behavior. Glow wire test. Flammability and extinction ability.
NF C 20-456. Fire hazard testing. Test methods. Needle flame test
NF C 20-902-1. Smoke opacity
NF C 26-212. Solid electrical insulation
NF C 27-300. Dielectric liquids
NF C 32-070. Insulated cables and flexible cords for installations. Classification tests on cables and cords with respect to their behavior to fire
NF C 32-071U. Electrical cables
NF C 32-072. Electrical cables
NF C 32-073-1. Electrical cables, smoke density
NF C 32-073-2. Electrical cables, smoke density
NF C 32-074. Acidity by pH and conductivity
NF C 32-130. Conductors and cables
NF C 32-131. Cables without halogens
NF C 32-310. Conductors and cables, fire resistant
NF C 32-323. Conductors and cables, without halogens

XP C 32-324. Cables, without halogens

NF C 68-106. Conduits, insulation

NF C 68-107. Conduits, insulation

NF C 68-108. Conduits

NF F 63-295. Conductors and cables

UTE C 20-450. Fire behavior of electrotechnical (electric, electromechanic, and electronic) materials, components, and equipment. General

UTE C 20-452. Test methods. Fire behavior. Determination of smoke opacity in closed chamber (similar to ASTM E 662).

UTE C 27-251U. Dielectric liquids

UTE C 32-071. Test electric cables under fire conditions. Guide stating the test method for checking small single insulated cores of the category C2

German standards relevant to fire safety and plastics include the following:

VDE 0304. Glow bar test

VDE 0318. Flame test, horizontal (similar to UL 94)

VDE 0340. Self-adhesive insulating tapes

VDE 0345. Flash-ignition temperature

VDE 0470. Hot mandrel test

VDE 0471. Glow wire test

VDE 0472. Part 804. Vertical cable test (similar to IEEE 383)

There is concern regarding the evolution of toxic combustion gases from electrical and electronic products.

In New York State, electrical wire insulations and synthetic electrical conduits permanently installed in buildings have to be tested for combustion toxicity and the data filed, in accordance with Article 15, Part 1120, of the New York State Fire Prevention and Building Code.

There is concern regarding the evolution of corrosive combustion gases from electrical and electronic products.

In a test developed by the Central Electricity Generating Board (CEGB) in the United Kingdom, designated as E/TSS/EX5/8056 Part 3, aqueous solutions of combustion products are tested for pH and ionic conductivity.

In a test developed by the Centre National d'Etudes et de Telecommunications (CNET) in France, designated as DEC/0611/C, a printed circuit board carrying electric current is used to determine the effect of corrosion by gases produced by burning.

A significant number of wire and cable applications restrict the HCl content of the gases produced from burning.

Specifications which require halogen-free materials include:

1. MIL-C-24640. General Specification for Lightweight Electrical Cable for Shipboard Use
2. MIL-C-24643. General Specification for Low Smoke Electrical Cable and Cord for Shipboard Use
3. PMS-400-881. General Specification for Special Purpose Lightweight Electric Cable for Shipboard Use

Nippon Telegraph and Telephone (NTT) in Japan and Telecom Australia have moved toward non-halogen and halogen-free materials in telecommunications applications.

SECTION 6.7. THE CONSUMER GOODS MARKET

The consumer goods market can be considered to consist of five areas, interior furnishings, apparel, appliances, housewares, and toys. This market may be considered to include parts of the electrical and electronics market.

Interior furnishings include furniture, curtains and draperies, and floor coverings. Wall and ceiling coverings, and cabinets, counters, and shelving, are sometimes considered interior furnishings, and sometimes considered part of the building.

The quantities of some plastics used in furniture and furnishings are shown in Table 6.21.

The furniture market can be divided into four areas, each a substantial market in itself:

1. The permanent live-in residential market, which includes single-family homes and apartments
2. The temporary live-in residential market, which includes hotels and motels
3. The live-in institutional market, which includes hospitals, nursing homes, and correctional institutions
4. The non-live-in business and institutional market, which includes offices

Furniture includes upholstered furniture, bedding, and occasional furniture.

The Code of Federal Regulations (CFR) is a codification of the general and permanent rules published in the Federal Register by the Executive departments and agencies of the Federal Government of the United States. The Code is divided into 50 titles which represent broad areas subject to Federal regulation. Each title is divided into chapters which usually bear the name of the issuing agency. Each chapter is further subdivided into parts covering specific regulatory areas. The outline of the Code of Federal Regulations, and some documents relevant to fire safety and plastics, are presented in Table 6.1.

In the Code of Federal Regulations, government requirements in the consumer goods market are organized as follows:

Title 16. Commercial Practices
  Chapter II. Consumer Product Safety Commission
  Subchapter D. Flammable Fabrics Act Regulations
    Part 1610. Clothing Textiles
    Part 1611. Vinyl Plastic Film
    Part 1615. Children's Sleepwear Sizes 0 through 6X
    Part 1616. Children's Sleepwear Sizes 7 through 14
    Part 1630. Carpets and Rugs
    Part 1631. Small Carpets and Rugs
    Part 1632. Mattresses and Mattress Pads

Tests for furniture can be divided into small-scale tests, such as those developed by the National Bureau of Standards (NBS), California Bureau of Home Furnishings and Thermal Insulation, and Upholstered Furniture Action Council (UFAC), and larger-scale tests, such as the California Technical Bulletin 133 test developed by the California Bureau of Home Furnishings and Thermal Insulation, and room fire tests.

There are no Federal flammability standards for upholstered furniture in force, although such standards have been proposed. The most important standards for upholstered furniture are California Technical Bulletins and voluntary standards developed by industry groups. There are some local standards, such as those of the Port Authority of New York and New Jersey (PANYNJ) and the City of Boston.

Voluntary standards have been developed for the permanent live-in residential market by the Upholstered Furniture Action Council (UFAC), by Underwriters Laboratories in UL 1056, and by the National Fire Protection Association in NFPA 260 and NFPA 261. These standards have been adopted as customer requirements to varying degrees.

Voluntary standards have been developed for the non-live-in business market by the Business and Institutional Furniture Manufacturers Association (BIFMA), and by Underwriters Laboratories in UL 1286. These standards have been adopted as customer requirements to varying degrees.

The State agency which is most active in developing flammability standards for furniture and furnishings is the California Bureau of Home Furnishings and Thermal Insulation, 3485 Orange Grove Avenue, North Highlands, California 95660-5595.

The California Technical Bulletins issued by the California Bureau of Home Furnishings and Thermal Insulation for furniture and furnishings include:

California Technical Bulletin 106. mattress
California Technical Bulletin 116. upholstered furniture

California Technical Bulletin 117. upholstered furniture
California Technical Bulletin 121. mattress
California Technical Bulletin 129. mattress
California Technical Bulletin 133. upholstered furniture

Mattresses and mattress pads sold in the United States must comply with the requirements of FF 4-72, contained in 16 CFR 1632 in the Code of Federal Regulations.

California Technical Bulletin 106 is similar to 16 CFR 1632 (FF 4-72).

Materials used in upholstered furniture sold in California must comply with the requirements of California Technical Bulletins 116 and 117. These requirements have been adopted in some places outside California.

Mattresses sold for public occupancies in some places in California must comply with the requirements of California Technical Bulletin 129. These requirements have been adopted in some places outside California.

California Technical Bulletin 129 is similar to the following standards:

ASTM E 1590
UL 1895
NFPA 267

Seating furniture sold for public occupancies in some places in California must comply with the requirements of California Technical Bulletin 133. These requirements have been adopted in some places outside California.

California Technical Bulletin 133 is similar to the following standards:

ASTM E 1537
UL 1056
NFPA 266
Boston BFD IX-10

Mattresses sold for high-risk occupancies such as correctional institutions in California must comply with the requirements of California Technical Bulletin 121.

Curtains and draperies are usually tested in the vertical position, the position in which they are most often used. Multi-layered fabrics are a subject of concern because the combination of fabrics may fail a test even though the individual fabrics pass the test. The most widely used tests are UL 214 and NFPA 701.

Floor coverings are usually tested in the horizontal position, the position in which they are most often used. The most widely used tests are ASTM E 648 and ASTM E 84.

Floor coverings sold in the United States must comply with the following requirements in the Code of Federal Regulations:

FF 1-70, 16 CFR 1630. Carpets and Rugs
FF 2-70, 16 CFR 1631. Small Carpets and Rugs

Furniture and furnishings sold in the United Kingdom must comply with the Furniture and Furnishings (Fire)(Safety) Regulations 1988, first laid before Parliament on 28 July 1988.

British standards (BS) issued by the British Standards Institution (BSI) for furniture and furnishings include:

BS 5852. Fire tests for furniture
    Part 1. Ignitability of upholstered assemblies by either a smoldering cigarette or a lighted match
    Part 2. Ignitability of upholstered composites by flaming sources
    Section 1. General and guidance
    Section 2. Smoldering ignition source
    Section 3. Flaming ignition sources
    Section 4. Methods of test for ignitability of upholstery composites
    Section 5. Methods of test for ignitability of complete items of furniture
BS 5867. Flammability of fabrics for curtains and drapes
BS 6307. Ignition of textile floor coverings (methenamine tablet test) (ISO 6925)
BS 6807. Methods of test for ignitability of mattresses
BS 7176. Resistance to ignition of upholstered furniture

Apparel sold in the United States must comply with the following requirements in the Code of Federal Regulations:

16 CFR 1610. Clothing Textiles
16 CFR 1611. Vinyl Plastic Film
16 CFR 1615. Children's Sleepwear Sizes 0 through 6X (FF 3-71)
16 CFR 1616. Children's Sleepwear Sizes 7 through 14 (FF 5-74)

Appliances consumed 1557 million lb. of plastics in 1994.

Flammability tests are described in some of the many standards for appliances developed by Underwriters Laboratories.

SECTION 6.8. THE INDUSTRIAL GOODS MARKET

The industrial goods market covers the areas of machinery, machine parts, and industrial coatings. and may be considered to include part of the packaging market and part of the electrical and electronics market. The areas of machinery, machine parts, and industrial coatings do not have plastics consumption data available on the same basis as the data presented for other markets.

The complexity of this market makes knowledge of all flammability requirements very difficult. In many cases, there are no flammability requirements other than those based on government or insurance requirements. Materials which are difficult to ignite appear to be generally acceptable if they meet other performance requirements.

Government requirements on the Federal level are largely based on the Federal government's position as a major purchaser, and generally come from the General Services Administration (GSA). Various government agencies indirectly influence requirements.

The Code of Federal Regulations (CFR) is a codification of the general and permanent rules published in the Federal Register by the Executive departments and agencies of the Federal Government of the United States. The Code is divided into 50 titles which represent broad areas subject to Federal regulation. Each title is divided into chapters which usually bear the name of the issuing agency. Each chapter is further subdivided into parts covering specific regulatory areas. The outline of the Code of Federal Regulations, and some documents relevant to fire safety and plastics, are presented in Table 6.1.

The Mine Safety and Health Administration requires conveyor belting and hose to pass flame resistance tests described in Section 18.65 of Title 30, Part 18.

The military specifications pertinent to the industrial goods market include those in the following groups in the Federal Supply Classification:

Group 28. Engines, Turbines, and Components
Group 29. Engine Accessories
Group 30. Mechanical Power Transmission Equipment
Group 31. Bearings
Group 32. Woodworking Machinery and Equipment
Group 34. Metalworking Machinery
Group 35. Service and Trade Equipment
Group 36. Special Industry Machinery
Group 37. Agricultural Machinery and Equipment
Group 38. Construction, Mining, Excavating, and Highway Maintenance Equipment
Group 39. Materials Handling Equipment
Group 40. Rope, Cable, Chain, and Fittings
Group 41. Refrigeration and Air Conditioning Equipment
Group 43. Pumps and Compressors
Group 44. Furnace, Steam Plant, and Drying Equipment; and Nuclear Reactors
Group 46. Water Purification and Sewage Treatment Equipment
Group 47. Pipe, Tubing, Hose, and Fittings

Group 48. Valves
Group 49. Maintenance and Repair Shop Equipment
Group 58. Communication, Detection, and Coherent Radiation Equipment
Group 59. Electrical and Electronic Equipment Components
Group 61. Electric Wire, and Power and Distribution Equipment
Group 66. Instruments and Laboratory Equipment
Group 67. Photographic Equipment
Group 64. Office Machines, Visible Record Equipment, and Data Processing Equipment
Group 81. Containers, Packaging, and Packing Supplies

Insurance requirements on specific products in this market have been developed by Underwriters Laboratories (UL) and Factory Mutual (FM).

SECTION 6.9. THE INTERNATIONAL MARKET

Plastics are produced and consumed in countries all over the world, and the market for plastics has become an international market. Many plastics material manufacturers in the United States have parent companies outside the United States, such as Akzo Nobel, Bayer, and Hoechst.

The international market can be divided into six regions:

1. North America: United States, Canada, and Mexico
2. South America: Argentina, Brazil, Chile, Colombia, Ecuador, Paraguay, Peru, Uruguay, Venezuela, and others
3. Asia and Oceania: Australia, China, Japan, New Zealand, South Korea, Taiwan, and others
4. Western Europe: Belgium, Denmark, France, Germany, Ireland, Italy, Netherlands, Norway, Portugal, Spain, Sweden, Switzerland, United Kingdom, and others
5. Eastern Europe: former Czechoslovakia, Hungary, Poland, Romania, former Soviet Union, and others
6. Africa and Mideast

Estimated plastics consumption by region and country is shown in Table 1.2. Estimated plastics production by region is shown in Table 1.3.

Each country is different from the United States and has its own culture and its own way of doing business. It has developed its own standards, influenced heavily by its national history and experience. In many cases, it has developed its own fire safety standards and its own flammability tests, influenced heavily by its particular experience, and is understandably reluctant to replace them with foreign or even consensus standards and tests, partly because of national pride and partly because of substantial investment in existing standards and tests.

The predominantly English-speaking countries have advantages in communication with each other because they share a common language, and to varying degrees, similar cultures and similar values. Their fire safety standards and flammability tests do not need to be carefully translated to be understood by each other.

The unification of Europe has been a theory and a dream for centuries, in concepts such as a United States of Europe. It was partly accomplished for brief periods through military conquest by Napoleon and Hitler. A United States of Europe would represent a market greater than that of the United States of America. Western Europe already represents a market comparable to that of the United States of America.

The cause of European unification was advanced by the formation of groups of countries with common interests.

The European Community (EC) was formed with twelve member countries: United Kingdom, Ireland, France, Belgium, Netherlands, Luxembourg, Denmark, West Germany, Portugal, Spain, Italy, and Greece.

The European Free Trade Association (EFTA) was formed with five member countries: Norway, Sweden, Finland, Austria, and Switzerland.

In 1992, the European Community became one unified internal market. A unified market requires uniform standards, and the development of uniform standards within the European Community was the task of the Committee for European Normalization (CEN).

Three flammability tests widely used within the European Community (EC) market were:

1. For ignitability, the French NF P 92-501 Epiradiateur test
2. For heat release, the German DIN 4102 test
3. For flame spread, the British BS 476 Part 7 test

The European Union (EU) has 15 member countries:

Austria
Belgium
Denmark
Finland
France
Germany
Greece
Ireland
Italy
Luxembourg
Netherlands
Portugal

Spain
Sweden
United Kingdom

With the inception of the then-named European Common Market (ECM), now the European Union (EU), countries in the Pacific area feared that they would be at a disadvantage in international standards organizations, primarily the International Standardization Organization (ISO) and the International Electrotechnical Commission (IEC). The Pacific Area Standards Congress (PASC) was formed and had 22 members in 1997:

Australia
Brunei Darussalam
Canada
Chile
People's Republic of China
Colombia
Commonwealth of Independent States
Fiji
Hong Kong
Indonesia
Japan
Republic of Korea
Malaysia
Mexico
New Zealand
Papua New Guinea
Philippines
Singapore
Thailand
South Africa
United States of America
Vietnam

The national standards bodies of the countries located within the American continent formed an association with complete autonomy and unlimited duration. The Comision Panamericana de Normas Tecnicas (COPANT) or Pan American Standards Commission, was founded in July 1949 as a committee and became operational in April 1961, and acquired its present name in 1964.

Active members are limited to the national standards bodies of the 24 countries located within the American continent:

Argentina
Barbados
Bolivia

Brazil
Canada
Central America
Chile
Colombia
Costa Rica
Cuba
Dominican Republic
Ecuador
Grenada
Guatemala
Guyana
Jamaica
Mexico
Panama
Paraguay
Peru
Trinidad and Tobago
United States
Uruguay
Venezuela

Adherent members are countries that are not located within the American continent, and currently include France, Italy, Portugal, and Spain.

The active and adherent members of COPANT are listed in Table 6.22.

In December 1994, the leaders of 34 Western Hemisphere countries met to launch the Free Trade Area of the Americas (FTAA) initiative and committed themselves to complete negotiations by the year 2005. This would be the culmination of over half a century of gradual movement toward this goal.

In December 1960, the Central America Common Market (CACM) was founded, and entered into force in 1973 with the Puntarenas Declaration. There were five member countries: Costa Rica, El Salvador, Guatemala, Honduras, and Nicaragua. The population in 1996 was 30.4 million, and total GDP in 1996 was $3.2 billion.

In 1969, the Andean Pact, a customs union, was founded with the Cartagena Agreement, and modified in May 1988 with the Quito Protocol. There were five member countries: Bolivia, Colombia, Ecuador, Peru, and Venezuela. The population in 1996 was 101.3 million, and total GDP in 1996 was $193.9 billion.

In July 1973, the Caribbean Common Market (CARICOM) was founded with the Treaty of Chaguaramas. There were 14 member countries:

Antigua and Barbuda

Bahamas
Barbados
Belize
Dominica
Grenada
Guyana
Jamaica
Montserrat
St. Kitts and Nevis
St. Lucia
St. Vincent and Grenadines
Suriname
Trinidad and Tobago

The population in 1996 was 5.9 million, and total GDP in 1996 was $28.6 billion.

In March 1991, Mercosur was founded with the Treaty of Asuncion, and entered into force in January 1995. There were four member countries: Argentina, Brasil, Paraguay, and Uruguay. The population in 1996 was 206.9 million, and total GDP in 1996 was $667 billion.

In December 1992, Canada, Mexico, and the United States signed the North American Free Trade Agreement (NAFTA), effective January 1994. The population in 1996 was 387.8 million, and total GDP in 1996 was $7,117 billion.

In June 1994, Colombia, Mexico, and Venezuela founded the Group of Three (G3) in Cartagena de Indias, effective January 1995. The population in 1996 was 149.9 million, and total GDP in 1996 was $384.6 billion.

If successful, the FTAA would result in the largest market in the world, with more than 750 million people (about 16 per cent of the world population) and a combined GDP of nearly $9 trillion (about 32 per cent of the global wealth).

The most prominent international standards organization is the International Standardization Organization (ISO), P.O. Box 56, CH-1211 Geneva 20, Switzerland. Some ISO standards relevant to fire safety and plastics are shown in Table 4.1.

The International Standardization Organization (ISO), with participation from standards organizations in 86 countries, has made progress toward a standardized system of tests. The countries participating in ISO are listed in Table 6.23.

All worldwide standards, except those of the United States, are in terms of metric units (versus inch/pound or a "soft conversion" of inch/pound units to metric), and ISO/IEC metric standards already exist. The need for "harmonization" is apparent.

A "soft conversion" refers to the practice of directly converting an inch/pound unit to a metric unit. For example, 0.5 inch is converted to 12.7 mm. A "hard conversion" would convert 0.5 in to 13.0 mm.

There are at least five standards organizations in the United States which issue standards relevant to fire safety and plastics:

1. American National Standards Institute (ANSI), 11 West 42nd Street, New York, New York 10036, Tel 212-642-8908, Fax 212-398-0023
2. American Society for Testing and Materials (ASTM), 100 Barr Harbor Drive, West Conshohocken, Pennsylvania 19428-2959, USA, Tel 610-832-9500, Fax 610-832-9555. Some ASTM standards relevant to fire safety and plastics are shown in Table 4.2.
3. Underwriters Laboratories (UL), 300 Pfingsten Road, Northbrook, Illinois 60062-2096, USA, Tel 847-272-8800. Some UL standards relevant to fire safety and plastics are shown in Table 4.3.
4. National Fire Protection Association (NFPA), 1 Batterymarch Park, P.O. Box 9101, Quincy, Massachusetts 02269-9101, USA, Tel 617-770-3000. Some NFPA standards relevant to fire safety and plastics are shown in Table 4.4.
5. International Conference of Building Officials (ICBO), 5360 South Workman Mill Road, Whittier, California 90601, USA, Tel 562-699-0541. Some Uniform Building Code (UBC) standards relevant to fire safety and plastics are shown in Table 4.5.

There are at least three standards organizations in Canada which issue standards relevant to fire safety and plastics:

1. Standards Council of Canada (SCC)
2. Canadian Standards Association (CSA), 178 Rexdale Boulevard, Etobicoke, Ontario M9W 1R3, Tel 416-747-4033, Fax 416-747-2475. Some CSA standards relevant to fire safety and plastics are shown in Table 4.6.
3. Underwriters Laboratories of Canada (ULC), 7 Crouse Road, Scarborough, Ontario M1R 3A9, Tel 416-757-3611, Fax 416-757-9540. Some ULC standards relevant to fire safety and plastics are shown in Table 4.7.

The standards organization in Australia is Standards Association of Australia (SAA). Some Australian standards (AS) standards relevant to fire safety and plastics are shown in Table 4.8.

The standards organization in New Zealand is Standards New Zealand (SNZ).

The standards organization in the United Kingdom is British Standards Institution (BSI), 2 Park Street, London W1A 2BS, United Kingdom. Some British standards (BS) relevant to fire safety and plastics are shown in Table 4.9.

The standards organization in Japan is Japanese Standards Association (JSA). Some Japanese Industrial Standards (JIS) relevant to fire safety and plastics are shown in Table 6.24.

There are at least two standards organizations in France which issue standards relevant to fire safety and plastics:

1. Association Francaise de Normalisation (AFNOR), Tour Europe, 92049 Paris La Defense Cedex, France, Tel +33 1 42 91 55 55, Fax +33 1 42 91 56 56. Some NF standards relevant to fire safety and plastics are shown in Table 6.25.
2. Union Technique de l'Electricite (UTE). Some UTE standards relevant to fire safety and plastics are shown in Table 6.25

The standards organization in Belgium is Institut Belge de Normalisation (IBN), Avenue de la Brabanconne, 29, B 1000 Bruxelles, Belgium, Tel 02 738 01 13, Fax 02 733 42 64.

The standards organization in the Netherlands is Netherlands National Institute (NNI).

There are at least two standards organizations in Germany which issue standards relevant to fire safety and plastics:

1. Deutsches Institut fur Normung (DIN). Some DIN standards relevant to fire safety and plastics are shown in Table 6.26.
2. Verband Deutscher Elektrotechniker (VDE). Some VDE standards relevant to fire safety and plastics are shown in Table 6.26.

The standards organization in Switzerland is Vereinigung Kantonaler Feuerverschingerungen (VKF), Bundesgasse 20, 3011 Bern, Switzerland, Tel 031 320 22 22, Fax 031 320 22 99.

The standards organization in Denmark is Dansk Institut for Provning og Justering (Dantest), Danish National Institute for Testing and Verification. Some DS standards relevant to fire safety and plastics are shown in Table 6.27.

There are at least two standards organizations in Norway which issue standards relevant to fire safety and plastics:

1. Norsk Allmennstandardisering (NAS), Drammensveien 145, Postboks 360, Skoyen 0212 Oslo, Norway, Tel 22 04 92 00, Fax 22 04 92 15
2. Norges Branntekniske Laboratorium (NBL), Norwegian Fire Research Laboratory

Some NS standards relevant to fire safety and plastics are shown in Table 6.28.

The standards organization in Sweden is Sveriges Provnings- och Forskningsinstitut (SP), Swedish National Testing and Research Institute, Box 857, S-501 15 Boras, Sweden, Tel +46 33 16 50 00, Fax +46 33 11 77 59. Some SP standards relevant to fire safety and plastics are shown in Table 6.29.

The standards organization in Finland is Suomen Standardisoimisliitto (SFS), Finnish Standards Association, P.O. Box 116, 00241 Helsinki (Maistraatinportti 2), Finland, Tel +358 9 149 9331, Fax +358 9 146 4925. Some SF standards relevant to fire safety and plastics are shown in Table 6.30.

The standards organization in the former Czechoslovakia was Vyzkumny Ustav Pozemnich Staveb (VUPS), Building Research Institute. Some CSN standards relevant to fire safety and plastics are shown in Table 6.31.

The standards organization in Slovenia is Standards and Metrology Institute of the Republic of Slovenia, Kotnikova 6, SI-1000 Ljubljana, Slovenia.

REFERENCES

Babrauskas, V., Krasny, J., "Fire Behavior of Upholstered Furniture", NBS Monograph 173, National Bureau of Standards, Gaithersburg, Maryland (November 1985)

Barile, P. W., "A Systematic Approach for Predicting Compliance with Technical Bulletin 133 for Untested Combinations of Chair Styles and Fabrics", Proceedings of the International Conference on Fire Safety, Vol. 18, 20-31 (January 1993)

Benisek, L., "Flammability of Upholstered Seating", Proceedings of the International Conference on Fire Safety, Vol. 12, 294-306 (January 1987)

Benisek, L., Myers, P. C., Palin, M. J., Woollin, R., "Latest Developments in Wool Aircraft Textile Furnishings", Proceedings of the International Conference on Fire Safety, Vol. 14, 153-167 (January 1989)

Benisek, L., Woollin, R., "Performance of Carpets and Critical Parameters in the NBS Flooring Radiant Panel Test", Proceedings of the International Conference on Fire Safety, Vol. 15, 1-10 (January 1990)

Benisek, L., "Environmentally Sound Aircraft Carpets", Proceedings of the International Conference on Fire Safety, Vol. 17, 180-197 (January 1992)

BOCA National Building Code, 1993 Edition, Building Officials and Code Administrators International, Country Club Hills, Illinois (1993)

Cahill, P. L., "Electrical Short Circuit and Current Overload Tests on Aircraft Wiring", DOT/FAA/CT-TN94/55, Federal Aviation Administration, Atlantic City, New Jersey (March 1995)

Cahill, P. L., "The Development of a Flammability Test Method for Aircraft Blankets", DOT/FAA/AR-96/15, Federal Aviation Administration, Atlantic City, New Jersey (March 1996)

Cahill, P. L., "Continued Fireworthiness Compliance of Aircraft Interior Materials", DOT/FAA/AR-TN96/25, Federal Aviation Administration, Atlantic City, New Jersey (April 1996)

Cahill, P. L., "Electrical Short Circuit and Current Overload Tests on Aircraft Wiring", Proceedings of the International Conference on Fire Safety, Vol. 22, 81-89 (July 1996)

Cahill, P. L., "Continued Fireworthiness Compliance of Interior Aircraft Materials", Proceedings of the International Conference on Fire Safety, Vol. 22, 114-118 (July 1996)

Cahill, P. L., "The Development of a Flammability Test Method for Aircraft Blankets", Proceedings of the International Conference on Fire Safety, Vol. 22, 137-168 (July 1996)

Cahill, P. L., "Evaluation of Fire Test Methods for Aircraft Thermal Acoustical Indulation", DOT/FAA/AR-97/58, Federal Aviation Administration, Atlantic City, New Jersey (September 1997)

Caudill, L. M., Hoover, J. R., "Fire Testing of Communications Cables", Proceedings of the International Conference on Fire Safety, Vol. 21, 206-212 (January 1996)

Cleary, T. G., Ohlemiller, T. J., Villa, K. M., "The Influence on Ignition Source on the Flaming Fire Hazard of Upholstered Furniture", NISTIR 4847, National Institute of Standards and Technology, Gaithersburg, Maryland (June 1992)

Coaker, A. W., Hirschler, M. M., "Fire Characteristics of Standard and Advanced PVC Wire and Cable Compounds", Proceedings of the International Conference on Fire Safety, Vol. 13, 397-416 (January 1988)

Creyf, H., "Development of Fire Legislation for Upholstered Furniture and Bedding within the European Community", Proceedings of the International Conference on Fire Safety, Vol. 15, 27-33 (January 1990)

Damant, G. H., McCormack, J. A., Mikami, J. F., Wortman, P. S., Hilado, C. J., "The California Technical Bulletin 133 Test: Some Background and Experience", Proceedings of the International Conference on Fire Safety, Vol. 14, 1-12 (January 1989)

Damant, G. H., Nurbakhsh, S., Hilado, C. J., "Flammability of Seating Furniture: California Technical Bulletin 133 Test History and Development", Proceedings of the International Conference on Fire Safety, Vol. 16, 7-25 (January 1991)

Damant, G. H., Nurbakhsh, S., Mikami, J., "Full Scale Heat Release Rate Tests on Bedding Systems", Proceedings of the International Conference on Fire Safety, Vol. 17, 38-63 (January 1992)

Damant, G. H., Nurbakhsh, S., "Developing a 'Code of Practice' for Technical Bulletin

133 Testing", Proceedings of the International Conference on Fire Safety, Vol. 18, 48-71 (January 1993)

Damant, G. H., "Cigarette Ignition of Upholstered Furniture", Journal of Fire Sciences, Vol. 13, No. 5, 337-349 (September/October 1995)

Damant, G. H., "Use of Barriers and Fire Blocking Layers to Comply with Full-Scale Fire Tests for Furnishings", Journal of Fire Sciences, Vol. 14, No. 1, 3-25 (January/February 1996)

Damant, G. H., Nurbakhsh, S., "Development of Furnishings Flammability Standards for Public Buildings and Private Residences", Journal of Fire Sciences, Vol. 13, No. 6, 417-433 (November/December 1995)

Damant, G. H., "Use of Barriers and Fire Blocking Layers to Comply with Full-Scale Fire Tests for Furnishings", Proceedings of the International Conference on Fire Safety, Vol. 24, 1-19 (July 1997)

Delaney, H., "Stacking the Deck in Europe", ASTM Standardization News, Vol. 24, No. 8, 34-37 (August 1996)

Diaz Portocarrero, M., "COPANT's Past, Present and Future", ASTM Standardization News, Vol. 25, No. 8, 26-29 (August 1997)

D'Silva, A. P., Sorensen, N., "The Flammability Aspects of Decorative Trimmings: Part 1 - Flammability of Trimmings Used on Upholstered Furniture", Journal of Fire Sciences, Vol. 14, No. 1, 26-49 (January/February 1996)

Ferm, D. J., "A Small Scale Chair Test for Evaluation of the Flammability of Flexible Polyurethane Seating Foam Containing Zinc or Ammonium Borates as Flame Retardant Additives", Proceedings of the International Conference on Fire Safety, Vol. 18, 241-257 (January 1993)

Fernandez-Pello, A. C., Hasegawa, H. K., Staggs, K., Lipska-Quinn, A., Alvares, N. J., "A Study of the Fire Performance of Electrical Cables", Proceedings of the 3rd International Symposium on Fire Safety Science, 237-247

Flisi, U., "Harmonisation of Fire Testing in the European Community", Proceedings of the International Conference on Fire Safety, Vol. 14, 41-52 (January 1989)

Forsberg, C. W., "Nonmetallic Electrical Tubing for the Plenum", Proceedings of the International Conference on Fire Safety, Vol. 15, 269-280 (January 1990)

Forsten, H. H., "Furniture Fire Blockers: A Route Toward Improved Fire Protections", Proceedings of the International Conference on Fire Safety, Vol. 15, 11-22 (January 1990)

Forsten, H. H., "Fabric/Foam/Blocker Systems: Evaluation of Relative Flammability Using a Cone Calorimeter", Proceedings of the International Conference on Fire Safety, Vol. 18, 104-120 (January 1993)

Friedman, R., "Review of Fire Safety in Spacecraft", Proceedings of the International Conference on Fire Safety, Vol. 20, 170-174 (January 1995)

Gallagher, R. M., Wadehra, I., "Comparison of Flame Propagation and Smoke Development from Cables Installed in a Simulated Plenum and Modified UL-910 Test", Proceedings of the International Conference on Fire Safety, Vol. 13, 256-272 (January 1988)

Grauers, K., Persson, B., "Measurements of Thermal Properties in Building Materials at High Temperatures", Nordtest Project No. 1028-02, SP Report 1994:09, Swedish National Testing and Research Institute, Boras, Sweden (1994)

Grayson, S. J., Hirschler, M. M., "National and International Developments in Standards for Buildings and Contents", Proceedings of the International Conference on Fire Safety, Vol. 19, 75-88 (January 1994)

Griffiths, T., "Combustion Modified Foams and the U.K. Furniture Regulations", Proceedings of the International Conference on Fire Safety, Vol. 14, 96-113 (January 1989)

Haslim, L. A., "Light Weight Fire Retardant Crashworthy Aircraft Seat Cushioning", Proceedings of the International Conference on Fire Safety, Vol. 18, 163-177 (January 1993)

Heckenliable, M. A., "Flammability Regulations and the Business Furniture Industry", Proceedings of the International Conference on Fire Safety, Vol. 22, 105-113 (July 1996)

Hilado, C. J., "Flammability Handbook for Electrical Insulation", Technomic Publishing Company, Westport, Connecticut (1982)

Hilado, C. J., "Flammability Handbook for Thermal Insulation", Technomic Publishing Company, Lancaster, Pennsylvania (1983)

Hilado, C. J., "Flammability Handbook for Plastics", 4th Ed., Technomic Publishing Co., Lancaster, Pennsylvania (1990)

Hill, R. G., "Aircraft Fire Material Testing and the Creation of an International Working Group", Proceedings of the International Conference on Fire Safety, Vol. 18, 178-182 (January 1993)

Hirschler, M. M., "Fire Hazard Assessment for Rail Transportation: Progress to Develop an ASTM Standard", Proceedings of the International Conference on Fire Safety, Vol. 20, 179-188 (January 1995)

Hirschler, M. M., "Analysis of Cone Calorimeter and Room-Scale Data on Fire Performance of Upholstered Furniture", Proceedings of the International Conference on Fire Safety, Vol. 23, 59-78 (January 1997)

Hirschler, M. M., "Testing Techniques Associated with Heat Release: the Cone Calorimeter (and its Applications) and Room/Furniture Scale Tests", Proceedings of the International Conference on Fire Safety, Vol. 23, 159-169 (January 1997)

Hoover, J. R., Caudill, L. M., Chapin, J. T., "Results of Full-Scale Fire Tests of Communications Cables Used in Concealed-Space Applications", Proceedings of the International Conference on Fire Safety, Vol. 23, 208-212 (January 1997)

Horak, E., "The Contribution of Man-Made Fibers to Fire Safety", Proceedings of the International Conference on Fire Safety, Vol. 14, 13-34 (January 1989)

Hovde, P. J., University of Trondheim, Norwegian Institute of Technology, private communication (August 18, 1989)

Innes, J. D., "Flame Retardants and Their Market Applications", Proceedings of the March 1996 FRCA Conference, 61-69 (1996)

Japanese Standards Association, "The Japanese Standards Association", ASTM Standardization News, Vol. 25, No. 10, 16-18 (October 1997)

Johnson, R. M., "Round-Robin Comparison of Heat Release Apparatus", DOT/FAA/CT-TN94/42, Federal Aviation Administration, Atlantic City, New Jersey (September 1994)

Kajastila, R., "Fire Tests on Coverings with a Substrate of Cellular Plastics", VTT Research Reports 632, Technical Research Centre of Finland, Espoo, Finland (July 1989)

Kaminski, A., "A Study of Smoke Emission of Materials Used in Passenger Rail Vehicles", Journal of Fire Sciences, Vol. 6, No. 4, 267-289 (July/August 1988)

Kaufman, S., "The 1987 National Electrical Code Fire Safety Requirements for Communications Cables", Proceedings of the International Conference on Fire Safety, Vol. 11, 109-127 (January 1986)

Keogh, M. J., "Polyolefin Cable Materials with Reduced Smoke Toxicity and Smoke Corrosivity", Proceedings of the International Conference on Fire Safety, Vol. 13, 388-396 (January 1988)

Keski-Rahkonen, O., Bjorkman, J., Farin, J., "Derating of Cables at High Temperatures", VTT Publications 302, Technical Research Centre of Finland, Espoo, Finland (1997)

Kolk, S. B., "The BIFMA Voluntary Flammability Standards for Non-Live-In Business and Institutional Furniture", Proceedings of the International Conference on Fire Safety, Vol. 10, 104-111 (January 1985)

Kushnier, G. W., "The Pacific Area Standards Congress (PASC)", ASTM Standardization News, Vol. 25, No. 10, 19-21 (October 1997)

Landrock, A. H., "Handbook of Plastics Flammability and Combustion Toxicology", Noyes Publications, Park Ridge, New Jersey (1983)

Litant, I., "Comparison of U.S. and French Regulations for Rail Transit Fire Safety", Proceedings of the International Conference on Fire Safety, Vol. 13, 203-217 (January 1988)

Londono, C., "A Free Trade Area of the Americas by the Year 2005?", ASTM Standardization News, Vol. 25, No. 6, 28-34 (June 1997)

Lyon, R. E., "Fire-Resistant Thermoset Polymers for Aircraft Interiors", Proceedings of the International Conference on Fire Safety, Vol. 20, 120-169 (January 1995)

Mansson, M., Blomqvist, P., Isaksson, I., Rosell, R., "Sampling and Chemical Analysis of Smoke Gas Components from the SP Industry Calorimeter", SP Report 1994:35, Swedish National Testing and Research Institute, Boras, Sweden (1994)

Marchant, R., "European Developments in Control of Furniture Fire Hazards", Proceedings of the International Conference on Fire Safety, Vol. 18, 32-37 (January 1993)

Marker, T. R., Sarkos, C. P., "Full Scale Test Evaluation of Aircraft Fuel Fire Burnthough Resistance Improvements", Proceedings of the International Conference on Fire Safety, Vol. 23, 96-109 (January 1997)

Marker, T. R., Sarkos, C. P., "Full-Scale Test Evaluation of Aircraft Fuel Fire Burn-Through Resistance Improvements", SAMPE Journal, Vol. 33, No. 4, 32-39 (July/August 1997)

McGarry, F. A., "Administration of New York State's Fire Gas Toxicity Program", Proceedings of the International Conference on Fire Safety, Vol. 13, 273-276 (January 1988)

National Electrical Code, 1996 Edition, National Fire Protection Association, Quincy, Massachusetts (1996)

Nishizawa, H., "Flame Retardant Properties of Vibration Damping Materials", Proceedings of the March 1995 FRCA Conference, 67-78 (1995)

Ohlemiller, T. J., Villa, K. M., "Furniture Flammability: An Investigation of the California Bulletin 133 Test. Part II: Characterization of the Ignition Source and a Comparable Gas Burner", NISTIR 4348, National Institute of Standards and Technology, Gaithersburg, Maryland (June 1990)

Ohlemiller, T. J., Villa, K., "Characterization of the California Technical Bulletin 133 Ignition Source and a Comparable Gas Burner", Fire Safety Journal, Vol. 18, 325-354 (1992)

Peterson, J. M., "Development of the Total Design", Proceedings of the International Conference on Fire Safety, Vol. 14, 146-152 (January 1989)

Reynolds, T. L., "Emerging Technology: Improvements in Aircraft Passenger Safety", Proceedings of the International Conference on Fire Safety, Vol. 17, 149-170 (January 1992)

Roman, S. W., "Present Day Material Usage in Mass Transit Rail Vehicles", Proceedings of the International Conference on Fire Safety, Vol. 12, 182-191 (January 1987)

Roux, H. J., "The NIBS SMOTOX WG Potential Toxic Hazard Test", presentation at the 14th International Conference on Fire Safety, Millbrae, California (January 12, 1989)

Ruikar, V. G., "Customized Tests for Detention and Correctional Facilities, ASTM Standardization News, Vol. 24, No. 5, 16-19 (May 1996)

Ryan, J. D., "Fire Hazard Assessment in Evaluatimg Cable Performance", Proceedings of the International Conference on Fire Safety, Vol. 12, 366-374 (January 1987)

Sarkos, C. P., "FAA's Cabin Fire Safety Test Program: Status and Recent Findings", Proceedings of the International Conference on Fire Safety, Vol. 15, 108-125 (January 1990)

Sarkos, C. P., "Application of Full-Scale Fire Tests to Characterize and Improve the Aircraft Postcrash Fire Environment", International Colloquium on Advances in Combustion Toxicology, Oklahoma City, Oklahoma (April 1995)

Sarkos, C. P., Hill, R. G., Johnson, R. M., "Implementation of Heat Release Measurements as a Regulatory Requirement for Commercial Aircraft Materials", 50th Calorimetry Conference, Gaithersburg, Maryland (July 1995)

Serkov, B. B., Asseva, R. M., "The Influence of Thermal Ageing on Fire Safety of Electrical and Electronic Products Based on Polyolefins", Proceedings of the International Conference on Fire Safety, Vol. 22, 63-80 (July 1996)

Serkov, B. B., "Some Aspects of Fire Regulation, Testing, and Certification Materials in Russia", Proceedings of the International Conference on Fire Safety, Vol. 22, 169-188 (July 1996)

SFS Catalogue 1997, Suomen Standardisoimisliitto, Helsinki, Finland (1997)

Slenski, G., Soloman, R. S., "New Insulation Constructions for Aircraft Wiring", Proceedings of the International Conference on Fire Safety, Vol. 15, 126-135 (January 1990)

Sonderbom, J., "Smoke Spread Experiments in Large Rooms. Experimental Results and Numerical Simulations", SP Report 1992:52, Swedish National Testing and Research Institute, Boras, Sweden (1992)

Sorathia, U., Lyon, R., Ohlemiller, T., Grenier, A., "A Review of Fire Test Methods and Criteria for Composites", SAMPE Journal, Vol. 33, No. 4, 23-31 (July/August 1997)

Troitzsch, J., "International Plastics Flammability Handbook", Hanser, Munich (1983)

Troitzsch, J., "Trends in Fire Safety and Consumer Electronics in Europe", Proceedings of the March 1997 FRCA Conference, 91-99 (1997)

UL Building Materials Directory 1997, Underwriters Laboratories, Northbrook, Illinois (1997)

UL Catalog of Standards 1996/1997, Underwriters Laboratories, Northbrook, Illinois (1996)

Uniform Building Code, 1997 Edition, International Conference of Building Officials, Whittier, California (1997)

Van Hees, P., Thureson, P., "Burning Behaviour of Cables - Modelling of Flame Spread", BRANDFORSK Project 725-942, SP Report 1996:30, Swedish National Testing and Research Institute, Boras, Sweden (1996)

Villa, K. M., Babrauskas, V., "Cone Calorimeter Rate of Heat Release Measurements for Upholstered Composites of Polyurethane Foams", NISTIR 4652, National Institute of Standards and Technology, Gaithersburg, Maryland (August 1991)

Walsh, P., "Reengineering the Standards Preparation Process Down Under", ASTM Standardization News, Vol. 25, No. 10, 14-15 (October 1997)

Watson, S. J., "Demand for Globalized Plastics Standards", ASTM Standardization News, Vol. 24, No. 9, 50-52 (September 1996)

Wilging, R. C., "Plastic Pipe Flammability", Proceedings of the International Conference on Fire Safety, Vol. 22, 1-9 (July 1996)

Williams, F. W., Beitel, J., Beyler, C., "The U.S. Navy's Passive Fire Safety Philosophy", Proceedings of the International Conference on Fire Safety, Vol. 24, 191-217 (July 1997)

Williams, S. S., "Flammability Regulations and Standards in the United States for Upholstered Furniture", Technomic Publishing Company, Lancaster, Pennsylvania (1985)

Youngs, R. W., "Flame Retardant Packaging for Increased Fire Safety in the Transportation and Warehousing of Hazardous Materials", Proceedings of the International Conference on Fire Safety, Vol. 19, 266-271 (January 1994)

**Table 6.1. The Code of Federal Regulations (CFR) and
Some Documents Relevant to Fire Safety and Plastics, CFR Documents.**

Title 1. General Provisions
Title 2. Reserved
Title 3. The President
Title 4. Accounts
Title 5. Administrative Personnel
Title 6. Reserved
Title 7. Agriculture
Title 8. Aliens and Nationality
Title 9. Animals and Animal Products
Title 10. Energy
Title 11. Federal Elections
Title 12. Banks and Banking
Title 13. Business Credit and Assistance
Title 14. Aeronautics and Space
  Chapter I. Federal Aviation Administration, Department of Transportation (Parts 1–199)
    Subchapter C. Aircraft
      Parts 23, 25, 27, 29. Airworthiness standards
      Part 23. Normal, utility, and commuter airplanes
      Part 25. Transport category airplanes
      Part 27. Normal category rotorcraft
      Part 29. Transport category rotorcraft
    14 CFR 23. Normal, utility, and commuter category airplanes
      FAR 23.853. Compartment interiors
    14 CFR 25. Transport category airplanes
      FAR 25.853. Compartment interiors
      FAR 25.855. Cargo and baggage compartments
    14 CFR 27. Normal category rotorcraft
      FAR 27.853. Compartment interiors
      FAR 27.855. Cargo and baggage compartments
    14 CFR 29. Transport category rotorcraft
      FAR 29.853. Compartment interiors
      FAR 29.855. Cargo and baggage compartments
    Subchapter G. Air carriers
      Part 121. Domestic, flag, and supplemental carriers
      Part 125. Airplanes with 20 or more passengers or 6000 lb. or more payload
    14 CFR 121. Domestic, flag, and supplemental air carriers
      FAR 121.215. Cabin interiors
    14 CFR 125. Airplanes of certain sizes and larger
      FAR 125.113. Cabin interiors
  Chapter V. National Aeronautics and Space Administration (Parts 1200–1299)
Title 15. Commerce and Foreign Trade
Title 16. Commercial Practices
  Chapter I. Federal Trade Commission (Parts 0–999)
  Chapter II. Consumer Product Safety Commission (Parts 1000–1799)
    Subchapter D. Flammable Fabrics Act regulations
      Part 1610. Clothing textiles
      Part 1611. Vinyl plastic film
      Part 1615. Children's sleepwear sizes 0 through 6X
      Part 1616. Children's sleepwear sizes 7 through 14
      Part 1630. Carpets and rugs
      Part 1631. Small carpets and rugs
      Part 1632. Mattresses and mattress pads
    16 CFR 1610. Clothing textiles

16 CFR 1611. Vinyl plastic film
16 CFR 1615, FF 3-71. Children's sleepwear sizes 0 through 6X
16 CFR 1616, FF 5-74. Children's sleepwear sizes 7 through 14
16 CFR 1630, FF 1-70. Carpets and rugs
16 CFR 1631, FF 2-70. Small carpets and rugs
Title 17. Commodity and Securities Exchanges
Title 18. Conservation of Power and Water Resources
Title 19. Customs Duties
Title 20. Employees' Benefits
Title 21. Food and Drugs
Title 22. Foreign Relations
Title.23. Highways
  Chapter I. Federal Highway Administration, Department of Transportation (Parts 1–999)
  Chapter II. National Highway Traffic Safety Administration and Federal Highway
    Administration, Department of Transportation (Parts 1200–1299)
  Chapter III. National Highway Traffic Safety Administration, Department of
    Transportation (Parts 1300–1399)
Title 24. Housing and Urban Development
Title 25. Indians
Title 26. Internal Revenue
Title 27. Alcohol, Tobacco Products, and Firearms
Title 28. Judicial Administration
Title 29. Labor
Title 30. Mineral Resources
  Chapter I. Mine Safety and Health Administration, Department of Labor (Parts 1–199)
    30 CFR 18.65. Flame test of conveyor belting and hose
  Chapter VI. Bureau of Mines, Department of the Interior (Parts 600–699)
Title 31. Money and Finance: Treasury
Title 32. National Defense
Title 33. Navigation and Navigable Waters
Title 34. Education
Title 35. Panama Canal
Title 36. Parks, Forests, and Public Property
Title 37. Patents, Trademarks, and Copyrights
Title 38. Pensions, Bonuses, and Veterans' Relief
Title 39. Postal Service
Title 40. Protection of Environment
Title 41. Public Contracts and Property Management
Title 42. Public Health
Title 43. Public Lands: Interior
Title 44. Emergency Management and Assistance
Title 45. Public Welfare
Title 46. Shipping
  Subchapter H. Passenger Vessels
  46 CFR 72.05. Structural fire protection
    72.05–5(g). Standard fire test
    72.05–10. Fire control bulkheads and decks
    72.05–15. Ceilings, linings, trims, and decorations in accomodation spaces and safety
     areas
    72.05–55. Furniture and furnishings
  Subchapter I. Cargo and miscellaneous vessels
  46 CFR 92.05. General fire protection

continued

Table 6.1 (continued). The Code of Federal Regulations (CFR) and Some Documents Relevant to Fire Safety and Plastics, CFR Documents.

Title 46. Shipping (continued)
  45 CFR 92.07. Structural fire protection
  92.07-5(a). Standard fire test
  92.07-10. Construction
Title 47. Telecommunication
Title 48. Federal Acquisitions Regulation System
Title 49. Transportation
  Chapter I. Research and Special Programs Administration, Department of
    Transportation (Parts 100–199)
  Chapter II. Federal Railroad Administration, Department of Transportation (Parts
    200–299)
  Chapter III. Federal Highway Administration, Department of Transportation (Parts
    300–399)
  Chapter IV. Coast Guard, Department of Transportation (Parts 400–499)
  Chapter V. National Highway Traffic Safety Administration, Department of Transportation
    (Parts 500–599)
    Part 571. Federal Motor Vehicle Safety Standards
  49 CFR 571.301
    Standard No. 301. Fuel system integrity
  49 CFR 571.302
    Standard No. 302. Flammability of interior materials
  Chapter VI. Federal Transit Administration, Department of Transportation (Parts 600–699)
  Chapter VII. National Railroad Passenger Corporation (AMTRAK) (Parts 700–799)
  Chapter VIII. National Highway Transportation Safety Board (Parts 800–999)
  Chapter IX. Surface Transportation Board, Department of Transportation (Parts
    1000–1399)
Title 50. Wildlife and Fisheries

Table 6.2. Some Plastics Used in Building and Construction.

United States	Million lb.	
	1995	1996
HDPE (corrugated pipe and conduit)	154	93
PVC (pipe and conduit)	4545	5360
PVC (siding)	1440	1740
PVC (windows and doors)	323	355
PVC (flooring, calendering)	217	235
PVC (flooring, coating)	207	225
PS (molding)	62	62
PS (extrusion)	80	78
PS foam (board)	185	196
Expandable polystyrene (EPS)	239	252
Polyurethane rigid foam (insulation)	630	647
ABS (pipe, extrusion)	130	125
ABS (pipe fittings, injection)	22	21
Acrylic (glazing)	100	105
Polycarbonate (glazing)	134	139
Acetals (plumbing, hardware)	49	52

Western Europe	Million lb.	
	1995	1996
LDPE (pipe and conduit)	264	261
LLDPE (pipe and conduit)	55	75
HDPE (pipe and conduit)	1049	1130
Rigid PVC (pipe and conduit)	3172	3238
Rigid PVC (profile extrusion)	2300	2335
Flexible PVC (flooring)	573	595
Flexible PVC (tubing, other profiles)	474	441
Expandable polystyrene (EPS)	997	970
ABS (pipe and fitting)	24	22
Polycarbonate (glazing)	152	132
Acetals (plumbing, hardware)	9	9

Japan	Million lb.	
	1995	1996
LDPE (pipe)	11	9
LLDPE (pipe)	57	57
HDPE (pipe)	176	172
Polystyrene foam	494	498
Reinforced polyester (construction)	90	88
Reinforced polyester (bldg. materials)	483	494

Reference: Modern Plastics, January 1997.

Table 6.3. Construction Classifications.

**1997 Uniform Building Code (UBC) Issued by
the International Conference of Building Officials (ICBO)**

Type I Fire resistive buildings (noncombustible walls
   and permanent partitions)
Type II Buildings
   Type II—F.R.
   Type II One-hour
   Type II—N
Type III Buildings
   Type III One-hour
Type IV Buildings
Type V Buildings
   Type V One-hour

**1993 BOCA National Building Code Issued by the
Building Officials and Code Administrators International**

Type 1 Noncombustible, protected, 1A and 1B
Type 2 Noncombustible
   Protected, 2A and 2B
   Unprotected, 2C
Type 3 Noncombustible/combustible
   Protected, 3A
   Unprotected, 3B
Type 4 Noncombustible/combustible, heavy timber
Type 5 Combustible
   Protected, 5A
   Unprotected, 5B

Table 6.4. Occupancy Classifications in the Uniform Building Code (UBC) Issued by the International Conference of Building Officials (ICBO).

Group and Division	Description of Occupancy
Group A	Assembly
A-1	Assembly room with an occupant load of 1,000 or more and a stage
A-2	Assembly room with an occupant load of less than 1,000 and a stage
A-2.1	Assembly room with an occupant load of 300 or more without a stage
A-3	Assembly room with an occupant load of less than 300 without a stage
A-4	Stadiums, reviewing stands, amusement park structures
Group B	Business. Business, office, professional, or service-type transactions, including storage of records and accounts; eating and drinking establishments with an occupant load of less than 30
Group E	Educational
E-1	Education through 12th grade, 50 or more persons
E-2	Education through 12th grade, less than 50 persons
E-3	Day care purposes, more than 6 persons
Group F	Factory and Industrial
F-1	Moderate hazard
F-2	Low hazard
Group H	Hazardous
H-1	High explosion hazard
H-2	Moderate explosion hazard or hazard from accelerated burning
H-3	High fire or physical hazard
H-4	Repair garages not classified as S-3
H-5	Aircraft repair hangars not classified as S-5
H-6	Hazardous production materials (HPM)
H-7	Health hazard
Group I	Institutional
I-1.1	Nurseries for children under age 6, up to 5 children; hospitals, sanitariums, nursing homes with nonambulatory patients, more than 5 patients
I-1.2	Health care centers with ambulatory patients, more than 5 patients
I-2	Nursing homes with ambulatory patients, more than 5 patients; homes for children age 6 or over, more than 5 children
I-3	Mental hospitals, mental sanitariums, jails, prisons, reformatories
Group M	Mercantile. Display and sale of merchandise, involving stocks of goods, wares, or merchandise, and accessible to the public
Group R	Residential
R-1	Hotels and apartment houses; congregate residences with more than 10 persons
R-3	Dwellings and lodging houses
Group S	Storage
S-1	Moderate hazard storage
S-2	Low hazard storage
S-3	Repair garages
S-4	Open parking garages
S-5	Aircraft hangars
Group U	Utility
U-1	Private garages, carports, sheds, agricultural buildings
U-2	Fences over 6 ft (1829 mm) high, tanks, towers

Table 6.5. Occupancy Classifications in the BOCA National Building Code Issued by the Building Officials and Code Administrators International.

Group and Division	Description of Occupancy
Group A	Assembly
A-1	With stages and platforms
A-2	Without stages, with viewing
A-3	Without stages, with platforms
A-4	Religious
A-5	Outdoor assembly
Group B	Business
Group E	Educational
Group F	Factory and Industrial
F-1	Moderate hazard
F-2	Low hazard
Group H	High Hazard
H-1	Detonation hazard
H-2	Deflagration (accelerated burning) hazard
H-3	Physical hazard
H-4	Health hazard
Group I	Institutional
I-1	6 or more persons in supervised environment
I-2	6 or more persons not capable of self-preservation
I-3	6 or more persons under restraint
Group M	Mercantile
Group R	Residential
R-1	Hotels, motels, boarding houses
R-2	Multiple family dwellings
R-3	One- or two-family dwelling units
R-4	Detached one- or two-family dwellings
Group S	Storage
S-1	Moderate hazard storage
S-2	Low hazard storage
Group U	Utility

Table 6.6. Occupancy Classifications in the Standard Building Code Issued by the Southern Building Code Congress International (SBCC).

Group	Description of Occupancy
Group A	Assembly
A-1	Large assembly
A-2	Small assembly
Group B	Business
Group E	Educational
Group F	Factory and Industrial
Group H	Hazardous
Group I	Institutional
Group M	Mercantile
Group R	Residential
Group S	Storage
Group U	Utility

Table 6.7. Flame Spread Classification for Interior Finish.

Uniform Building Code (UBC) Issued by the International Conference of Building Officials (ICBO)	
Class	Flame Spread Index
I	0–25
II	26–75
III	76–200

BOCA National Building Code Issued by the Building Officials and Code Administrators International	
Class	Flame Spread Index
I	0–25
II	26–75
III	76–200

Standard Building Code Issued by the Southern Building Code Congress International (SBCC)		
Class	Flame Spread Index	Smoke Developed
A	0–25	0–450
B	26–75	0–450
C	76–200	0–450

Table 6.8. Maximum Flame Spread Class for Interior Finish.

1997 Uniform Building Code (UBC) Issued by the International Conference of Building Officials (ICBO)

Occupancy Group	Enclosed Vertical Exitways	Other Exitways	Rooms or Areas
A	I	II	II*
B	I	II	III
E	I	II	II
F	II	III	II
H	I	II	III*
I-1.1, I-1.2, I-2	I	I*	II*
I-3	I	I*	I*
M	I	II	III
R-1	I	II	III
R-3	III	III	III*
S-1, S-2	II	II	III
S-3, S-4, S-5	I	II	III
U		no restrictions	

1993 BOCA National Building Code Issued by the Building Officials and Code Administrators International

Occupancy Group	Required Vertical Exits and Passageways	Corridors Providing Exit Access	Room or Enclosed Spaces
A-1, A-2, A-3	I	I*	II*
A-4, B, E, F	I	II	III
H	I	II	III*
I-1	I	II	III
I-2	I*	I*	I*
I-3	I	I	III
M walls	I	II	III
M ceilings	I	II	II*
R-1, R-2	I	II	III
R-3	III	III	III
S	II	II	III

*Refer to Building Code for additional information.

276

Table 6.9. Some Plastics Used in Transportation.

United States	Million lb.	
	1995	1996
Polyurethane flexible foam	500	545
Polyurethane rigid foam	82	89
Polyurethane RIM elastomers	159	170
Polypropylene	368	408
Acrylonitrile-butadiene-styrene	295	285
Acetals	31	33
Nylon	320	351
Polycarbonate	99	107
Polycarbonate/ABS blends	93	102
PBT incl. PC/PBT blends	120	127
Acrylic (automotive)	50	50

Western Europe	Million lb.	
	1995	1996
Acrylonitrile-butadiene-styrene	269	260
Nylon	277	293
Polycarbonate	15	17
Polycarbonate/ABS blends	73	79
PBT and PC/PBT blends	66	70
Acetals	99	99
Polystyrene (automotive)	13	13

Japan	Million lb.	
	1995	1996
ABS (vehicles)	152	150
Nylon 6 (automotive)	45	49
Nylon 66 (automotive)	68	71
Polyacetal	49	51
Polycarbonate	62	64
Polybutylene terephthalate	46	49
Reinforced polyester (vehicles)	49	46

Reference: Modern Plastics, January 1997.

Table 6.10. Some Fire Test Requirements for Passenger Rail Vehicles.

Test	Material	Requirement
ASTM E 162	Window, light diffuser	100 max I_s
	Wall, ceiling panels	35 max I_s
	End cap roof housings	35 max I_s
	Interior, exterior boxes	35 max I_s
	Seat and mattress frames	35 max I_s
	Floor covering	25 max I_s
	Thermal, acoustic insulation	25 max I_s
ASTM D 3675	Cushions and mattresses	25 max I_s
ASTM E 648	Floor covering	5 kW/m² min CRF
ASTM C 542	Window gaskets, diaphragms	Pass
ASTM E 119	Structural flooring	15 min min
ASTM E 662	Cushions and mattresses	100 max D_s at 1.5 min
		175 max D_s at 4.0 min
	Panels and flooring, elastomers	100 max D_s at 1.5 min
		200 max D_s at 4.0 min
	End cap roof housings, interior exterior boxes	100 max D_s at 1.5 min
		200 max D_s at 4.0 min
	Thermal, acoustic insulation	100 max D_s at 4.0 min

278

Table 6.11. Some Fire Test Requirements for Aircraft Materials Issued by Federal Aviation Administration (FAA).

Test	Requirement
Vertical bunsen burner test for cabin and cargo compartment materials (FAR 25.853 and FAR 25.855) (ASTM F 501)	15 sec max flame time 3 or 5 sec max drip flame time 6 or 8 in max burn length
45-degree bunsen burner test for cargo compartment liners and waste stowage compartment materials (FAR 25)	15 sec max flame time no flame penetration 10 sec max glow time
Horizontal bunsen burner test for cabin, cargo compartment, and miscellaneous materials (FAR 25.853) (ASTM F 776)	2.5 or 4 in/min max burn rate
60-degree bunsen burner test for electric wire (FAR 25.869) (ASTM F 777)	30 sec max exting. time 3 sec max drip exting. time
Heat release rate test for cabin materials (FAR 25.853)	65 kW/m² max HRR during 5 min 65 kW-min/m² max total HR in 2 min
Smoke test for cabin materials (FAR 25.853)	200 max D_m during 4 min
Oil burner test for seat cushions (FAR 25.853)	17 in max burn length 10% max weight loss
Oil burner test for cargo liners (FAR 25.855)	No burn through in 5 min 400°F max backside temperature
Radiant heat test for evacuation slides, ramps, and rafts (TSO C69A) (ASTM F 828)	No failure in 90 sec
Fire containment test for combustible waste containers, carts, and compartments (FAR 25.853)	Able to contain fire
Fire test for aircraft thermal acoustical insulation	15 sec max flame time 6 in max burn length

Table 6.12. Some Fire Test Requirements for Aircraft Materials Issued by Boeing Airplane Company in Specification D6-51377.

Test	Requirement
BSS 7230—vertical bunsen burner test for cabin and cargo compartment materials (FAR 25.853 and FAR 25.855) (ASTM F 501)	15 sec max flame time 3 or 5 sec max drip flame time 6 or 8 in max burn length
BSS 7230—45-degree bunsen burner test for cargo compartment liners and waste stowage compartment materials (FAR 25)	15 sec max flame time No flame penetration 10 sec max glow time
BSS 7230—horizontal bunsen burner test for cabin, cargo compartment, and miscellaneous materials (FAR 25.853) (ASTM F 776)	2.5 or 4 in/min max burn rate
BSS 7238—smoke test for cabin materials (FAR 25.853)	200 max D_m during 4 min 150 max D_s after 4 min panels 100 max D_s after 4 min cargo liners 300 max D_s after 4 min carpet 150 max D_s after 4 min other textile components 200 max D_s after 3 min windows 200 max D_s after 4 min thermoplastic and elastomeric parts 50 max D_s after 4 min fuselage insulation 50 max D_s after 4 min electric wire and cable insulation
BSS 7239—toxicity of gases from smoke chamber	3500 ppm max CO 150 ppm max HCN 200 ppm max HF 500 ppm max HCl 100 ppm max SO_2 100 ppm max NO_x
BSS 7303—oil burner test for seat cushions (FAR 25.853)	17 in max burn length 10% max weight loss
BSS 7304—radiant panel test (ASTM E 162)	25 max I_s
BSS 7322—heat release rate test for cabin materials (FAR 25.853)	65 kW/m² max HRR during 5 min 65 kW-min/m² max total HR in 2 min
BSS 7323—oil burner test for cargo liners (FAR 25.855)	No burn through in 5 min 400°F max backside temperature
BSS 7324—60-degree bunsen burner test for electric wire (FAR 25.869) (ASTM F 777)	30 sec max exting. time 3 sec max drip exting. time
BSS 7357—fire test for aircraft thermal acoustical insulation	15 sec max flame time 6 in max burn length
Radiant heat test for evacuation—slides, ramps, and rafts (TSO C69A) (ASTM F 828)	No failure in 90 sec
Fire containment test for combustible waste containers, carts, and compartments (FAR 25.853)	Able to contain fire

Table 6.13. Some Fire Test Requirements for Aircraft Materials Issued by Airbus Industrie in Specification ABD-0031.

Test	Requirement
AITM 2.002—vertical burner test for cabin and cargo compartment materials (JAR/FAR 25.853)	15 sec max flame time 3 or 5 sec max drip flame time 6 or 8 in max burn length
AITM 2.0003—horizontal burner test for cabin and cargo compartment materials (JAR/FAR 25.853)	2.5 or 4 in/min max burn rate
AITM 2.0004—45-degree burner test for cargo compartment liners (JAR/FAR 25.855)	15 sec max flame time No flame penetration 10 sec max glow time
AITM 2.0005—60-degree burner test for electric wire (JAR/FAR 25.1359)	3 in max burn length 30 sec max exting. time 3 sec max drip exting. time
AITM 2.0006—heat release rate test (JAR/FAR 25.853)	65 kW/m² max HRR during 5 min 65 kW-min/m² max total HR in 2 min
AITM 2.007—smoke density test (JAR/FAR 25.853)	150 max D_m within 4 min 150 max D_m within 4 min panels 100 max D_m within 4 min cargo liners 200 max D_m within 4 min textile covered panels 250 max D_m within 4 min carpet 200 max D_m within 4 min other textile components 200 max D_m within 4 min thermoplastic and elastomeric parts 100 max D_m within 4 min fuselage insulation and airducting
AITM 2.0008—smoke density of electrical wire/cable insulation	20 max D_m in 16 min
AITM 2.009—oil burner test for seat cushions (JAR/FAR 25.853)	17 in max burn length 10% max weight loss
AITM 2.0010—oil burner test for cargo liners (FAR 25.855)	No burn through in 5 min 400°F max backside temperature
AITM 2.0038—heat shrinkable tubings, 60-degree burner test	3 in max burn length 30 sec max exting. time 3 sec max drip exting. time
AITM 3.0005—toxicity of gases from smoke chamber	1000 ppm max CO 150 ppm max HCN 100 ppm max HF 150 ppm max HCl 100 ppm max SO_2/H_2S 100 ppm max NO/NO_2

Table 6.14. *Some Fire and Toxicity Test Requirements in Military Standard MIL-STD-2031 (SH) for Composite Material Systems Inside Naval Submarines.*

Fire Test Characteristic	Requirement
ASTM E 162	
Flame spread (index)	20 maximum
ASTM E 662	
Smoke obscuration	
D_{max}	200 maximum
ASTM E 1354	
Ignitability (seconds)	
100 kW/m² irradiance	60 minimum
75 kW/m² irradiance	90 minimum
50 kW/m² irradiance	150 minimum
25 kW/m² irradiance	300 minimum
Heat release (kW/m²)	
100 kW/m² irradiance	
Peak	150 maximum
Average 300 seconds	120 maximum
75 kW/m² irradiance	
Peak	100 maximum
Average 300 seconds	100 maximum
50 kW/m² irradiance	
Peak	65 maximum
Average 300 seconds	50 maximum
25 kW/m² irradiance	
Peak	50 maximum
Average 300 seconds	50 maximum
Combustion gas generation	
25 kW/m² irradiance	
CO (p/m²)	200 maximum
CO_2 (% by volume)	4 maximum
HCN (p/m)	30 maximum
HCl (p/m)	100 maximum

Table 6.15. *Estimated Plastics Demand in Flexible Packaging.*

	Million lb.	
	1993	1998
LDPE	3630	3720
LLDPE	1560	2500
HDPE	865	1195
Polypropylene (PP)	735	850
Polyvinyl chloride (PVC)	250	225
Polystyrene (PS)	250	315
Thermoplastic polyester (TPE)	250	400
EVA	80	90
Nylon	85	100
Polyvinyl acetate	50	65
Epoxy	40	45
Other	200	240
Total	7995	9745

Reference: Modern Plastics Encyclopedia, 1996.

Table 6.16. *Estimated Plastics Demand in Rigid Packaging.*

	Million lb.	
	1993	1998
HDPE	4625	5520
Polystyrene (PS)	1710	1950
Thermoplastic polyester (TPE)	1580	2430
Polypropylene (PP)	1085	1340
LLDPE	280	420
Polyvinyl chloride (PVC)	210	175
LDPE	80	70
Polyurethane	120	145
Other	430	520
Total	10,120	12,570

Reference: Modern Plastics Encyclopedia, 1996.

Table 6.17. Some Plastics Used in Wire and Cable.

United States	Million lb.	
	1995	1996
LDPE and PE copolymers	196	190
LLDPE	168	160
HDPE	127	130
Polypropylene	22	11
Polyvinyl chloride	416	450
Nylons	51	54

Western Europe	Million lb.	
	1995	1996
LDPE	440	419
LLDPE	44	55
HDPE	106	103
Polyvinyl chloride	987	936

Japan	Million lb.	
	1995	1996
LDPE	176	170
LLDPE	7	11
Polyvinyl chloride	706	701

Reference: Modern Plastics, January 1997.

Table 6.18. Some Plastics Used in Electrical/Electronics.

United States	Million lb.	
	1995	1996
Acetals	20	21
Nylons	110	110
Polycarbonate	34	34
Polybutylene terephthalate	46	49
Polystyrene (radio/TV/stereo)	193	205
Polystyrene (cassettes, reels)	299	309
ABS (appliances)	263	252
ABS (business machines)	86	84
ABS (consumer electronics)	27	22
ABS (telecommunications)	25	24

Western Europe	Million lb.	
	1995	1996
Acrylonitrile-butadiene-styrene	291	276
Nylons	202	205
Polycarbonate	95	99
Polycarbonate/ABS blends	48	48
Acetals	24	24
Polystyrene (TV/radio etc.)	587	600

Japan	Million lb.	
	1995	1996
Polystyrene (electrical/industrial)	639	635
Polybutylene terephthalate	55	53
Nylon 6	20	18
Nylon 66	35	37
Polyacetal	75	75
Polycarbonate	90	95
ABS (electrical appliances)	249	245

Reference: Modern Plastics, January 1997.

Table 6.19. Types of Conductor Insulation in the National Electrical Code.

FEP	Fluorinated ethylene propylene
FEPB	Fluorinated ethylene propylene
MI	Mineral insulation (metal sheathed)
MTW	Flame-retardant, moisture-, heat- and oil-resistant thermoplastic
PFA	Perfluoroalkoxy
PFAH	Perfluoroalkoxy
RH	Flame-retardant thermoset
RHH	Flame-retardant thermoset
RHW	Flame-retardant, moisture-resistant thermoset
RHW-2	Flame-retardant, moisture-resistant thermoset
SA	Silicone
SIS	Flame-retardant thermoset
TBS	Thermoplastic and fibrous outer braid
TFE	Extended polytetrafluoroethylene
THHN	Flame-retardant, heat-resistant thermoplastic
THHW	Flame-retardant, moisture- and heat-resistant thermoplastic
THW	Flame-retardant, moisture- and heat-resistant thermoplastic
THWN	Flame-retardant, moisture- and heat-resistant thermoplastic
TW	Flame-retardant, moisture-resistant thermoplastic
UF	Underground feeder and branch-circuit cable, single conductor
USE	Underground service entrance cable, single conductor
XHH	Flame-retardant thermoset
XHHW	Flame-retardant, moisture-resistant thermoset
XHHW-2	Flame-retardant, moisture-resistant thermoset
Z	Modified ethylene tetrafluoethylene
ZW	Modified ethylene tetrafluoethylene

Table 6.20. Some International Standards Relevant to Fire Safety and Plastics Issued by International Electrotechnical Commission (IEC), IEC Standards.

IEC 79-1	Explosive gas mixtures
IEC 112	Tracking indices, solid insulating materials
IEC 197	High voltage connecting wire with fire retarding insulation for television receivers
IEC 249-2-3	Epoxide cellulose paper-clad laminated sheet
IEC 249-2-5	Epoxide-woven glass fabric copper-clad laminated sheet
IEC 249-2-6	Phenolic cellulose paper copper-clad laminated sheet
IEC 249-2-7	Phenolic cellulose paper copper-clad laminated sheet
IEC 249-2-9	Paper core, epoxide glass cloth surface copper-clad laminated sheet
IEC 249-2-10	Epoxide nonwoven/woven glass reinforced copper-clad laminated sheet
IEC 249-2-12	Thin epoxide woven glass fabric copper-clad laminated sheet
IEC 249-2-14	Phenolic cellulose paper copper-clad laminated sheet
IEC 249-2-15	Flexible copper clad polyimide film
IEC 249-2-16	Polyimide woven glass fabric copper-clad laminated sheet
IEC 249-2-17	Thin polyimide woven glass fabric copper-clad laminated sheet
IEC 249-2-18	Epoxide woven glass fabric copper-clad laminated sheet
IEC 249-2-19	Epoxide woven glass fabric copper-clad laminated sheet
IEC 331	Electric cables, fire resisting characteristics
IEC 332-1	Electric cables, single vertical insulated wire or cable
IEC 332-2	Electric cables, single small vertical insulated wire
IEC 332-3	Electric cables, bunched wires or cables
IEC 371-3-8	Mica paper tapes
IEC 684-3-209	Heat shrinkable sleeving, general purpose
IEC 684-3-246	Heat shrinkable sleeving, dual wall
IEC 695-1-1	Guidance for assessing fire hazard
IEC 695-1-2	Guidance for preparation
IEC 695-1-3	Guidance for preparation
IEC 695-2-1	Glow-wire test
IEC 695-2-2	Needle-flame tests
IEC 695-2-3	Bad connection test with heaters
IEC 695-2-4/0	Diffusion type and premixed type flame test
IEC 695-2-4/1	1 kW nominal premixed test flame
IEC-695-2-4/2	500 W nominal test flames
IEC 695-3-1	Fire hazard assessment
IEC 695-4	Fire test terminology
IEC 695-5-1	Potential corrosion damage by fire effluent
IEC 695-5-2	Potential corrosion damage by fire effluent
IEC 695-6-30	Obscuration hazards of vision
IEC 695-7-1	Toxic hazards due to fires
IEC 695-7-4	Toxic hazards due to fires
IEC 707	Flammability, solid insulating materials
IEC 1100	Insulating liquids, fire point
IEC 1197	Insulating liquids, linear flame propagation
IEC 1300-2-36	Passive components

Table 6.21. Some Plastics Used in Furniture and Furnishings.

United States	Million lb.	
	1995	1996
Polyurethane flexible foam (bedding)	246	258
Polyurethane flexible foam (furniture)	777	790
Polystyrene (molding)	55	59
Polystyrene (extrusion)	32	32
ABS (furniture)	10	10

Western Europe	Million lb.	
	1995	1996
Polystyrene (furniture)	232	236
ABS (furniture)	20	20

Reference: Modern Plastics, January 1997.

Table 6.22. Comision Panamericana de Normas Technicas (COPANT) (Pan American Standards Commission).

Active Members	
ABNT	Associacao Brasileira de Normas Tecnicas (Brazil)
ANSI	American National Standards Institute (USA)
BNSI	Barbados National Standards Institute (Barbados)
COGUANOR	Comision Guatemalteca de Normas (Guatemala)
COVENIN	Comision Venezolana de Normas Industriales (Venezuela)
DGN	Direccion General de Normas (Mexico)
DIGENOR	Direccion General de Normas y Sistemas de Calidad (Dominican Republic)
GNBS	Guyana National Bureau of Standards (Guyana)
GRENADA	Grenada Bureau of Standards (Grenada)
IBNORCA	Instituto Boliviano de Normalizacion y Calidad (Bolivia)
ICAITI	Instituto Centroamericano de Investigaciones y Tecnologia (Central America)
ICONTEC	Instituto Colombiano de Normas Technicas y Certificacion (Colombia)
INDECOPI	Instituto Nacional de Defensa de la Competencia y de la Proteccion a la Propiedad Intelectual (Peru)
INEN	Instituto Ecuatoriano de Normalizacion (Ecuador)
INN	Instituto Nacional de Normalizacion (Chile)
INTECO	Instituto de Normas Tecnicas de Costa Rica (Costa Rica)
INTN	Instituto Nacional de Tecnologia y Normalizacion (Paraguay)
IRAM	Instituto Argentino de Normalizacion (Argentina)
JBS	Jamaica Bureau of Standards (Jamaica)
NC	Oficina Nacional de Normalizacion (Cuba)
SCC	Standards Council of Canada (Canada)
TTBS	Trinidad and Tobago Bureau of Standards (Trinidad and Tobago)
UNIT	Instituto Uruguayo de Normas Tecnicas (Uruguay)
Adherent Members	
AENOR	Asociacion Espanola de Normalizacion y Certificacion (Spain)
AFNOR	Association Francaise de Normalisation (France)
IPQ	Instituto Potugues da Qualidade (Portugal)
UNI	Ente Nazionale Italiano de Unificazione (Italy)

Table 6.23. *International Standardization Organization (ISO) Participants.*

Albania	DSC	Korea, South	KNITQ
Algeria	NAPI	Libya	LNCSM
Argentina	IRAM	Macedonia	ZSM
Armenia	SARM	Malaysia	DSM
Australia	SAA	Mauritius	MSB
Austria	ON	Mexico	DGN
Bangladesh	BSTI	Mongolia	MNCSM
Belarus	BELST	Morocco	SNIMA
Belgium	IBN	Netherlands	NNI
Bosnia/Herzegovina	BASMP	New Zealand	SNZ
Brazil	ABNT	Nigeria	SON
Bulgaria	BDS	Norway	NSF
Canada	SCC	Pakistan	PSI
Chile	INN	Panama	COPANIT
China	CSBTS	Philippines	BPS
Colombia	ICONTEC	Poland	PKN
Costa Rica	INTECO	Portugal	IPQ
Croatia	DZNM	Romania	IRS
Cuba	NC	Russia	GOST R
Cyprus	CYS	Saudi Arabia	SASO
Czech Republic	COSMT	Singapore	PSB
Denmark	DS	Slovakia	UNMS
Ecuador	INEN	Slovenia	SMIS
Egypt	EOS	South Africa	SABS
Ethiopia	EAS	Spain	AENOR
Finland	SFS	Sri Lanka	SLSI
France	AFNOR	Sweden	SIS
Germany	DIN	Switzerland	SNV
Ghana	GSB	Syria	SASMO
Greece	ELOT	Tanzania	TBS
Hungary	MSZT	Thailand	TISI
Iceland	STRI	Trinidad/Tobago	TTBS
India	BIS	Tunisia	INNORPI
Indonesia	DSN	Turkey	TSE
Iran	ISIRI	Ukraine	DSTU
Ireland	NSAI	United Kingdom	BSI
Israel	SII	United States	ANSI
Italy	UNI	Uruguay	UNIT
Jamaica	JBS	Uzbekistan	UZGOST
Japan	JISC	Venezuela	CONVENIN
Kazakhstan	KAZMEMST	Vietnam	TCVN
Kenya	KEBS	Yugoslavia	SZS
Korea, North	CSK	Zimbabwe	SAZ

Table 6.24. Some Japanese Standards Relevant to Fire Safety and Plastics Issued by Japanese Standards Association (JSA), Japanese Industrial Standards (JIS).

A. Civil Engineering and Architecture
JIS A 1301. Method of fire test for wooden structural parts of buildings
JIS A 1302. Method of fire test for noncombustible structural parts of buildings
JIS A 1304. Method of fire resistance test for structural parts of buildings
JIS A 1306. Measuring method of smoke density using light extinction method
JIS A 1311. Method of fire protecting test of fire door of buildings
JIS A 1312. Method of fire test for roof of buildings
JIS A 1321. Testing method for incombustibility of interior finish material and procedure of buildings Announcement No. 1231. Gas toxicity test
JIS A 1322. Testing method of incombustibility of thin materials for buildings
JIS A 1323. Flame retardant testing method for spark droplets of welding and gas cutting on fabric sheets in construction works
C. Electrical Engineering
JIS C 0060. Fire hazard testing, Part 2. Test methods, glow-wire test and guidance
JIS C 0061. Fire hazard testing, Part 2. Needle flame test
JIS C 3521. Flame test method for flame retardant sheath of telecommunication cables
D. Automotive Engineering
JIS D 1201. Test method for flammability of organic interior materials for automobiles (similar to MVSS 302)
K. Chemical Engineering
JIS K 7201. Testing method for flammability of polymeric materials using the oxygen index method (similar to ASTM D 2863)
JIS K 7217. Analytical method for determining gases evolved from burning plastics
JIS K 7228. Method of measuring smoke density and concentration of gases evolved by incineration or decomposition of plastics
L. Textile Engineering
JIS L 1091. Testing methods for flammability of clothes
Z. Fundamental and General
JIS Z 2120. Method of inflammability test for wood
JIS Z 2150. Method of flame test for materials

Table 6.25. Some French Standards Relevant to Fire Safety and Plastics.

Issued by Association Francaise de Normalisation (AFNOR)

NF C 17-300. Dielectric liquids

NF C 20-453. Basic environmental testing procedures. Test methods. Conventional determination of corrosiveness of smoke

NF C 20-454. Basic environmental testing procedures. Test methods. Fire behavior. Analysis and titrations of gases evolved during pyrolysis or combustion of materials used in electrotechnics. Exposure to abnormal heat or fire. Tube furnace method

NF C 20-455. Test methods. Fire behavior. Glow wire test. Flammability and extinction ability.

NF C 20-456. Fire hazard testing. Test methods. Needle flame test

NF C 20-902-1. Smoke opacity

XP C 20-90221. Smoke opacity

XP C 20-903. Acidity by pH and conductivity

NF C 20-940. Terminology of fire tests.

NF C 26-212. Solid electrical insulation

NF C 27-300. Dielectric liquids

NF C 32-070. Insulated cables and flexible cords for installations. Classification tests on cables and cords with respect to their behavior to fire.

NF C 32-071U. Electrical cables

NF C 32-072. Electrical cables

NF C 32-073-1. Electrical cables, smoke density

NF C 32-073-2. Electrical cables, smoke density

NF C 32-074. Acidity by pH and conductivity

NF C 32-130. Conductors and cables

NF C 32-131. Cables without halogens

NF C 32-310. Conductors and cables, fire resistant

NF C 32-323. Conductors and cables, without halogens

XP C 32-324. Cables, without halogens

NF C 68-106. Conduits, insulation

NF C 68-107. Conduits, insulation

NF C 68-108. Conduits

NF F 16-101. Fire behavior

NF F 16-102. Fire behavior

NF F 16-201. Fire test

NF F 63-295. Conductors and cables

NF G 00-100. Textiles

NF G 07-128. Textiles, fire behavior

NF G 07-180. Textiles, fire behavior

NF G 07-182. Textiles, flame spread, 45°

NF G 07-184. Textiles, fire behavior

NF G 35-027. Textiles, fire behavior

NF P 92-501. Safety against fire. Building materials. Reaction to fire tests. Radiation test used for rigid materials, or for materials on rigid substrates (flooring and finishes) of all thicknesses, and for flexible materials thicker than 5 mm (Epiradiateur test)

NF P 92-503. Safety against fire. Building materials. Reaction to fire tests. Electrical burner test used for flexible materials 5 mm thick or less

NF P 92-504. Safety against fire. Building materials. Reaction to fire tests. Speed of the spread of flame test used for the materials which are not intended to be glued on a rigid substrate. Complementary test

NF P 92-505. Safety against fire. Building materials. Reaction to fire tests. Dripping test with electrical radiator, used for melting materials. Complementary test

NF P 92-506. Safety against fire. Building materials. Reaction to fire tests. Radiant panel test for flooring

continued

Table 6.25 (continued). Some French Standards Relevant to Fire Safety and Plastics.

Issued by Association Francaise de Normalisation (AFNOR)

NF P 92-507. Fire behavior

NF P 92-508. Safety against fire. Building materials. Reaction to fire tests. Setting of the electrical burner

NF P 92-509. Safety against fire. Building materials. Reaction to fire tests. Calibration of the radiator

NF P 92-510. Safety against fire. Building materials. Reaction to fire tests. Determination of upper calorific potential

NF P 92-511. Fire tests

NF P 92-512. Fire tests

NF P 92-701. Fire behavior

NF P 92-702. Fire behavior

NF P 92-703. Fire resistance

NF P 92-704. Fire behavior

NF Q 64-024. Fire tests

NF R 18-501. Determination of combustion

NF T 51-071. Plastics. Determination of flammability by oxygen index (similar to ASTM D 2863)

NF T 51-072. Plastics. Flammability test. Behavior of plastics in the form of bars (similar to UL 94)

NF T 51-073. Plastics. Reaction to fire. Test method to measure the density of smoke

NF T 51-074. Plastics. Fire behavior. Hot wire test

NF T 51-075. Plastics. Fire behavior. Needle burner test

NF T 51-077. Plastics, fire tests

NF T 51-079. Plastics, fire tests

NF T 56-125. Rubber and plastics, fire behavior, horizontal

NF T 57-102. Reinforced plastics

NF T 57-501. Plastics

NF X 10-702. Fire test methods. Smoke emission. Test for measuring the specific optical density of smoke emitted by the combustion or pyrolysis of solid materials, Parts 1, 2, 3, 4, and 5.

NF X 41-027. Adhesive tape. Inflammability test

NF X 65-010. Fire tests

NF X 70-100. Fire tests. Analysis of pyrolysis and combustion gases. Tube furnace method

NF X 70-101. Safety against fire. Fire behavior tests. Analysis of gases resulting from combustion or pyrolysis. Test chamber method.

Issued by Union Technique de l'Electricite (UTE) UTE Standards

UTE C 20-450. Fire behavior of electrotechnical (electric, electromechanic, and electronic) materials, components, and equipment. General

UTE C 20-452. Test methods. Fire behavior. Determination of smoke opacity in closed chamber (similar to ASTM E 662)

UTE C 27-251U. Dielectric liquids

UTE C 32-071. Test electric cables under fire conditions. Guide stating the test method for checking small single insulated cores of the category C2

Table 6.26. Some German Standards Relevant to Fire Safety and Plastics.

Issued by Deutsches Institut fur Normung (DIN) DIN Standards

DIN 4102. Fire performance of building materials; Part 1. Building materials; Part 2. Building components; Part 3. Fire walls and exterior walls; Part 4. Grouping of materials, components, and elements

DIN 4842. Heat protective clothing

DIN 4845. Flameproof fabrics of cellulose fibers

DIN 22118. Conveyor belts for coal mining

DIN 23320. Fireproof protective clothing for mining

DIN 23325. Flameproof protective clothing

DIN 53436. Toxicity of thermal decomposition products

DIN 53438. Fabrics, vertical

DIN 54331. Fabrics, semi-circle test

DIN 54332. Textile floor coverings

DIN 54333. Fabrics, horizontal

DIN 54334. Fabrics, ignition time

DIN 54335. Fabrics, 45° method

DIN 54336. Fabrics, vertical method

DIN 66080. Classification of textile products

DIN 66081. Classification of textile floor coverings

DIN 66082. Classification of drapery and curtain materials

DIN 66090. Textile floor coverings, burning behavior

DIN 75200. Interior materials in motor vehicles

Issued by Verband Deutscher Elektrotechniker (VDE) VDE Standards

VDE 0304. Glow bar test

VDE 0318. Flame test, horizontal (similar to UL 94)

VDE 0340. Self-adhesive insulating tapes

VDE 0345. Flash-ignition temperature

VDE 0470. Hot mandrel test

VDE 0471. Glow wire test

VDE 0472. Part 804. Vertical cable test (similar to IEEE 383)

Table 6.27. Some Danish Standards Relevant to Fire Safety and Plastics—Issued by Dansk Institut for Provning og Justering (Dantest), DS Standards.

DS 1058.1	Building products, ignitability—DS/INSTA 410, NT FIRE 002
DS 1058.2	Building products, heat release, smoke—DS/INSTA 412, NT FIRE 004
DS 1059.1	Roofings, fire spread—DS/INSTA 413, NT FIRE 006
DS 1059.2	Floorings, fire spread, smoke—DS/INSTA 414, NT FIRE 007
DS 1060.1	Coverings, fire protection ability—DS/INSTA 411, NT FIRE 003

Table 6.28. Some Norwegian Standards Relevant to Fire Safety and Plastics Issued by Norges Branntekniske Laboratorium (NBL) Norwegian Fire Research Laboratory.

NS Standards	
NS 3904	Elements of building construction—ISO 834
NS 3907	Door and shutter assemblies—ISO 3008
NS 3908	Glazed elements, fire resistance—ISO 3009

IMO Standards	
IMO A 214 (VII)	Primary deck coverings
IMO A 471	Textiles, foils
IMO A 564 (14)	Bulkhead and deck finish materials
IMO A 652 (16)	Upholstered furniture, cigarette and flame
IMO A 653 (16)	Floor covering, ships
IMO A 687 (17)	Primary deck coverings
IMO A 688 (17)	Bedding components, cigarette and flame
IMO A 753 (18)	Plastic pipes
IMO A 754 (18)	Bulkheads and decks

NT FIRE Standards	
NT FIRE 001	Building materials, non-combustibility—ISO 1182
NT FIRE 002	Building products, ignitability—INSTA 410
NT FIRE 003	Coverings, fire protection ability—INSTA 411
NT FIRE 004	Building products, heat release, smoke—INSTA 412
NT FIRE 005	Elements of building construction—ISO 834
NT FIRE 006	Roofings, fire spread—INSTA 413
NT FIRE 007	Floorings, fire spread, smoke—INSTA 414
NT FIRE 008	Door and shutter assemblies—ISO 3008
NT FIRE 009	Glazed elements, fire resistance—ISO 3009
NT FIRE 010	Fire dampers
NT FIRE 011	Dust clouds, minimum explosible dust concentration
NT FIRE 012	Solid materials, smoke—ANSI/ASTM E 662
NT FIRE 013	Plastics, oxygen index—ISO 4589
NT FIRE 014	Upholstered furniture, ignitability—BS 5852
NT FIRE 015	Textiles and films, vertical—IMO A 471
NT FIRE 017	Filing cabinets and data cabinets
NT FIRE 018	Textile floor coverings, tablet test—ISO 6925
NT FIRE 019	Multilayer textiles, tablet test
NT FIRE 021	Insulation of steel structures
NT FIRE 022	Plastics and textiles, flammability—BS 2782-1
NT FIRE 025	Surface layer, building coverings and components
NT FIRE 027	Textile fabrics, ignition, vertical—ISO 6940
NT FIRE 028	Textile fabrics, flame spread, vertical—ISO 6941
NT FIRE 029	Textile fabrics, ignition, flame spread—ASTM D 1230
NT FIRE 030	Building products, fire spread, smoke, full scale
NT FIRE 032	Upholstered furniture, full scale
NT FIRE 033	Building products, ignitability—ISO 5657
NT FIRE 034	Ventilation ducts, fire resistance
NT FIRE 035	Loose-fill thermal insulation, flame, smoldering
NT FIRE 036	Pipe insulation, fire spread, smoke, full scale
NT FIRE 037	Bedding components, cigarette and flame—BS 6807
NT FIRE 038	Building material, combustible content
NT FIRE 046	Atrium roof constructions

Table 6.29. Some Swedish Standards Relevant to Fire Safety and Plastics Issued by Sveriges Provnings- och Forskningsinstitut (SP).

	SP Standards
SP 1514	Coverings—NT FIRE 003
SP 1515	Element of building construction—ISO 834, NT FIRE 005
SP 1516	Chimney—ISO 4736, NT FIRE 020
SP 1517	Refuse cabinet—SP A4 104
SP 1518	Flexible duct
SP 1519	Facade claddings, balcony, window—SP FIRE 105
SP 1522	Fire ventilators
SP 1523	Doors and shutters—ISO 3008, NT FIRE 008
SP 1524	Glazed elements—ISO 3009, NT FIRE 009
SP 1525	Light element of roofs—SP A04 020
SP 1526	Roofs—SP A04 021
SP 1527	Roofs—SP A04 022
SP 1528	Ventilation unit—SP A04 023
SP 1529	Fire dampers—NT FIRE 010
SP 1530	Bulkheads and decks—IMO A 754 (18)
SP 1531	Door closer
SP 1532	Vertical elements or details—SP A4 117
SP 1533	Fixings
SP 1534	Vertical and horizontal elements or details
SP 1535	Atrium roof constructions—NT FIRE 046
SP 1536	Insulation of steel structures—NT FIRE 021
SP 1537	Filing cabinets and data cabinets—NT FIRE 017
SP 1538	Joint sealing compound
SP 1539	Insulation for ventilation ducts
SP 1540	Cable and pipe penetration seals
SP 1541	Lift landing doors
SP 1542	Walls—DIN 4102 teil 3
SP 1544	Insulation, covering—SS-ISO 1182
SP 1545	Insulation, covering
SP 1546	Surface layer, building coverings, components—SS 02 48 23
SP 1547	Roof covering, flame spread—SS 02 48 24
SP 1548	Floor covering, flame spread, smoke
SP 1549	Textiles, insulation and foils, ignitability—SIS 65 00 82
SP 1550	Materials, heating thermal stability
SP 1551	Aircraft interior material, vert.—FAR 25.853 F(b) (4)
SP 1552	Textile floor coverings, ignitability—ISO 6925
SP 1553	Components for electrical equipment
SP 1554	Building material, smoke production—NT FIRE 012
SP 1555	Motor vehicle interior material, flame—SS-ISO 3795
SP 1556	Textiles, foils, ignitability—IMO A 471
SP 1557	Textile floor coverings, burning rate—NT FIRE 029
SP 1558	Building material in plane form—DIN 4102, teil 1
SP 1559	Electric cables, ignitability—SS-IEC 332-1
SP 1560	Building materials, calorific value—ISO 1716
SS 1561	Sleeping bags, ignitability—NT FIRE 019
SS 1563	Rubber foil, ignition, small flame—SS 16 22 22
SP 1564	Handtools, ignition, small flame—IEC 900
SP 1566	Components for electrical business equipment
SP 1567	Surface layer, building coverings, components—ISO 9705
SP 1568	Surface layer, building coverings, components—NT FIRE 030
SP 1569	Upholstered furniture, heat, smoke, toxic gas—NT FIRE 032

continued

	SP Standards (cont.)
SP 1570	Pipe insulation, heat release, spread, smoke—NT FIRE 036
SP 1571	Upholstered furniture, cigarette and flame—CAL TB 117
SP 1573	Plastic components in equipment, ignitability—ANSI/UL-94
SP 1574	Mattresses, ignitability—BS 6807
SP 1575	Plastic pipes, flame spread, endurance—IMO A 753 (18)
SP 1576	Building material, ignitability—ISO 5657
SP 1577	Ship, cables, furniture, heat release—ISO 5660-1
SP 1578	Bedding components, cigarette and flame—NT FIRE 037
SP 1579	Building material, combustible content—NT FIRE 038
SP 1580	Upholstered furniture, cigarette—ISO 8191-1
SP 1581	Upholstered furniture, small flame—ISO 8191-2
SP 1582	Upholstered furniture, flame and cribs—BS 5852-2
SP 1583	Curtains, draperies—NT FIRE 043
SP 1584	Floor covering, ships, flame, heat—IMO A 653 (16)
SP 1585	Primary deck coverings, flame, heat—IMO A 687 (17)
SP 1586	Surface layer, flame spread—ISO 5658-2
SP 1587	Floor covering, flame spread, smoke—DIN 4102 teil 14
SP 1588	Electric cables, flame spread—IEC 332-3
SP 1659	Motor vehicle interior material, flame spread—FMVSS 302
SP 1660	Aircraft interior material, horiz.—FAR 25.853 F(b) (5)
SP 1664	Glazed elements and doors—DIN 4102 teil 2 + teil 5
SP 1665	Pipe insulation, electrical, oxygen index—ASTM D 2863
SP 1691	Upholstered furniture, small flame—SS-EN 1021-2
SP 1699	Motor vehicle interior material, flame—Volvo 5031,1
SP 1700	Motor vehicle material, ignitability—Volvo 1027, 5158
SP 1701	Motor vehicle cable harness ignitability—Volvo 7611,133
SP 1702	Motor vehicle electric cables ignitability—Volvo 7611,131
SP 1703	Motor vehicle protecting hose ignitability—Volvo 7411,171
SP 1704	Motor vehicle cable splices ignitability—Volvo 7411,1715
SP 1705	Motor vehicle cable sleeves ignitability—Volvo 7621,2
SP 1706	Motor vehicle interior material, flame—SAAB 232
SP 1707	Motor vehicle cables, PVC—SAAB 3305
SP 1708	Motor vehicle cables, not PVC—SAAB 3526
SP 1711	Upholstered furniture, cigarette—SS-EN 1021-1
SP 1712	Upholstered furniture, cigarette to cribs—BS 5852
SP 1716	Inset to filing cabinets and data cabinets
SP 2118	Pipe insulation, electrical, oxygen index—ISO 4589
SP 2205	Fabric for tents
SP 2207	Ship, cables, furniture, heat release, smoke—ASTM E 1354
SP 2208	Upholstered furniture, heat release, smoke—ASTM E 1474
SP 2209	Bedding components, cigarette and flame—IMO A 688 (17)
SP 2210	Bedding components, smoldering cigarette
SP 2211	Bedding components, small flame
SP 2212	Electric cables, ignitability—SS-IEC 332-2
SP 2213	Electric cables, ignitability
SP 2214	Electric cables, flame spread
SP 2215	Upholstered furniture, cigarette, flame—BS 5852-1
SP 2216	Upholstered furniture, cigarette, flame—IMO A 652 (16)
SP 2217	Surface layer, building coverings, components—NT FIRE 025
SP 2219	Components for electrical household equipment
SP 2120	Plastic pipes, flame spread, endurance, smoke

Table 6.29 (continued). Some Swedish Standards Relevant to Fire Safety and Plastics Issued by Sveriges Provnings- och Forskningsinstitut (SP).

SP Standards (cont.)	
SP 2284	Aircraft interior material, vert.—JAR 25.853 F(b) (4)
SP 2285	Aircraft interior material, horiz.—JAR 25.853 F(b) (5)
SP 2286	Combustible material, edge ignition—DIN 53 438-2
SP 2287	Combustible material, surface ignition—DIN 53 438-3
SP 2288	Ship, cables, smoke, toxicity—ISO 5659-2
SP 2290	Electrical equipment, flameproof enclosure
SP 2312	Electrical equipment, flameproof enclosure
SP 2341	Motor vehicle cables, ignitability—ISO 6722-1

NT FIRE Standards	
NT FIRE 001	Building materials, non-combustibility—ISO 1182
NT FIRE 002	Building products, ignitability—INSTA 410
NT FIRE 003	Coverings, fire protection ability—INSTA 411
NT FIRE 004	Building products, heat release, smoke—INSTA 412
NT FIRE 005	Elements of building construction—ISO 834
NT FIRE 006	Roofings, fire spread—INSTA 413
NT FIRE 007	Floorings, fire spread, smoke—INSTA 414
NT FIRE 008	Door and shutter assemblies—ISO 3008
NT FIRE 009	Glazed elements, fire resistance—ISO 3009
NT FIRE 010	Fire dampers
NT FIRE 011	Dust clouds, minimum explosible dust concentration
NT FIRE 012	Solid materials, smoke—ANSI/ASTM E 662
NT FIRE 013	Plastics, oxygen index—ISO 4589
NT FIRE 014	Upholstered furniture, ignitability—BS 5852
NT FIRE 015	Textiles and films, vertical—IMO A 471
NT FIRE 017	Filing cabinets and data cabinets
NT FIRE 018	Textile floor coverings, tablet test—ISO 6925
NT FIRE 019	Multilayer textiles, tablet test
NT FIRE 021	Insulation of steel structures
NT FIRE 022	Plastics and textiles, flammability—BS 2782-1
NT FIRE 025	Surface layer, building coverings and components
NT FIRE 027	Textile fabrics, ignition, vertical—ISO 6940
NT FIRE 028	Textile fabrics, flame spread, vertical—ISO 6941
NT FIRE 029	Textile fabrics, ignition, flame spread—ASTM D 1230
NT FIRE 030	Building products, fire spread, smoke, full scale
NT FIRE 032	Upholstered furniture, full scale
NT FIRE 033	Building products, ignitability—ISO 5657
NT FIRE 034	Ventilation ducts, fire resistance
NT FIRE 035	Loose-fill thermal insulation, flame, smoldering
NT FIRE 036	Pipe insulation, fire spread, smoke, full scale
NT FIRE 037	Bedding components, cigarette and flame—BS 6807
NT FIRE 038	Building material, combustible content
NT FIRE 046	Atrium roof constructions

Table 6.30. Some Finnish Standards Relevant to Fire Safety and Plastics Issued by Suomen Standardisoimisliitto (SFS) Finnish Standards Association, SFS Standards.

SFS 3187	Textile floor coverings, flammability, pill test
SFS 3723	Plastics, decomposition temperature—ISO 871-1
SFS 3829	Textile glass reinforced plastics, ignition—ISO 1172
SFS 4190	Building products, ignitability—INSTA 410
SFS 4191	Coverings, fire protection ability—INSTA 411
SFS 4192	Building products, heat release, smoke—INSTA 412
SFS 4193	Fire resistance, elements of building construction
SFS 4194	Roofings, fire spread—INSTA 413
SFS 4195	Floorings, fire spread, smoke—INSTA 414
SFS 4814	Building materials, calorific potential
SFS 4871	Plastics, thin PVC sheeting, flammability
SFS 4878	Solid materials, smoke density
SFS 4902	Plastics, oxygen index
SFS 5099	Rigid cellular plastics, vertical position
SFS 5222	Floating equipment, inflammability
SFS-ISO 181	Plastics, flammability, incandescent rod
SFS-ISO 1182	Building materials, non-combustibility
SFS-ISO 6940	Textile fabrics, ignition, vertical specimens
SFS-ISO 6941	Textile fabrics, flame spread, vertical specimens
SFS-EN 71-2	Toys, flammability
SFS-EN 597-1	Mattresses and upholstered bed bases, cigarette
SFS-EN 597-2	Mattresses and upholstered bed bases, match flame
SFS-EN 1021-1	Upholstered furniture, cigarette
SFS-EN 1021-2	Upholstered furniture, match flame
SFS-EN 20340	Conveyor belts
SFS-EN 28030	Rubber and plastics hoses for underground mining
SFS-EN ISO 4589-3	Plastics, oxygen index

Table 6.31. Some Czechoslovak Standards Relevant to Fire Safety and Plastics Issued by Vyzkumny Ustav Pozemnich Staveb (VUPS) Building Research Institute, CSN Standards.

CSN 64 0146	Heat stability, fire resistance, hot mandrel test
CSN 64 0149	Ignitability of materials
CSN 64 0752	Incandescence resistance of plastics
CSN 64 0755	Flammability of plastics
CSN 64 0756	Plastics, low oxygen index method
CSN 64 0757	Flammability of plastics in the form of film
CSN 64 0758	Hydrogen cyanide in combustion product of plastics
CSN 64 5464	Cellular plastics, cellular rubber, horizontal burning characteristics
CSN 73 0861	Combustibility of building materials
CSN 73 0862	Flammability of building materials
CSN 73 0863	Surface flame propagation of building materials
CSN 73 0864	Heating value of combustible solids under fire conditions
CSN 73 0865	Fire protection, materials drainage of soffits of ceilings and roofs

CHAPTER 7

Flammability and Product Liability

Product liability litigation has become one of the major economic factors in product safety. The increasing number of claims, the increasing size of awards made by juries, the increasing cost of litigation, and the increasing cost of settlements, all together have multiplied the cost of this type of litigation to manufacturers and their insurance companies. The insurance companies have increased the premiums for liability insurance accordingly, and the cost of increased premiums becomes part of the price of the finished product when sold in commerce.

There are at least two opposing philosophies regarding product liability litigation. Opponents of such litigation believe that the substantial resources expended in litigation could be better spent in making the product safer. Proponents of such litigation believe that only the economic penalty of prohibitive jury awards or costly settlements is effective in forcing manufacturers to improve the safety of their products.

SECTION 7.1. HISTORY OF PRODUCT LIABILITY LITIGATION

Product liability litigation has developed gradually over the years from a relatively simple matter into a sophisticated technology. Under the rule of privity, a seller was liable for injury by his product only to the buyer to whom he contracted to sell the product, and this rule was involved in the case of Winterbottom v Wright in 1842. In the case of Brown v Kendall in 1850, it was held that the plaintiff had to prove negligence by the defendant in order to impose liability for accidental injury.

Two exceptions to the rule of privity began to decrease the protection which the privity rule afforded manufacturers. In the case of Thomas v Winchester in 1852, products considered imminently dangerous because of the manufacturer's negligence made the seller liable for injury regardless of whether there was a contract. In the case of Huset v J. I. Case in 1903, a manufacturer who delivered a product without giving notice of a known dangerous condition was held to be liable to anyone injured using that product.

For most practical purposes, the rule of privity was essentially destroyed in 1916 in the case of McPherson v Buick Motor Company, when it was held that the

299

manufacturer was liable if there was negligence in the assembly of the product, and was liable to persons, other than the purchaser, who used the product.

An express warranty theoretically limits the manufacturer's liability to replacement of the defective part or assembly. In the case of Henningen v Bloomfield Motors in 1960, it was held that each product, by being put on the market for sale, bears an implied warranty that it is reasonably safe for use. This 1960 decision in considered by some to be the modern beginning of the principle of strict liability, which was established in 1963 by the California Supreme Court in the case of Greenman v Yuba Power Products. Under this doctrine, a manufacturer is strictly liable when the product proves to have a defect that causes injury to a human being, and the plaintiff does not need to prove negligence.

In the case of Noel v United Aircraft in 1964, it was ruled that the supplier of a product can be found liable by failing to make a newly developed safety device available to owners of the product, or by failing to warn them of a dangerous condition discovered after the product was delivered.

In the case of Bart v B.F. Goodrich Tire Company, it was ruled that contributory negligence of the plaintiff or others is no defense in a strict liability action. The degree of the plaintiff's negligence, generally less than 50 per cent, is used to pro-rate the liabilities.

In the case of Elmore v American Motors in 1969, it was ruled that even a bystander could sue if injured, and could sue not just the operator of a car but even the car manufacturer. When the third-party suit is used against third parties capable of paying off large claims, either in judgements or settlements, the philosophy is sometimes called the deep-pocket principle.

In the case of Thomas v General Motors in 1970, the manufacturer was held liable for any and all foreseeable intended uses, misuses, and abuses of the product and even abnormal uses which were foreseeable and could have been safeguarded against.

In the case of Cronin v J. B. E. Olson Corporation in 1972, it was ruled that a product being defective and resulting in an injury is enough to impose strict liability on the manufacturer, and that the plaintiff's awareness of the defect is no defense.

In the case of Balido v Improved Machinery in 1973, it was ruled that warnings and directors do not absolve the manufacturer of liability if there is a defect in design or manufacture. In the case of Glass v Ford Motor Company, also in 1973, it was ruled that the burden of determining whether the defective part was due to a manufacturing or design defect is no longer imposed on the plaintiff.

Under the theory of enterprise liability or industry-wide liability, suggested in the 1972 case of Hall v E. I. du Pont de Nemours and Company, if the defendants, essentially all the companies in an industry and its trade association, had adhered to an industry-wide standard concerning the safety features, manufacture, and design of the product, if the plaintiffs could trace the product to one of the defendants, then the burden of proof as to causation would shift to all the defendants, and each industry member would have contributed to the plaintiff's injury.

Under the theory of market share liability, adopted in a ruling of the California Supreme Court in 1980 in the case of Sindell v Abbott Laboratories, each defendant is held liable for a percentage of the judgement, as determined by that defendant's market share, unless a particular defendant can demonstrate that it could not have made the product that injured the plaintiff.

The contingency fee basis used in many product liability cases makes many more claims possible than would be the case if injured parties had to rely only on their own resources. In a contingency fee situation, the plaintiff's lawyers are entitled to a previously agreed upon fraction of any awards made, instead of legal fees, and in effect are advancing funds to finance the litigation.

Personal injury lawyers believe that, without such contingency agreements, injured parties generally could not afford the legal fees involved. Opponents of this system believe that the contingency fee basis increases the number of cases and the amount of each claim, and therefore the total cost of liability litigation.

SECTION 7.2. FIRE LOSSES AND LITIGATION

The fire losses which result in litigation fall into two classes: personal injury and death, and property and business losses.

The litigation involving personal injury and death covers the following:

1. Compensation for loss of life, which may include the estimated loss of earnings to support a family
2. Compensation for hospital costs and other medical expenses
3. Compensation for estimated future costs of special care and support
4. Compensation for pain and suffering

The litigation involving property and business losses covers the following:

1. Compensation for loss of the structure, which may range from the amount of insurance to full replacement cost
2. Compensation for loss of contents of the structure, which may include the value of the contents, not only on the date of loss, but even on a later date at which the contents could have been sold for higher prices had it not been for loss of structure and contents
3. Compensation for losses due to business interruption

Fire litigation appears to have established almost standard procedures and strategies. There is no dispute regarding the number of deaths and the number of injuries. In the case of death, dispute may begin over the estimated loss of earnings. In the case of injury, dispute may begin over extent of injuries and estimated future costs of special care and support. In the case of property and business loss, dispute may begin over everything.

Attorneys representing the families, heirs, and estates of the dead and the injured survivors, the plaintiff attorneys, usually filed lawsuits against the owners of the structure in which they were occupants, and against other parties. That structure may have been a building, a motor vehicle, an aircraft, or a ship. The other parties which were sued, in efforts to determine and assign liability and the financial responsibility for liability, included product manufacturers, component manufacturers, material suppliers, and raw material suppliers.

The owners of a building filed lawsuits against the architects who designed the building, the contractors who built the building, and the suppliers and manufacturers of the products in the building.

The attorneys for the parties which were sued, the defense attorneys, contested the suit. The plaintiff attorneys and the defense attorneys engaged in extensive and expensive activities before and during trial to establish liability. The parties involved extracted information from the other parties involved in a process called discovery. Experts were retained by the parties involved to provide information as consultants, and to testify in depositions before trial and at trial.

In many cases, the lawsuits did not go to trial or were terminated before trial was completed, because the defendants settled out of court, individually or as groups, for amounts which were negotiated and which were kept secret by agreement.

The defendants usually sought to recover their liabilities, including their attorney fees, by filing lawsuits against other parties to obtain indemnity. Insurance companies for defendants filed lawsuits to recover their losses in a process called subrogation.

In efforts to distribute the liability and the cost of liability among as many parties as possible, and thereby reduce their own liability and their own cost of liability, building owners filed lawsuits against product manufacturers, who in turn filed lawsuits against component manufacturers, who in turn filed lawsuits against material suppliers, who in turn filed lawsuits against raw material suppliers.

In tort law, there are no special rules governing fire litigation. The manufacturers, distributors, and users of flammable products face the same liability as manufacturers, distributors, and users of other products.

Negligence is a prime recovery vehicle in product liability suits. Under negligence theory, the supplier of the product is obliged to keep abreast of scientific knowledge and advances within the product field. The manufacturer is chargeable with the knowledge of risk known to science during the period during which a plaintiff would be exposed to the product. Only excusable ignorance is a potential defense in an action for negligent marketing.

With regard to warnings, the manufacturer or supplier runs the risk of negligent conduct by giving an inadequate warning not commensurate with the potential danger. The manufacturer or supplier must identify the severity of the risk, describe the nature of the hazard, and provide the user sufficient information to help avoid the hazard. Within a chain of distribution, a manufacturer may itself become liable for damages if the warning is not communicated to the ultimate user. The manufacturer's duties may be found to extend to the ultimate user or consumer. A duty to warn of hazards is not discharged because others in the processing or distribution chain have equal or greater knowledge about inherent hazards.

The doctrine of strict product liability holds that sellers engaged in the business of selling defective products or components become strictly liable for damages if the product is in a defective condition or unreasonably dangerous to the consumer or user and causes injury to person or property. Liability attaches only where the defective condition of the product makes it unreasonably dangerous. The seller remains liable even though it has exercised all possible care in the sale and preparation of the product. In California, once a plaintiff shows that an injury was proximately caused by a product's design, the burden shifts to the manufacturer to prove that the product was not defective.

The liability of manufacturers and distributors of products has been extended under emerging tort theories characterized as "concert of action", "alternative" liability, and "market share" liability. Such theories are applied where an injured plaintiff faces difficulty in tracing a product to a specific manufacturer. Such theories of relief come to the fore where there is a showing by injured plaintiffs of "consciously parallel conduct" or other concerted action among manufacturers of similar products with regard to their testing, manufacture, and marketing.

The policies underlying strict product liability favored extension of liability where the charged parties were able to pass along the "hidden costs" of the defective product through pricing or the purchase of liability insurance.

Negligence is conduct-oriented, asking whether a defendant's actions were reasonable. Strict liability is product-oriented, asking whether the product was reasonably safe for its foreseeable purposes. The "state of the art" defense (what was the state of scientific knowledge about the hazard at the time the product was sold) and the "industry practice" defense (what was industry custom and practice at the time the product was sold) are used to demonstrate the reasonable character of the defendant's conduct. They are basically negligence defenses, and may be less effective as strict product liability defenses.

SECTION 7.3. MAJOR FIRE LITIGATION

On May 28, 1977, a fire in the Beverly Hills Supper Club in South Gate, Kentucky, near Cincinnati, Ohio, took the lives of 164 people. Most of the victims were in the Cabaret Room and were reported to have died of smoke inhalation and carbon monoxide poisoning.

On November 21, 1980, a fire in the MGM Grand Hotel in Las Vegas, Nevada, took the lives of 85 people. Many of the victims were in the upper 10 floors of he 26-story hotel and were reported to have died of carbon monoxide poisoning.

On December 31, 1986, a fire in the Dupont Plaza Hotel in San Juan, Puerto Rico, took the lives of 97 people. Most of the victims were in the casino.

These three major fires can be considered the "big three" of fire litigation in terms of size, extensiveness, and expense. They were characterized by large loss of life which multiplied the amount of money involved, extensive and detailed documentation which permitted thorough investigation and scientific analysis, and aggressiveness and skill on the part of both plaintiffs and defendants.

SECTION 7.4. VULNERABILITY OF PLASTICS TO FIRE LITIGATION

The factors affecting the relative vulnerability of a particular plastic to fire litigation are the following:

1. Overall availability. Large-volume commodity plastics are more likely to be involved in a fire than small-volume specialty plastics.
2. Market share of a particular application. Plastics which comprise a large portion of a particular application are more likely to be involved in a particular fire than plastics which have only a small market share.
3. Relative vulnerability of the application. A specific plastic used in a low-risk application is less likely to be involved in a fire than the same plastic used in a high-risk application. For example, a plastic used in fuel tanks would be more likely to be involved in a fire than a plastic used in ice-cream containers.
4. Relative amount contained in a single product item. A product item containing a large quantity of plastic is more likely to play a significant role in a fire than a small product item. For example, the plastic in a handbag would play a much less significant role in a fire than the plastic in the wall covering on all four walls of a room.
5. Relative flammability of the plastic. A particular plastic formulated for greater fire resistance would be less likely to be involved in a fire than a less fire resistant formulation. For example, a polyurethane flexible foam cushion in furniture made to be sold in California would be expected to be less likely to be involved in a fire than a polyurethane flexible foam cushion in furniture made to be sold where fire-resistance requirements were less demanding.
6. Relative smoke evolution. A plastic material which tends to produce more dense

smoke would be more likely to be involved in fire litigation than one which tends to produce relatively little smoke.

7. <u>Relative toxic gas evolution</u>. A plastic which has been extensively studied, and publicized, for toxic gas evolution would be more likely to be involved in fire litigation than one which has received less attention.

8. <u>Relative corrosive gas evolution</u>. A plastic which has been extensively studied, and publicized, for corrosive gas evolution would be more likely to be involved in fire litigation than one which has received less attention.

9. <u>Relative size and sophistication of applicators</u>. A plastic which can be installed by an applicator with relatively little investment or training would be more likely to be involved in fire litigation than one which can be installed only with substantial investment and significant training. For example, many fire cases involving rigid polyurethane foam resulted from one-of-a-kind applications and relatively small applications by small contractors with minimal equipment and minimal training.

10. <u>Relative control by manufacturer over ultimate use</u>. A manufacturer which has substantial knowledge and some measure of control over the ultimate application would be less likely to be involved in fire litigation than one which has no knowledge of intended use once the product has been delivered to a distributor. For example, neoprene flexible foam, which tends to be used in well-defined applications such as mass-transit seating and shipboard mattresses, has been much less involved in fire litigation than polyurethane flexible foam, which is sold as slabs, blocks, and even scrap by small outlets to any customer with few if any questions asked.

REFERENCES

Best, R. L., "Investigative Report on the MGM Grand Hotel Fire", National Fire Protection Association, Quincy, Massachusetts (1981)

Heinlein, C. A., Hill, C. G., "Practical Concerns in Fire Hazard Assessment and Litigation", Proceedings of the International Conference on Fire Safety, Vol. 13, 337-346 (January 1988)

Hilado, C. J., Huttlinger, P. A., "The Economic Importance of Potential Litigation in the Selection of Fire-Safe Materials", Proceedings of the International Conference on Fire Safety, Vol. 8, 268-271 (January 1983)

Hilado, C. J., "Flammability Handbook for Plastics", 4th Ed., Technomic Publishing Co., Lancaster, Pennsylvania (1990)

Hill, R. G., "Investigation and Characteristics of Major Fire-Related Accidents in Civil Air Transports over the Past Ten Years", Proceedings of the International Conference on Fire Safety, Vol. 15, 136-150 (January 1990)

Kemp, W., "Modern Trends in Fire Litigation", Proceedings of the International Conference on Fire Safety, Vol. 16, 104-114 (January 1991)

Klem, T. J., "Investigation Report on the Dupont Plaza Hotel Fire", National Fire Protection Association, Quincy, Massachusetts (1987)

McHugh, J. A., "The Materials Revolution: A Liability Overview for Makers, Distributors, and Professional Users of High Risk Flammable Products", Proceedings of the International Conference on Fire Safety, Vol. 8, 41-71 (January 1983)

Millar, S. A., "Legal Issues of Codes and Standards", Proceedings of the International Conference on Fire Safety, Vol. 14, 261-268 (January 1989)

Nelson, G. L., "Combustion Product Toxicity and Fire Litigation", Proceedings of the International Conference on Fire Safety, Vol. 23, 139-155 (January 1997)

Smith, E. E., "Fire Tests and Fire Litigation", Proceedings of the International Conference on Fire Safety, Vol. 9, 238-241 (January 1984)

CHAPTER 8

Flammability and Environmental Concerns

Concerns about the environment have increased in recent years, and those concerns affect plastics and flammability in varied and sometimes unexpected ways.

Concern about the atmospheric ozone layer is resulting in the reduction in the use of chlorofluorocarbons and their replacement with alternate materials. Chlorofluorocarbons, however, contribute to fire resistance, and the alternate materials may contribute less to fire resistance.

Waste disposal is a major environmental concern, plastics are a significant component of solid waste, and incineration is an important method of waste disposal. Plastics designed to be more difficult to burn unintentionally may be more difficult to burn intentionally.

Plastics which exhibit little or no fire resistance burn to produce relatively clean combustion products such as carbon dioxide, carbon monoxide, and water, and are associated with corrosion no more than are natural materials such as wood, cotton, and paper. Plastics which exhibit fire resistance include those which contain chlorine, bromine, phosphorus, nitrogen, and other elements which could appear in combustion products corrosive to metals.

SECTION 8.1. CHLOROFLUOROCARBONS AND FLAMMABILITY

There has been great urgency in replacing fully halogenated chlorofluorocarbons (CFC) with other materials because of environmental concerns, primarily the protection of the global atmospheric ozone layer. It is believed that fully halogenated chlorofluorocarbons do not degrade in the lower atmosphere, and reach the upper atmosphere where they may be responsible for depletion of the ozone layer.

In 1987, nearly 90 countries including the United States agreed to phase out CFC use by the year 2000. Known as the Montreal Protocol, this agreement was amended several times after 1987, with the result that the timetable for replacement of CFCs was moved up to 1996.

In 1994, the production of CFCs ceased in developed countries, including the United States.

Chlorofluorocarbons have their greatest use as refrigerants. They have been rapidly replaced with other materials.

Chlorofluorocarbons have two other uses because they are nonflammable, and which therefore have impact on flammability: blowing agents and fire extinguishing agents.

Blowing agents are an essential factor in cellular plastics used as thermal insulation, packaging, and cushioning materials because of their role in producing a cellular structure and in providing low thermal conductivity and desired shock absorption. Chlorofluorocarbons have been important blowing agents in polyurethane and polyisocyanurate foams, and in extruded polystyrene foams.

The most widely used chlorofluorocarbon blowing agents have been:

CFC-11, trichloromonofluoromethane
CFC-12
CFC-114
CFC-22

CFC-11 has been used in polyurethane and polyisocyanurate foams. CFC-11, CFC-12, and CFC-114 have been used in extruded polystyrene foam board 1 to 4 inches thick, and CFC-22 has been used in extruded polystyrene foam sheet.

The most widely used chlorofluorocarbon blowing agent has been trichloromonofluoromethane, known as CFC-11, and considerable efforts were made to find alternatives for this material. Chlorofluorocarbons which are not fully halogenated are believed to degrade more readily in the lower atmosphere before reaching the ozone layer. Two promising alternative blowing agents have been dichlorotrifluoroethane, known as CFC-123 or HCFC-123, and dichloromonofluoroethane, known as CFC-141b or HCFC-141b. The latter is flammable within certain limits.

Plastics foam producers are permitted to use these alternative blowing agents which may themselves be damaging to the ozone layer, although much less so than the banned CFCs. Replacements such as HCFC-141b are themselves scheduled for phase-out by the year 2006.

Other alternative blowing agents, petrochemical blowing agents such as cyclopentane, can be used in polyurethane foam. Pentane is widely used in expanded polystyrene. There are, however, environmental problems associated with the release of hydrocarbons into the atmosphere.

Fire extinguishing agents are important in fire protection, and CFCs have been widely used as fire extinguishing agents. Water, carbon dioxide, and dry chemicals are used as fire extinguishing agents in ground applications, but normally not in aviation applications.

CFCs, called halons in aviation, are important in aircraft fire protection and airport firefighting capability. The selection of halons for aviation applications is based on a number of considerations, most notably extinguishment effectiveness per unit weight, effectiveness over a wide range of operational conditions, low toxicity, low corrosivity, and virtually no cleanup.

Halon 1301 (CF_3Br) and Halon 1211 (CF_2BrCl) are used extensively in commercial aircraft and by airport firefighters. In commercial aircraft, Halon 1301 is used in cargo compartments, powerplants, and lavatory trash receptacles, and Halon 1211 is required in portable extinguishers for use against passenger cabin fires. At airports, Halon 1211 is used in flight line extinguishers and crash fire rescue vehicles.

The Federal Aviation Administration (FAA) continues to use halon to maintain the current level of aircraft fire protection and airport firefighting capability, but has programs underway to evaluate replacement and alternate agents/systems for aircraft and airport applications.

SECTION 8.2. RECYCLING OF PLASTICS

The recycling of plastics has encountered economic and institutional problems. Major technological problems are associated with initial separation of plastics in the waste stream, segregation of plastic wastes by resin types, and cleaning the waste plastic.

The melt-processability of thermoplastics is a characteristic that gives than an advantage over thermosets in recycling, because they can be remelted and reprocessed into satisfactory, usable products. Nearly all producers of commodity thermoplastic resins recycle scrap parts. Scrap parts are commonly reground and blended with virgin resin at up to 25% loading ratios for injection molding resins. Most of these recycling operations are internal, with typical plastics processors regrinding and reusing their own scrap.

Some plastics used in post-consumer recyclate in the United States are shown in Table 8.1.

Flexible polyurethane foam production scrap is used in rebonded carpet underlay, and some post consumer waste is finding this use.

Pulverizing, or regrind, is one option for recycling polyurethane foam. In this process, the recycled foam is ground into very fine particles and used as a filler in the preparation of new foam. Either cryogenic grinding, or mechanical grinding under ambient conditions, can be used.

The market for recycled plastics in the United States continues to increase and evolve because of increasing recycled content legislation, expanding collection networks, improvements in recycling technologies, and further product and applications development. Polyethylene terephthalate (PET) and high-density polyethylene (HDPE) together account for 70% of all recycled plastics. Recycling focuses on PET and HDPE bottles culled from municipal solid waste (MSW).

Packaging is the largest market for recycled resins, based on consumer demands for environmentally-friendly packaging, and more cost-efficient technologies that allow closed-loop recycling of materials in the food and beverage industries.

Sources of recycled plastics in Europe include industrial packaging, consumers, agriculture, construction, and motor vehicles.

SECTION 8.3. DEPOLYMERIZATION OF PLASTICS

The most common form of recycling, mechanical recycling, is limited by inherent problems: costly sorting of the different types of plastics, concerns about the quality of recycled material, and finding reliable markets for products made from recycled plastics.

Other recycling processes, collectively known as advanced recycling technologies (ART) reclaim chemical or feedstock value from plastics through depolymerization. Advanced recycling technologies can be divided into two groups, based on whether the depolymerization results in monomer, or in hydrocarbon feedstock.

The back-to-monomer processes in general require the plastics to be sorted but can tolerate contaminants like dirt, additives, fillers, and colorants. The processes which produce feedstock have the ability to recycle commingled, contaminated plastics.

Plastics such as polyesters, polyamides, and polyurethanes are created via reversible reactions. It is feasible to convert these plastics back to their immediate starting materials, monomers, which are most valuable and can be used to remake the same plastic.

Glycolysis of polyurethane foam can lead to new polyols.

Plastics such as high-density polyethylene (HDPE), polypropylene (PP), and polyvinyl chloride (PVC) are created via irreversible reactions. These plastics can only be converted into basic petrochemical components through pyrolysis. Further refining and purification isolate monomers, monomer precursors, other petrochemicals, and fuel ingredients.

SECTION 8.4. INCINERATION OF PLASTICS

Paper and paper products dominate the contents of municipal solid waste (MSW), contributing about 35.6% by weight, while plastics contribute about 7.3% by weight. Plastics, however, have a much higher energy content, as high as 19,500 Btu/lb for polyolefins, compared to the average of about 5,000 Btu/lb for the total. While contributing only about 7% of the weight, plastics contribute about 25% of the energy content of the waste and are a significant component of the energy economics of resource recovery plants. Removal of plastics from municipal solid waste would significantly reduce energy content.

The majority of plastics in the waste, about 90%, consist of polyethylene, polypropylene, polystyrene, and PET, which contain carbon, hydrogen, and oxygen, and when incinerated, yield carbon dioxide and water. Other plastics such as polyvinyl chloride (PVC), polyamides or nylons, polyurethanes, and polysulfides, when incinerated, may give off, in addition to carbon dioxide and water, compounds such as HCl from PVC, HCN and NO_x from nylons and polyurethanes, and H_2S and SO_2 from polysulfides. These other plastics comprise less than 1% of the waste. Many non-plastic components of municipal solid waste, such as wood, paper, and leather, will also give off minor amounts of these additional gases, which are readily removed by existing scrubbing systems.

Energy recovery can be achieved by co-combustion of mixed plastics waste and municipal solid waste, and by incineration of mixed plastics with coal. In both processes, the contribution of acid gases and heavy metals from plastics is a matter of concern.

REFERENCES

Association of Plastics Manufacturers in Europe, "Plastics Recovery in Perspective", Association of Plastics Manufacturers in Europe, Brussels (1993)

Association of Plastics Manufacturers in Europe, "Energy Recovery Through Co-Combustion of Mixed Plastics Domestic Waste and Municipal Solid Waste", Association of Plastics Manufacturers in Europe, Brussels (1994)

Association of Plastics Manufacturers in Europe, "Weighing Up the Options: A Comparative Study of Recovery and Disposal Routes", Association of Plastics Manufacturers in Europe, Brussels (1994)

Carroll, W. F., "Is PVC in House Fires the Great Unknown Source of Dioxin?", Proceedings of the International Conference on Fire Safety, Vol. 20, 268-278 (January 1995)

Frankenhaeuser, M., "Packaging Derived Fuel (PDF) as a Source of Energy", European Centre for Plastics in the Environment, Brussels (1993)

Frankenhaeuser, M., Kojo, I., Manninen, H., Ruuskanen, J., Vesterinen, R., Virkki, J., "Mixed Plastics with Coal Incineration", European Centre for Plastics in the Environment, Brussels (1991)

Freiesleben, W., "Plastics in Municipal Incineration", European Centre for Plastics in the Environment, Brussels (1992)

Gibala, D., Cain, R. J., Salsamendi, M. C., "Effect of Post-Consumer Automotive Seating Granulate on TDI-Based Flexible Foam Properties", Proceedings of the Polyurethanes 1995 Conference, 261-269 (September 1995)

Hicks, D. A., Hemel, C. B., Kirk, A. C., Stapleton, R. J., Thompson, A. R., "Recycling and Recycled Content for Polyurethane Foam", Proceedings of the Polyurethanes 1995 Conference, 279-286 (September 1995)

Hilado, C. J., "Flammability Handbook for Plastics", 4th Ed., Technomic Publishing Co., Lancaster, Pennsylvania (1990)

Hilado, C. J., Furst, A., "Safety and Toxicity Aspects of Chlorofluorocarbons and Other Blowing Agents", Proceedings of the International Conference on Thermal Insulation, Vol. 5, 61-73 (March 1989)

Magee, R. S., "Plastics in Municipal Solid Waste Incineration: A Literature Study", Hazardous Substance Management Research Center, New Jersey Institute of Technology, Newark, New Jersey (January 1989)

Mark, F. E., "The Role of Plastics in Municipal Solid Waste Combustion", European Centre for Plastics in the Environment, Brussels (1993)

Mark, F. E., "Energy Recovery Through Co-Combustion of Mixed Plastics Waste and Municipal Solid Waste", Association of Plastics Manufacturers in Europe, Brussels (June 1994)

Mischutin, V., "Brominated Flame Retardants and the Environment", Proceedings of the International Conference on Fire Safety, Vol. 13, 347-356 (January 1988)

Modern Plastics Encyclopedia '96, McGraw-Hill, New York (1996)

Nelson, G. L., "Recycling of Plastics: A New Challenge for Flame Retardants", Proceedings of the International Conference on Fire Safety, Vol. 20, 189-199 (January 1995)

Sarkos, C. P., "Status of Halon Replacement Agent Evaluation in Commercial Aircraft and Airport Applications", Proceedings of the 3rd International Fire and Materials Conference, 189-197 (October 1994)

Stone, H., Lichvar, S., Sweet, F., "Commercial Potential for Recycling of Finley Ground Foam in Flexible Polyurethane Foam", Proceedings of the Polyurethanes 1995 Conference, 244-252 (September 1995)

Table 8.1. Some Plastics Used in Post-Consumer Recyclate.

United States	Million lb.	
	1995	1996
Polyethylene terephthalate (PET)	630	640
HDPE	650	680
LDPE	170	185
Polypropylene	255	270
Polystyrene	50	55
Polyvinyl chloride	10	10
All others	70	75
Total	1835	1915

Reference: Modern Plastics, January 1997.

CHAPTER 9

Commercial Fire, Smoke, and Smolder Retardants

Some plastics are inherently, to a relative degree, fire retardant or smoke retardant or smolder retardant, and their fire performance is acceptable for certain applications. Most plastic materials, however, do not inherently have acceptable fire performance for those applications with demanding requirements.

For many plastic materials, the most feasible method of improving their fire performance is the incorporation of commercially available retardants. In previous years, these retardants were considered primarily fire retardants; many are now viewed as smoke retardants and smolder retardants because smoke evolution and smoldering are fire response characteristics of concern in some applications.

Some materials used as fire retardants have been found to increase smoke under certain conditions, some other materials used as fire retardants have been found to increase smoldering under certain conditions, and yet other materials have been found to be effective with regard to two or even all three characteristics under certain conditions. Their extent of effectiveness is very much a function of the material, the application involved, and the test used, and it is not possible at this time to classify retardants on the basis of the characteristics on which they have the most desirable impact.

Retardants are generally recommended by manufacturers and distributors for use with those plastics with which they are expected to be effective in improving fire performance without excessive loss of other important performance characteristics. Some retardants have been found to be effective with certain plastics but at an unacceptable cost in other performance characteristics.

The products listed in Table 9.1 represent a spectrum of available commercial retardants. This listing was prepared from information provided to the author by companies in response to requests for product information. It is not intended to be a complete listing of all or even the majority of available commercial products. The chemical analyses came from the information provided, and the applications listed for each product are those recommended or suggested in the information provided. Each proposed retardant should be evaluated with the products with which it is intended to be used, in the application for which it is intended, in order to determine its effectiveness in each specific situation.

REFERENCES

Akzo Nobel Chemicals Inc., Dobbs Ferry, New York, product literature received November 1997

Albemarle Corporation, Baton Rouge, Louisiana, product literature received February and October 1997

Albright & Wilson Americas Inc., Glen Allen, Virginia, product literature received September 1997

Alcoa Industrial Chemicals, Bauxite, Arkansas 72011, product literature received August 1997

Buckman Laboratories, Memphis, Tennessee, product literature received September 1997

FMC Corporation, Philadelphia, Pennsylvania, product literature received November 1997

Great Lakes Chemical Corporation, West Lafayette, Indiana, product literature received November 1997

Laurel Industries, Cleveland, Ohio, product literature received August and November 1997

Martin Marietta Magnesia Specialties, Baltimore, Maryland, product literature received November 1997

Sherwin-Williams Chemicals, Coffeyville, Kansas, product literature received November 1997

U.S. Borax Inc., Valencia, California, product literature received June 1997

Table 9.1. Commercial Fire, Smoke, and Smolder Retardants.

Akzo Nobel Chemicals Inc., 5 Livingstone Avenue, Dobbs Ferry, New York 10522-3407

Product Name	Chemical Composition	Possible Application
Fyrol FR-2	Tri(1,3-dichloroisopropyl) phosphate 7.1% P 49.0% Cl	Flexible PU foam, epoxy, phenolic, polyester
Fyrol FR-2LV	Modified tri(1,3-dichloroisopropyl) phosphate 7.3% P 47.0% Cl	Flexible PU foam, epoxy, phenolic, polyester
Fyrol 38	Modified tri(1,3-dichloroisopropyl) phosphate 7.0% P 49.0% Cl	Flexible PU foam
Fyrol CEF	Tri(2-chloroethyl) phosphate 10.8% P 36.7% Cl	Rigid and flexible PU foam, epoxy, phenolic, PMMA, polyester
Fyrol PCF	Tri(2-chloroisopropyl) phosphate 9.5% P 32.5% Cl	Rigid PU foam, phenolic, polyester
Fyrol 25	Chloroalkyl phosphate/oligomeric phosphonate 10.2% P 37.0% Cl	Flexible PU foam
Fyrol 99	Oligomeric chloroalkyl phosphate 14.0% P 26.0% Cl	Rigid PU foam, phenolic, epoxy, polyester
Fyrol PBR	Brominated FR based on pentabromo diphenyl ether 2.2% P 51.5% Br	Flexible PU foam
Fyrol DMMP	Dimethyl methyl phosphonate 25.0% P	Rigid PU foam, polyester
Fyrol 6	Diethyl N,N bis(2-hydroxyethyl) amino methyl phosphonate 12.0%	Rigid PU foam
Fyrol 51	Oligomeric phosphonate 20.5% P	Cellulosic, PET, phenolic
Fyrol 58	Fyrol 51 and an organic solvent 17.1% P	Cellulosic
Fyrol 42	Proprietary customized additive 5.3% P	Flexible PU foam
Phosflex 31P	Isopropylated triaryl phosphate 7.9% P	PVC
Phosflex 41P	Isopropylated triaryl phosphate 7.9% P	PVC
Phosflex 11P	Isopropylated triaryl phosphate 7.9% P	PVC
Phosflex 21P	Isopropylated triaryl phosphate 7.9% P	PVC
Phosflex 61B	t-Butylated triphenyl phosphate ester mixture 8.0% P	PVC, phenolic
Phosflex 71B	t-Butylphenyl diphenyl phosphate 8.5% P	PVC, PVA, cellulosic
Phosflex 72B	Proprietary blend based on a t-butylated triphenyl phosphate 5.1% P	
Phosflex 362	2-Ethylhexyl diphenyl phosphate 8.5% P	PVC, cellulosic
Phosflex 370	Blend of alkyl diaryl/triaryl phosphate esters 8.3% P	PVC, cellulosic
Phosflex 390	Isodecyl diphenyl phosphate 7.9% P	PVC, PVA
Phosflex 179A	Trixylyl phosphate 7.8% P	PVC
Lindol	Tricresyl phosphate 8.4% P	PVC, phenolic
Phosflex TPP	Triphenyl phosphate 9.5% P	Cellulosic
Phosflex 4	Tributyl phosphate 11.7% P	Cellulosic
Phosflex T-BEP	Tributoxyethyl phosphate 7.8% P	
Fyrolflex RDP	Resorcinol bis-(diphenyl phosphate) 11.0% P	PPO, PC
Phosflex CDP	Cresyl diphenyl phosphate 9.1% P	PVC

continued

Table 9.1 (continued). Commercial Fire, Smoke, and Smolder Retardants.

Albemarle Corporation, 451 Florida Street, Baton Rouge, Louisiana 70801

Product Name	Chemical Composition	Possible Application
Saytex 102E	Decabromodiphenyl oxide 83.3% Br	ABS, HIPS, nylon, TPE, PC, PP, PE, silicone, PVC, epoxy, phenolic
Saytex 120	Tetradecabromodiphenoxy benzene 81.8% Br	ABS, HIPS, nylon, TPE, PC, PP, PE, silicone, EPDM
Saytex 600L	Hexabromocyclododecane 74.7% Br	PP, EPS
Saytex 8010	Brominated aromatic (prop.) 82.3% Br	ABS, HIPS, nylon, TPE, PC, PP, PE, silicone, PVC, epoxy, phenolic
Saytex BC-48	Tetrabromocyclo-octane 74.7% Br	EPS, silicone, epoxy
Saytex BC-56HS	Brominated aliphatic (prop.) 56.0% Br	PP, SAN, PS, EPS
Saytex BCL-462	Dibromoethyl-dibromocyclohexane 74.7% Br	EPS, silicone, epoxy
Saytex BN-451	Ethylene-bis-dibromonorbornane-dicarboximide 47.6% Br	Nylon, PP
Saytex BT-93	Ethylene-bis-tetrabromophthalimide 67.2% Br	ABS, HIPS, TPE, PC, PP, PE, silicone, EPDM, epoxy, phenolic
Saytex BT-93W	Ethylene-bis-tetrabromophthalimide 67.2% Br	ABS, HIPS, TPE, PC, PP, PE, silicone, EPDM, epoxy, phenolic
Saytex HBCD-LM	Hexabromocyclododecane 74.7% Br	HIPS, PP, EPS, silicone, epoxy
Saytex HBCD-SF	Hexabromocyclododecane blend 70.2% Br	HIPS, PP, EPS, silicone, epoxy
Saytex HP-7010 P/G	Polymeric FR 69% Br	Nylon, TPE, PC
Saytex HP-800	Tetrabromobisphenol-A-bis-(2,3-dibromopropyl ether) 68% Br	SAN, EPS, PS
Saytex RB-49	Tetrabromophthalic anhydride 68.9%Br	Epoxy, phenolic
Saytex RB-100	Tetrabromobisphenol-A 59% Br	ABS, HIPS, epoxy, phenolic
Saytex RB-79/7980	RB-79/phosphate ester blend 36% Br	RPU

Albright & Wilson Americas Inc., P.O. Box 4439, Glen Allen, Virginia 23058-4439

Amgard MC	Ammonium polyphosphate 30–30.5% P 15.9% N	
Amgard NH	Melamine phosphate based 13.4% P 38.6% N	PE, PP, PS, PMMA, PU foam
Amgard NK	19.6% P	PE, PP, EVA, TPE, TPU, epoxy
Amgard NP	15.7% P	PE, PP, TPE, TPU, epoxy
Antiblaze 519	Triaryl phosphate ester 8.3% P	PVC
Antiblaze 521	Triaryl phosphate ester 8.1% P	PVC
Antiblaze 524	Triaryl phosphate ester 7.7% P	PVC
Antiblaze 1045	Phosphorus based 20.8% P	Nylon, epoxy

Alcoa Industrial Chemicals, P.O. Box 300,4701 Alcoa Road, Bauxite, Arkansas 72011

Hydral 705	Alumina trihydrate (ATH) 64.7% Al_2O_3	PP, PVC, EPDM, polyester, epoxy
Hydral 710	Alumina trihydrate (ATH) 65.1% Al_2O_3	PP, PVC, EPDM, polyester, epoxy
Hydral 7025	Alumina trihydrate (ATH) 64.7% Al_2O_3	PP, PVC, EPDM, polyester, epoxy
Hydral PGA	Alumina trihydrate (ATH) 65.0% Al_2O_3	PP, PVC, EPDM, polyester, epoxy
SpaceRite S-3	Aluminum trihydroxide 99.0% $Al(OH)_3$	
SpaceRite S-11	Aluminum trihydroxide 99.0% $Al(OH)_3$	
SpaceRite S-23	Aluminum trihydroxide 98.0% $Al(OH)_3$	
Onyx Classica 1500	Alumina trihydrate (ATH) 65.0% Al_2O_3	
BayGranite 17	Alumina trihydrate (ATH) 65.0% Al_2O_3	
BAO	Basic aluminum oxalate 25.7% Al	PP, nylon, PBT

Table 9.1 (continued). *Commercial Fire, Smoke, and Smolder Retardants.*

Buckman Laboratories Inc., 1265 North McLean Boulevard, Memphis, Tennessee 38108-1241

Bulab Flamebloc	Hydrated borate	PVC, neoprene, SBR

FMC Corporation, 1735 Market Street, Philadelphia, Pennsylvania 19103

Kronitex TCP	Tricresyl phosphate 8.4% P	Cellulosic, NBR, SBR
Kronitex TXP	Trixylenyl phosphate 7.8% P	PVC
Reoflam PB-370	Brominated phosphate 3% P 70% Br	PP, PC, ABS
Reoflam FG-372		PP
Reofos 50	Triaryl phosphate 8.3% P	PVC, PUR, cellulosic
Reofos 65	Triaryl phosphate 7.9% P	PVC, PUR
Reofos 95	Triaryl phosphate 7.4% P	PVC, cellulosic
Reofos 1884	Emulsifiable triaryl phosphate based on Reofos 65	PVC
Reofos 3600	Phosphate ester plasticizer 8.3% P	PVC
Reofos RDP	Resorcinol bis-(diphenyl phosphate) 10.8% P	PPO, PC/ABS, PUR
Reomol TBP	Tributyl phosphate	PUR, cellulosic PVC
Reomol TOF	Tris(2-ethylhexyl)phosphate	PVC, cellulosic
Reomol 249	Triaryl phosphate ester	PVC
Reomol 293	Triaryl phosphate ester	PVC
KP-140	Tributoxyethyl phosphate	PUR, cellulosic

Great Lakes Chemical Corporation, P.O. Box 2200, West Lafayette, Indiana 47906-0200

Product Name	Chemical Composition	Possible Application
BA-59P	Tetrabromobisphenol A 58.8% Br	ABS, HIPS, PC, epoxy, polyester
BA-50P	Tetrabromobisphenol A bis (2-hydroxyethyl ether) 50.6% Br	PBT, TPE, polyester
PE-68	Tetrabromobisphenol A bis (2,3-dibromopropyl ether) 67.7% Br	PP
BE-51	Tetrabromobisphenol A bis(allyl ether) 51.2% Br	PS foam, polyester
BC-52	Phenoxy-terminated carbonate oligomer of tetrabromobisphenol A 51.3% Br	ABS, PC, PBT
BC-58	Phenoxy-terminated carbonate oligomer of tetrabromobisphenol A 58.7% Br	ABS, PC, PBT
PH-73	2,4,6-tribromophenol 72.5% Br	
PHE-65	Tribromophenol allyl ether 64.2% Br	PS foam
PO-64P	Poly-dibromophenylene oxide 62.0% Br	PC, PA, PET, PBT, TPE
FF-680	Bis(tribromophenoxy) ethane 70.0% Br	ABS, HIPS, PC
DE-83	Decabromodiphenyl oxide 83.3% Br	ABS, HIPS, PP, PC, PA, PET, PBT, PVC, TPE, epoxy, polyester, PUR
DE-83R	Decabromodiphenyl oxide 83.3% Br	ABS, HIPS, PP, PC, PA, PET, PBT, PVC, TPE, epoxy, polyester, PUR
DE-79	Octabromodiphenyl oxide 79.8% Br	ASB, HIPS, PP, PA, TPE
DE-71	Pentabromodiphenyl oxide 70.8% Br	TPE, epoxy, polyester, PUR foam
DE-60F	Pentabromodiphenyl oxide blend 52.0% Br	TPE, epoxy, polyester, PUR foam
DE-61	Pentabromodiphenyl oxide blend 51.0% Br	PUR foam

continued

Table 9.1 (continued). Commercial Fire, Smoke, and Smolder Retardants.

Great Lakes Chemical Corporation, P.O. Box 2200, West Lafayette, Indiana 47906-0200 (cont.)

Product Name	Chemical Composition	Possible Application
DE-62	Pentabromodiphenyl oxide blend 51.0% Br	PUR foam
DBS	Dibromostyrene 59.0% Br	ABS, HIPS, TPE, PUR, epoxy, polyester
PDBS-80	Poly-(dibromostyrene) 59.0% Br	ABS, HIPS, PC, PA, PET, PBT, TPE
GPP-36	Polypropylene-dibromostyrene graft copolymer 36.0% Br	PP, PET
GPP-39	Proprietary derivative of GPP-36 39.0% Br	PP, PET
PHT4	Tetrabromophthalic anhydride 68.2% Br	Polyester
PHT4-DIOL	Tetrabromophthalate diol 46.0% Br	PUR foam
DP-45	Tetrabromophthalate ester 45.0% Br	PVC, TPE
FR-756	Disodium salt of tetrabromophthalic anhydride 81.0% Br	Textiles
CD-75P	Hexabromocyclododecane 74.7% Br	PS foam
SP-75	Stabilized hexabromocyclododecane 72.0% Br	HIPS, PP, PS foam
FB-72	Proprietary brominated flame retardant 72.0% Br	HIPS, PP
Firemaster HP-36	Halogenated phosphate ester 44.5% halogen 7.5% P	PS foam, polyester, PUR foam
Firemaster 836	Halogenated phosphate ester 44.5% halogen 7.5% P	PUR foam
Firemaster 642	Halogenated phosphate ester 49.2% halogen 6.5% P	PUR foam
NH-1197	Intumescent flame retardant 17.0% P	TPE, epoxy, polyester
NH-1511	Intumescent flame retardant 15.0% P	PP, TPE, epoxy, polyester

Laurel Industries, 30195 Chagrin Boulevard, Cleveland, Ohio 44124-5794

FireShield H FireShield L Ultrafine II FireShield HPM FireShield HPM-UF	Antimony trioxide Sb_2O_3	PE, PP, PS, ABS, PVC EPDM, PU, epoxy
LSFR	Proprietary	
Dechlorane Plus	Chlorine-containing cycloaliphatic 65.1% Cl	PE, PP, HIPS, ABS, PBT, nylon, epoxy, polyester, PUR, TPE

Martin Marietta Magnesia Specialties Inc., P.O. Box 15470, Baltimore, Maryland 21220-0470

Product Name	Chemical Composition	Possible Application
MagShield S	Magnesium hydroxide 98.5% $Mg(OH)_2$	PP, PS, PVC, PPO, EPDM
MagShield M	Magnesium hydroxide 98.5% $Mg(OH)_2$	PP, PS, PVC, PPO, EPDM

Table 9.1 (continued). Commercial Fire, Smoke, and Smolder Retardants.

Sherwin-Williams Chemicals, P.O. Box 1028, Coffeyville, Kansas 67337		
Kemgard 911A	Calcium zinc molybdate 12% ZnO	PVC
Kemgard 911A-S	Stabilized calcium zinc molybdate 8.8% Zn	PVC
Kemgard 911B	Basic zinc molybdate 75% ZnO 25% ZnMoO$_4$	PVC
Kemgard 911C	Zinc molybdate/magnesium silicate complex 7.2% Zn	PVC
Kemgard 911C-S	Stabilized zinc molybdate/magnesium silicate complex 8.6% MoO$_3$ 7.2% Zn	PVC
Kemgard 981	Phosphated zinc oxide 68.4% Zn	PVC
Kemgard 425	Calcium zinc molybdate 8% Zn	PVC
Kemgard X501	Experimental calcium molybdate complex 9.3% MoO$_3$	PVC
U.S. Borax Inc., 26877 Tourney Road, Valencia, California 91355-1847		
Firebrake ZB	Zinc borate, 2ZnO·3B$_2$O$_3$·3·5H$_2$O, 48.05% B$_2$O$_3$, 37.45% ZnO, 14.50% H$_2$O	PVC, epoxy, nylon
Firebrake 415	4ZnO·B$_2$O$_3$·H$_2$O	Nylon, PVC
Firebrake 500	Anhydrous zinc borate 2ZnO·3B$_2$O$_3$	

continued

Table 9.1 (continued). Commercial Fire, Smoke, and Smolder Retardants.

Information on the products of the following manufacturers was not available at the time this book went to press.

Ameribrom, Inc. 52 Vanderbilt Avenue New York, New York 10017	Bromine-containing compounds
Amspec Chemical Corporation 751 Water Street Gloucester City, New Jersey 08030	Sb_2O_3
Anzon (now part of Great Lakes) 2545 Aramingo Avenue Philadelphia, Pennsylvania 19125	Sb_2O_3
Clariant Corporation 5200 77 Center Drive Charlotte, North Carolina 28217	Phosphorus-containing compounds
Climax Molybdenum Company P.O. Box 980407 Ysilanti, Michigan 48198	Molybdic oxide
Elf Atochem North America Inc. 2000 Market Street Philadelphia, Pennsylvania 19103	Sb_2O_3 and bromine-containing compounds
Ferro Corporation 3000 Sheffield Avenue Hammond, Indiana 46327	Bromine-containing compounds
J. M. Huber Norcross, Georgia	ATH
Lonza Inc. 17-17 Route 208 Fair Lawn, New Jersey 07410	ATH and $Mg(OH)_2$
Melamine Chemicals P.O. Box 748 Donaldsonville, Louisiana 70346	Melamine
PQ Corporation P.O. Box 840 Valley Forge, Pennsylvania 19482	Sb_2O_5
Solutia Inc. (formerly Monsanto) 10300 Olive Street Boulevard St. Louis, Missouri 63166	Phosphorus-containing compounds

322

T - #0173 - 101024 - C0 - 229/152/19 [21] - CB - 9781566766517 - Gloss Lamination